T0309722

VEHICLE DYNAMICS

Automotive Series

Series Editor: Thomas Kurfess

VEHICLE DYNAMICS

Martin Meywerk

Helmut-Schmidt-University (University of the Federal Armed Forces Hamburg), Germany

This edition first published 2015
© 2015 John Wiley & Sons Ltd

Registered office
John Wiley & Sons Ltd, The Atrium, Southern Gate, Chichester, West Sussex, PO19 8SQ, United Kingdom

For details of our global editorial offices, for customer services and for information about how to apply for permission to reuse the copyright material in this book please see our website at www.wiley.com.

Library of Congress Cataloging-in-Publication Data Applied for.

ISBN: 9781118971352

A catalogue record for this book is available from the British Library.

Set in 11/13pt Times by Laserwords Private Limited, Chennai, India

1 2015

For my wife Annette
and my children Sophia, Aljoscha, Indira and Felicia

Contents

Foreword

This book is an extract of lectures on vehicle dynamics and mechatronic systems in vehicles held at the Helmut-Schmidt-University, University of the Federal Armed Forces, Hamburg, Germany. The lectures have been held since 2002 (Vehicle Dynamics) and 2009 (Vehicle Mechatronics). The book is an introduction to the field of vehicle dynamics and most parts of the book should be comprehensible to undergraduate students with a knowledge of basic mathematics and engineering mechanics at the end of their Bachelor studies in mechanical engineering. However, some parts require advanced methods which are taught in graduate studies (Master programme in mechanical engineering).

I wish to thank Mrs Martina Gerds for converting the pictures to Corel Draw with LaTeX labels and for typing Chapter 9. My thanks go to Mr Darrel Fernandes, B.Sc., for the pre-translation of my German scripts. I especially wish to thank Mr Colin Hawkins for checking and correcting the final version of the book with respect to the English language. My scientific assistants, especially Dr Winfried Tomaske and Dipl.-Ing. Tobias Hellberg, I thank for proofreading, especially with regard to the technical aspects. Special thanks for assistance in preparing a number of Solid Works constructions for pictures of suspensions and transmissions as well for help in preparing some MATLAB diagrams go to Mr Hellberg. Last but not the least, my thanks go to my family, my wife, Dr Annette Nicolay, and my children, Sophia, Aljoscha, Indira and Felicia, for their patience and for giving me a lot of time to prepare this book.

Series Preface

The automobile is a critical element of any society, and the dynamic performance of the vehicle is a key aspect regarding its value proposition. Furthermore, vehicle dynamics have been studied for many years, and provide a plethora of opportunities for the instructor to teach her students a wide variety of concepts. Not only are these dynamics fundamental to the transportation sector, they are quite elegant in nature linking various aspects of kinematics, dynamics and physics, and form the basis of some of the most impressive machines that have ever been engineered.

Vehicle Dynamics is a comprehensive text of the dynamics, modeling and control of not only the entire vehicle system, but also key elements of the vehicle such as transmissions, and hybrid systems integration. The text provides a comprehensive overview of key classical elements of the vehicle, as well as modern twenty-first century concepts that have only recently been implemented on the most modern commercial vehicles. The topics covered in this text range from basic vehicle rigid body kinematics and wheel dynamic analysis, to advanced concepts in cruise control, hybrid powertrain design and analysis and multi-body systems. This text is part of the *Automotive Series* whose primary goal is to publish practical and topical books for researchers and practitioners in industry, and post-graduate/advanced undergraduates in automotive engineering. The series addresses new and emerging technologies in automotive engineering supporting the development of next generation transportation systems. The series covers a wide range of topics, including design, modelling and manufacturing, and it provides a source of relevant information that will be of interest and benefit to people working in the field of automotive engineering.

Vehicle Dynamics presents a number of different designs, analysis and implementation considerations related to automobiles including power requirements, converters, performance, fuel consumption and vehicle dynamic models. The text is written from a very pragmatic perspective, based on the author's extensive experience. The book is written such that it is useful for both undergraduate and post-graduate courses, and is also an excellent reference text for those practicing automotive systems design and engineering, in the field. The text spans a wide spectrum of concepts that are critical to the understanding of vehicle performance, making this book welcome addition to the *Automotive Series*.

Thomas Kurfess
October 2014

Preface

This books covers the main parts of vehicle dynamics, which is divided into three topics: longitudinal, vertical and lateral dynamics. It also explains some applications, especially those with a mechatronic background, and outlines some components.

Figure 1 provides an overview of the chapters of the book. The main parts (longitudinal, vertical and lateral) as well as applications and component chapters are grouped together. Many principal aspects of dynamics are explained by using simple mechanical models (e.g. quarter-vehicle model and single-track model). As the virtual development process with very complex multi-body systems (MBS) is used in the design of modern cars, this simulation technique is described very briefly in the last chapter. Although these MBS models are able to predict many details, the user of such models should understand the principles of how vehicles behave, and the main theory behind dynamic behaviour. It is therefore important to learn the basic dynamic behaviour using the simple models described in this book.

Chapter 1 contains some general data for vehicles. These remarks are followed by an introduction to some of the basics of frames and axis systems. This introduction should be read by everyone. Following this are the three groups of longitudinal, vertical and lateral dynamics, which are largely independent. The longitudinal and the vertical parts are completely independent of the other parts and may be read and understood without any knowledge of the other parts. The third group, containing the lateral dynamics part, includes a number of aspects that may be difficult to understand without first reading the longitudinal or the vertical part.

The application chapters can only partly be understood without reading the corresponding theory chapter. Readers are therefore recommended to start with the basic parts: the basics for Chapters 7 (Hybrid Powertrains), 8 (Adaptive Cruise Control) and 17 (Torque and Speed Converters) can be found in the longitudinal dynamics chapters, while lateral dynamics is important for Chapter 16 (Suspension Systems), in the case of Chapter 18 (Shock Absorbers, Springs and Brakes) a knowledge is required of vertical dynamics as well as some aspects of longitudinal and lateral dynamics. Chapter 19 (Active Longitudinal and Lateral Systems), as the name reveals, involves longitudinal and lateral aspects. Chapter 20 is nearly independent of the theoretical considerations.

Figure 1 includes the letters B and M, which stand for Bachelor and Master, behind every chapter. Chapters of level B should be comprehensible for undergraduate

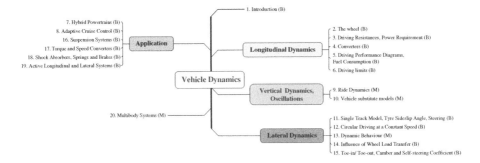

Figure 1   Chapters of the book

Figure 2   Bloom's taxonomy of learning

students with a knowledge of engineering mechanics and mathematics at the end of their Bachelor studies in mechanical engineering. Topics covered are: algebra; trigonometric functions; differential calculus; linear algebra; vectors; coordinate systems; force, torque, equilibrium; mass, centre of mass, moment of inertia; method of sections, friction, Newton's laws, Lagrange's equation. In chapters followed by level M, an advanced knowledge is useful, as is usually taught to graduate students: ordinary differential equations (ODE), stability of ODEs, Laplace transformation, Fourier transformation, stochastic description of uneven roads and spectral densities.

At the end of nearly each chapter, you will find some questions and exercises. These are for monitoring learning progress or for applying the material learned to some small problems. For this reason, the questions and tasks are arranged in classes according to Bloom's taxonomy of learning (cf. Figure 2).

The simplest class is *Remembering*, which means that you only have to remember the correct content (e.g. a definition or a formula). You should be able to answer the questions of the second class *Understanding* if you have understood the content. The tasks of the third class *Application* involve applying the content to some unknown problem. The remaining three classes *Analysing*, *Evaluating* and *Creating* are more suited to extended student works, such as Bachelor or Master theses, and are therefore rarely included in this book.

List of Abbreviations and Symbols

The tables on the following pages summarize the mathematical symbols and abbreviations used in this book. In most cases (but not in all), the indices used with the symbols indicate the following:

- v vehicle
- b body
- w wheel
- t tyre
- x, y, z: $\vec{e}_{x*}$, $\vec{e}_{y*}$, $\vec{e}_{z*}$

Sometimes a symbol, which is needed only in a very local part of the book, could be used in another meaning than it is described in the following tabular, as well as sometimes the units can differ from those given in the tabular. Symbols that occur only in a small part of the book are not listed in the tabular.

Table 1 List of symbols

Symbol	Description	Units	Page
a	Acceleration of the vehicle $a = \ddot{x}_v$	m/s^2	84
A	Aerodynamic area	m^2	27
α	Tyre slip angle (or locally used angle)	rad	177
ABS	Anti-lock braking system	–	283
A_c	Clothoid parameter	m	171
ACC	Adaptive cruise control	–	107
α_g	Angle of inclination of the road	rad	11, 29
α_{gz}	Progression ratio of a transmission $\alpha_{gz} = i_{z-1}/i_z, z = 2, \ldots, N_{z\max}$	1	48
α_j	Tyre slip angle ($j = 1$ front; $j = 2$ rear)	rad	186
$\bar{\alpha}_j$	Average tyre slip angle ($j = 1$ front; $j = 2$ rear)	rad	219
α_{ji}	Inner tyre slip angle ($j = 1$ front; $j = 2$ rear)	rad	217
α_{jo}	Outer tyre slip angle ($j = 1$ front; $j = 2$ rear)	rad	217
ASF	Active front steering	–	297
ASR	Anti-slip regulation	–	293
β	Vehicle sideslip angle	rad	170
γ	Camber angle	rad	229
c_α	Cornering stiffness	N/rad	180
$\bar{c}_\alpha$	Mean cornering stiffness	N/rad	219
c_d	Aerodynamic drag coefficient	1	27
c_{lj}	Aerodynamic lift coefficient ($j = 1$ front; $j = 2$ rear)	1	76
CPVA	Centrifugal pendulum vibration absorber	–	144
c_y	Aerodynamic crosswind coefficient	1	214
DAE	Differential algebraic equations	–	315
δ_{10}	Toe-in $\delta_{10} > 0$, toe-out $\delta_{10} < 0$	rad	229
ΔF_{zji}	Wheel load change at the inner wheel j; ($j = 1$ front; $j = 2$ rear)	N	217
ΔF_{zjo}	Wheel load change at the outer wheel j; ($j = 1$ front; $j = 2$ rear)	N	217
$(O, \vec{e}_{ix}, \vec{e}_{iy}, \vec{e}_{iz})$	Inertial reference frame	–	6
η_d	Efficiency of the differential	1	49
η_e	Efficiency of the engine	1	67
$\bar{\eta}_e$	Mean efficiency of the engine	1	63
η_t	Efficiency of the drivetrain (transmission and differential)	1	58
$\bar{\eta}_t$	Mean efficiency of the drivetrain (speed and torque converter, differential)	1	63
η_z	Efficiency of the zth gear of the transmission	1	49
ESP	Electronic stability programme	–	294

Table 2 List of symbols

Symbol	Description	Units	Page
$(S_{cp}, \vec{e}_{tx}, \vec{e}_{ty}, \vec{e}_{tz})$	Tyre reference frame	–	7
$(S_{cm}, \vec{e}_{vx}, \vec{e}_{vy}, \vec{e}_{vz})$	Vehicle reference frame	–	6
e_w	Eccentricity at the wheel	m	12
e_{wj}	Eccentricity at the wheels of axle j ($j = 1$ front; $j = 2$ rear)	m	74
$(S_{cmw}, \vec{e}_{wx}, \vec{e}_{wy}, \vec{e}_{wz})$	Wheel reference frame, $\vec{e}_{wz}$ perpendicular to the road plane	–	7
F_a	Aerodynamic drag in the longitudinal direction for simplified models in longitudinal dynamics	N	27
F_{ax}	Aerodynamic drag force in the $\vec{e}_{vx}$-direction for single track model and wheel load transfer	N	172
F_{ay}	Aerodynamic drag force in the $\vec{e}_{vy}$-direction	N	172
F_{az}	Aerodynamic lift force in the $\vec{e}_{vz}$-direction	N	76
F_{basic}	Basic demand of tractive force: $F_{\text{basic}} = F_r + F_a$	N	35
f_e	Frequency of the received signal	Hz	112
F_g	Gradient resistance	N	29
$F_{g/i}$	Combined gradient and inertial resistance for $y = p + \lambda \ddot{x}_v/g$	N	35
F_{ideal}	Ideal (demand) characteristic map of tractive force	N	37
F_i	Acceleration or inertial resistance	N	33
FMCW	Frequency modulated continuous wave	–	112
f_r	Rolling resistance coefficient	1	15
f_s	Frequency of the transmitted signal	Hz	112
f_{ri}	Coefficients for f_r approximation ($i = 0, 1, 4$)	1	16
F_r	Rolling resistance	N	15
F_{tot}	Total tractive force demand: $F_{\text{tot}} = F_a + F_i + F_g + F_r$	N	34
F_{wsz}	Force supplied at the wheel from the powertrain for gear z	N	50
F_x	Section force for tyre-road	N	13
F_{xj}	Section force for tyre-road ($j = 1$ front; $j = 2$ rear)	N	32
F_{zj}	Section force (wheel load) ($j = 1$ front; $j = 2$ rear)	N	74
$F_{zj\,\text{aero}}$	Wheel load aerodynamic portion ($j = 1$ front; $j = 2$ rear)	N	76
$F_{zj\,\text{stat}}$	Wheel load static portion ($j = 1$ front; $j = 2$ rear)	N	75

Table 3 List of symbols

Symbol	Description	Units	Page
$F_{zj\,\mathrm{dyn}}$	Wheel load dynamic portion ($j = 1$ front; $j = 2$ rear)	N	78
F_z	Wheel or axle load in the $\vec{e}_{wz}$-direction	N	12
g	Gravitational acceleration	m/s^2	13
G_{aj}	Weight of the axle j ($j = 1$ front; $j = 2$ rear)	N	75
G_b	Weight of body (sprung mass)	N	75
h	Distance: centre of mass S_{cm} – road	m	86
h_b	Distance: centre of mass of the body – road	m	75
h_{cm}	Distance: centre of mass S_{cm} – road	m	77
HP	Pump	–	297
h_{pp}	Distance: Centre of pressure S_{pp} – road for air flow in the $\vec{e}_{vx}$-direction	m	76
HSV	High-pressure selector valve	–	297
i_d	Transmission ratio of the differential (final drive)	1	44
i_g	Transmission ratio of the gearbox; for a stepped transmission: $i_g = i_z$	1	30
i_t	Total transmission ratio $i_t = i_z i_d$	1	58
i_z	Transmission ratio of gear z of the transmission, $z = 1, \ldots, N_{z\,\max}$	1	48
J_{aj}	Moment of inertia of the axle j	kg m^2	30
J_c	Moment of inertia of gear, differential, Cardan shaft	kg m^2	30
J_e	Moment of inertia of engine, clutch	kg m^2	30
J_z	Moment of inertia of the vehicle with respect to the $\vec{e}_z$-axis	kg m^2	172
κ	Angle of rotation of the body of the vehicle	rad	222
κ_{cc}	Instantaneous curvature ($\kappa_{\mathrm{cc}} = 1/\rho_{\mathrm{cc}}$) of the vehicle path	1/m	171
κ_w	Wavenumber of an uneven road	rad/m	164
ℓ	Wheelbase; distance between front and rear axle	m	75
ℓ_1	Distance in the $\vec{e}_{vx}$ direction between front axle centre of mass and centre of mass S_{cm} of the vehicle	m	75
ℓ_2	Distance in the $\vec{e}_{vx}$ direction between rear axle centre of mass and centre of mass S_{cm} of the vehicle	m	75
λ	Rotational mass factor	1	33
λ_e	Eigenvalue with respect to time	1/s	207
ℓ_{cm}	Distance: centre of gravity S_{cm} – centre of pressure S_{pp} in the $\vec{e}_{vx}$ direction	m	172

Table 4 List of symbols

Symbol	Description	Units	Page
$\mu = \mu(S)$	Tyre longitudinal force coefficient	1	21
M_a	Aerodynamic moment	Nm	74
μ_a	Coefficient of adhesion	1	21
m_{aj}	Mass of the axle ($j = 1$ front; $j = 2$ rear)	kg	32
M_{aj}	Section moment at the axle j ($j = 1$ front; $j = 2$ rear)	Nm	32
m_b	Mass of the body or sprung mass of the vehicle	kg	31
MBS	Multi-body systems	–	4
M_{cc}	Centre of curvature	–	170
M_{cr}	Instantaneous centre of rotation	–	173
M_e	Torque supplied from the engine	N	58
M_{100}	Full load moment of the engine	Nm	37
M_i	Input moment (e.g. at input of transmission or clutch)	Nm	45
$M(P_{\max})$	Moment where the power of the engines reaches a maximum	Nm	37
M_l	Torque loss from the engine	N	59
$M_{\max}$	Maximum torque of the engine	Nm	55
$M(n_{\max})$	Full load moment of the engine at $n_{\max}$	Nm	37
$M(n_{\min})$	Full load moment of the engine at $n_{\min}$	Nm	37
M_o	Output moment (e.g. at input of transmission or clutch)	Nm	45
μ_s	Coefficient of pure sliding	1	22
m_{tot}	Total mass (sprung and unsprung mass)	kg	29
M_{ws}	Torque supplied at the wheel from the powertrain	N	58
n_c	Total caster trail $n_c = n_{kc} + n_{tc}$	m	188
n_e	Engine speed (revolutions)	rad/s	44
n_i	Input speed or revolutions (e.g. at input of transmission or clutch)	rev/s	45
n_{iz}	Input speed (revolutions) of transmission at gear z	rad/s	48
n_{kc}	Kinematic caster trail	m	182
$n_{\max}$	Maximum speed of the engine	rpm	37
$n_{\min}$	Minimum speed of the engine	rpm	37
n_o	Output speed or revolutions (e.g. at input of transmission or clutch)	rev/s	45
$n(P_{\max})$	Engine speed where the power of the engines reaches a maximum	rpm	37
$n_{w\,\max}$	Maximum revolutions per minute of the wheel	rpm	44
n_{oz}	Output speed (revolutions) of transmission at gear z	rad/s	48

Table 5 List of symbols

Symbol	Description	Units	Page
n_{tc}	Tyre caster trail	m	177
n_w	Wheel speed (revolutions)	rad/s	44
$N_{z\,\max}$	Number of gears in a transmission	1	48
ODE	Ordinary differential equation	–	5
OEM	Original equipment manufacturer	–	121
ω_i	Input angular velocities (e.g. at input of transmission or clutch)	rad/s	45
ω_o	Output angular velocities (e.g. at input of transmission or clutch)	rad/s	45
p	Gradient (inclination) of a road $p = \tan\alpha_g$	1	29
ψ	Yaw angle	rad	7
φ	Roll angle	rad	7
P_{100}	Full load power of the engine	W = Nm/s	37
P_a	Power of aerodynamic drag force ($S = 0$: $P_a = F_a v_v$)	W = Nm/s	35
φ_{aj}	Angle of rotation of the axle j ($j = 1$ front; $j = 2$ rear)	rad	32
φ_b	Pitch angle of the body	rad	159
P_{basic}	Basic demand of power: $P_{\mathrm{basic}} = P_r + P_a$	N	35
$\dot{\varphi}_c$	Angular velocity gear, differential, Cardan shaft	rad/s	30
P_e	Power supplied from the engine	W = Nm/s	46
$\dot{\varphi}_e$	Angular velocity engine, clutch	rad/s	30
$P_{g/i}$	Power of combined gradient and inertial resistance	W = Nm/s	35
P_g	Power of gradient resistance	W = Nm/s	35
Φ_h	Spectral density of the stochastic road surface irregularity	m³	131
$\Phi_h(\Omega_0)$	Coefficient of roughness	m³	131
P_{basic}	Total power demand: $P_{\mathrm{basic}} = P_r + P_a$	N	34
P_{ideal}	Ideal (demand) characteristic map of power at the wheel	N	37
P_i	Input power (e.g. at input of transmission) in Chapter 4	W = Nm/s	45
P_i	Power of inertia forces in Chapter 3	W = Nm/s	35
$P_{\max}$	Maximum power of the engine reaches a maximum	W = Nm/s	37
P_o	Output power (e.g. at input of transmission)	W = Nm/s	45
P_r	Power of rolling resistance ($S = 0$: $P_r = F_r v_v$)	W = Nm/s	34
P_{tot}	Total power demand: $P_{\mathrm{tot}} = P_r + P_a + P_g + P_i$	N	34
φ_w	Rotational angle of the wheel w.r.t the $\vec{e}_{wy}$-axis	rad	14

Table 6 List of symbols

Symbol	Description	Units	Page
P_{wsi}	Power supplied at the wheel from the powertrain for gear i	$W = Nm/s$	50
P_w	Power at the wheel	$W = Nm/s$	63
ρ_a	Mass density of air	kg/m^3	27
ρ_{cc}	Instantaneous radius of curvature of the vehicle path	m	171
r_k	Scrub radius: distance between the intersection of the steering axis with road and the centre of the contact patch	m	183
r_σ	Kingpin offset between the wheel centre and the steering axis	m	184
R_{w0}	Dynamic rolling radius	m	20
R_{w0j}	Dynamic rolling radius ($j = 1$ front; $j = 2$ rear)	m	32
r_{wst}	Static radius of a wheel	m	14
r_{wstj}	Static radius of the wheels of the axle j ($j = 1$ front; $j = 2$ rear)	m	32
σ	Inclination angle of the steering axis; angle from the $\vec{e}_{Vz}$ direction to the projection of the steering axis on to the $\vec{e}_{Vz}$-$\vec{e}_{Vy}$-plane	rad	181
S_j	Slip at wheels of the axle j ($j = 1$ front; $j = 2$ rear)	1	32
s_j	Track of the axle j ($j = 1$ front; $j = 2$ rear)	m	219
S_{cm}	Centre of mass of the vehicle (sprung and unsprung mass)	–	6
S_{cmw}	Centre of mass of a wheel	–	7
SOV	Switch over valve	–	297
S_{pp}	Centre of pressure	–	76
S_{cp}	Centre point of the contact patch	–	7
σ_n	Normal stress distribution in the contact patch	N/m^2	12
SSF	Static stability factor	–	220
τ	Caster angle; angle from the $\vec{e}_{vz}$ direction to the projection of the steering axis on to the $\vec{e}_{vz}$-$\vec{e}_{vx}$-plane	rad	181
t_b	Pressure build-up time	s	80
t_{fb}	Foot force build-up time	s	80
t_f	Duration of full braking process	s	82
t_r	Reaction time	s	79
t_t	Transmission time	s	80
ϑ	Pitch angle	rad	7
v	Absolute value of $\vec{v}_v$: $v = |\vec{v}_v| = v_v$	m/s	27
v_a	Velocity of wind	m/s	27
$\vec{v}_a$	Velocity vector of wind	m/s	214

Table 7   List of symbols

Symbol	Description	Units	Page
v_c	Circumferential velocity of wheel	m/s	20
v_{ch}^2	Characteristic velocity squared	m^2/s^2	197
v_{crit}	Critical velocity	m/s	209
$\vec{v}_r$	Resultant velocity vector (wind and vehicle)	m/s	214
$\vec{v}_v$	Velocity of the vehicle at S_{cm}; $$v_v = |\vec{v}_v|; \vec{v} = \vec{v}_v$$ $$\vec{v} = (v_{vx}, v_{vy}, v_{vz}) \cdot (\vec{e}_{ix}, \vec{e}_{iy}, \vec{e}_{iz})^T$$	m/s	214
v_w	Velocity of wheel	m/s	179
w	Waviness of an uneven road	1	131
W_w	Work at the wheel	$J = Nm$	63
X	Section force between wheel and body	N	13
X_j	Section force axle-vehicle at the axle j ($j = 1$ front; $j = 2$ rear)	N	33
$\ddot{x}_v$	Acceleration of the vehicle	m/s^2	30
x_{aj}	Coordinate of the axle j ($j = 1$ front; $j = 2$ rear)	m	32
x_v, y_v, z_v	S_{cm} vehicle coordinates w.r.t. $(O, \vec{e}_{ix}, \vec{e}_{iy}, \vec{e}_{iz})$	m	6
x_w, y_w, z_w	S_{cmw} wheel coordinates w.r.t. $(S_{cm}, \vec{e}_{vx}, \vec{e}_{vy}, \vec{e}_{vz})$	m	14
Y	Section force between wheel and body	N	178
Z	Section force between wheel and body	N	13
$\mathcal{Z}$	Braking ration $\mathcal{Z} = -a/g$	1	84
z_1	Displacement of the wheel	m	264
z_2	Displacement of the body	m	264
z_3	Displacement of the seat	m	136
z_b	Displacement of the body	m	156
z_w	Displacement of the wheel	m	156

1

Introduction

Automobiles have been used for over 100 years for the transportation of people and goods. Despite this long period, essential elements of an automobile have in principle remained the same, i.e. four wheels and an internal combustion engine with a torque converter drive. However, the technical details of an automobile have changed a great deal, and the complexity has increased substantially. This has partly gone hand in hand with general technical progress, on the one hand, and increasing customer demands, on the other. Legal requirements have also led to distinct changes in automobiles.

The importance of automobiles becomes evident when we look at the graphs in Figures 1.1–1.4. You should bear in mind that the abscissas of most graphs are partitioned logarithmically. The quantity, the distances travelled and the distances travelled per capita are at a very high level, or these values are increasing at a high rate. If we look at some European countries or the United States of America, we can recognize stagnation at a high level, whereas emerging economies exhibit high rates of growth. The need to develop new, economic and ecological vehicles is evident. In order to do this, engineers should be familiar with the basic properties of automobiles. As the automobile is something which moves and which not only moves forward at a constant velocity, but also dynamic behaviour depends on these basic properties. Consequently, the basic dynamic properties form the main topic of this book.

The ecological aspect could be a dramatic limiting factor in the development of vehicles throughout the world. If the number of cars per 1000 inhabitants in China and Hong Kong grows from 22 in the year 2007 to 816, which is the number in the USA, then this represents a factor of 40. If we now multiply the CO_2 emissions of the USA from the year 2007 by 40, we obtain around 57 000 Mt, which is 12 times the world CO_2 emissions from fuel combustion in road transport for the year 2007. This seems to be very high (or perhaps too high), and vehicles with lower fuel consumption or hybrid or electric powertrains will have to be developed and improved in the coming decades.

Vehicle Dynamics, First Edition. Martin Meywerk.
© 2015 John Wiley & Sons, Ltd. Published 2015 by John Wiley & Sons, Ltd.
Companion Website: www.wiley.com/go/meywerk/vehicle

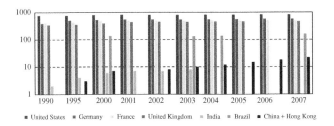

Figure 1.1 Passenger cars (and light trucks in US) per 1000 inhabitants (data from OECD 2014)

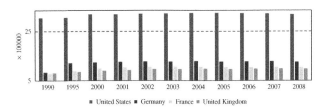

Figure 1.2 Road passenger km (million pkm) (data from OECD 2014)

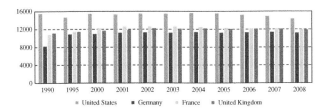

Figure 1.3 Passenger km/capita (data from OECD 2014)

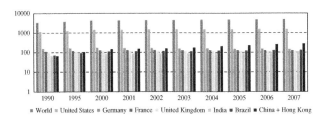

Figure 1.4 International Energy Agency (IEA) CO_2 from fuel combustion (Mt) in Road Transport (data from OECD 2014)

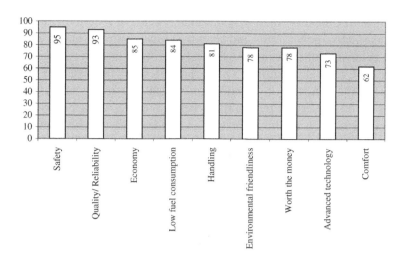

Figure 1.5 Importance of purchase criteria (Braess and Seiffert 2001)

The presentation of the most important buying criteria in Figure 1.5 highlights the ecological and economic aspects as well as safety, handling behaviour and comfort. The last three points, namely safety, handling behaviour and comfort, are strongly linked with the driving dynamics and suspension, making these aspects of particular importance in the automotive industry. Safety is generally subdivided into active safety (active safety systems help to avoid accidents) and passive safety (passive safety systems protect the occupants during an accident).

It is evident that the dynamics of the vehicle is of crucial importance because of the impact on active safety; handling behaviour and comfort are also closely associated with the properties of vehicle dynamics. For this reason, particular emphasis is placed on the aspect of dynamics in this course.

The aim of this course is to define and identify the basic concepts and relationships that are necessary for understanding the dynamics of a motor vehicle.

The content of this textbook is limited to the essentials, and the course closely follows the monograph of Mitschke and Wallentowitz 2004 (German). Further recommended reading can be found in the bibliography at the end of this book, e.g. Heissing and Ersoy 2011, Dukkipati et al. 2008, Gillespie 1992, Jazar 2014, or Reimpell et al. 2001.

1.1 Introductory Remarks

The content of this book is divided into four parts: longitudinal dynamics, vertical dynamics, lateral dynamics and structural design of vehicle components and automotive mechatronic systems. Longitudinal dynamics is included in Chapters 2–6, which discuss the process of acceleration and braking. Key importance here is given to the total running resistance, the demand and supply of power and the driving state diagrams. In Chapters 7 and 8, additional systems of longitudinal dynamics are described: alternative powertrains and adaptive cruise control systems. In Chapters 9 and 10, the behaviour of the vehicle when driving on an uneven surface is explained in the context of vertical dynamics. These chapters study the basics of the theory of oscillations and the influence of vibrations on humans. Lateral dynamics, the contents of Chapters 11–15, describes the handling behaviour of a vehicle during cornering. Important concepts such as slip, oversteer and understeer, toe and camber angle are explained. It deals with the influence of wheel load on the handling behaviour.

Chapters 16–19 highlight the engineering design (structural) aspects of an automobile. In addition to speed and torque converters, they also discuss brakes and chassis elements of active safety systems, such as anti-lock braking system (ABS), anti-slip regulation (ASR) and electronic stability programme (ESP). In Chapter 20, multi-body systems (MBS) are explained. MBS are computational models which allow more precise calculations of the dynamic behaviour of vehicles.

1.2   Motion of the Vehicle

To describe the dynamics of motor vehicles, we use, as in any other branch of engineering, models with a greater or lesser degree of detail. The complexity of the models depends on the questions under investigation. Today the MBS models are most commonly used in both science and research as well as in the development departments of the automotive industry. Multi-body systems consist of one or more rigid bodies which are interconnected by springs and/or shock absorbers and joints.

Figure 1.6 shows an MBS model of a vehicle. This model is taken from the commercial MBS programme ADAMS. Another example of a McPherson front axle is shown in Figure 1.7. These MBS models allow high accuracy in the simulation of dynamic behaviour. A lot of details can be incorporated into these models, even flexible parts can be considered. However, the detailed simulation yields a large number of effects in the calculated results and the engineer has to interpret and understand these results. As an example, an engineer has to distinguish between main effects and numerical phenomena. For this purpose, it is helpful to understand the basic dynamics and to know simple models for calculating the behaviour of a vehicle in order to interpret or even to check the MBS results. This book therefore takes vehicle dynamic behaviour and simple models as its main topics.

In a simplified view of the motor vehicle, a model could consist of five rigid bodies: the four wheels and the body structure. These are interconnected by springs, shock absorbers and rigid body suspensions with joints. A rigid body has six degrees of freedom. This simple model would therefore have $5 \times 6 = 30$ degrees of freedom[1].

---

[1] We may argue that the suspension between wheel carrier and body of the vehicle locks five degrees of freedom, the wheel bearing will unlock one degree of freedom, which results all together in only two degrees of freedom for one wheel. The sum for the whole vehicle will then be 14. That is correct under the assumption that there are no compliances in the suspension. Since modern cars have these compliances, the number of 30 is correct.

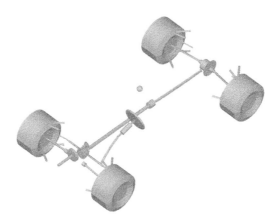

Figure 1.6 MBS model of a four-wheel-drive vehicle (example of the MBS programme ADAMS)

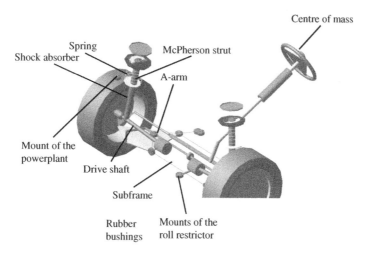

Figure 1.7 McPherson front axle with driven wheels (from MBS programme ADAMS)

It is evident that a description of even this simple model with 30 degrees of freedom will require 30 equations of motion of second order (with respect to time).

Equations of motion are ordinary differential equations that describe the motion of (rigid) bodies. A simple example of the equation of motion for a single mass oscillator is given below (mass m, stiffness of spring k, displacement z):

$$m\ddot{z} + kz = 0 \ . \tag{1.1}$$

Equations of motion are often second-order differential equations with respect to time. For specific problems, these models are therefore reduced to a few masses with limited motion options. Hence, we in turn limit ourselves to certain specific questions. This approach will be used in this book. For this reason, we first introduce the terminology and coordinates to describe the possible motions of a vehicle.

Frame system: A quadruple $(A, \vec{e}_x, \vec{e}_y, \vec{e}_z)$ is a frame system of an affine space. Here, A is a point (the origin) and $\vec{e}_x, \vec{e}_y, \vec{e}_z$ is a Cartesian tripod (the axis system). To describe the position of a point P with respect to A, three coordinates x, y, z are sufficient:

$$\overrightarrow{AP} = x\vec{e}_x + y\vec{e}_y + z\vec{e}_z \; . \tag{1.2}$$

The point A can be defined as fixed in space (or in an inertial frame). This is called an inertial frame system (sometimes called earth or world coordinate system). If the point A and the tripod $\vec{e}_x, \vec{e}_y, \vec{e}_z$ are fixed to a body and continues to be firmly connected to the body then the result is called a body-fixed frame system.

We introduce several frames. The first one is an inertial frame $(O, \vec{e}_{ix}, \vec{e}_{iy}, \vec{e}_{iz})$ which is fixed to the earth (or the world)[2]. To describe the motion of a point in this inertial frame, three Cartesian coordinates x, y, z are necessary, in the case of the centre of mass, S_{cm}, of the vehicle we introduce x_v, y_v, z_v. This point S_{cm} is the origin for two other, fixed body frame systems for the vehicle:

1. $(S_{\mathrm{cm}}, \vec{e}_{vx}, \vec{e}_{vy}, \vec{e}_{vz})$: vehicle frame system
2. $(S_{\mathrm{cm}}, \vec{e}_x, \vec{e}_y, \vec{e}_z)$: intermediate frame system

The first one is completely fixed to the body of the vehicle, i.e. all three vectors $\vec{e}_{vx}$, $\vec{e}_{vy}, \vec{e}_{vz}$ move together with the vehicle. The origin of the second is also fixed to the vehicle. To define the intermediate frame system, we assume only a rotation about the $\vec{e}_{iz}$ direction, this means that $\vec{e}_{iz} = \vec{e}_z$. Then the vector $\vec{e}_x$ is the vector $\vec{e}_{ix}$ rotated by the so-called yaw angle, ψ, about the $\vec{e}_{iz}$ direction. The vector $\vec{e}_y$ is oriented to the left side of the vehicle, perpendicular to $\vec{e}_x$ and parallel to the $\vec{e}_{ix} - \vec{e}_{iy}$ plane. The vector $\vec{e}_z = \vec{e}_x \times \vec{e}_y$ is the vector or cross product[3].

In order to define the orientation of the vehicle and the orientation of the axis system, $\vec{e}_{vx}, \vec{e}_{vy}, \vec{e}_{vz}$, with respect to the inertial axis system, $\vec{e}_{ix}, \vec{e}_{iy}, \vec{e}_{iz}$, three angles are necessary. There are different ways to use these three angles: here we use the Euler (see footnote) convention. This means that we first rotate about the $\vec{e}_{iz}$-axis; the angle for this first rotation is the yaw angle, ψ. After this, we rotate about the new $\vec{e}_{iy}'$-axis (which is the rotated $\vec{e}_{iy}$-axis from the first rotation); the angle for this second rotation is the pitch angle, ϑ. The third rotation is about the new $\vec{e}_{ix}''$-axis. The $\vec{e}_{ix}''$-axis is the

---

[2] In some MBS software tools this coordinate system is called world system. Strictly speaking, an earth frame system, i.e. a coordinate system which is fixed to the earth, is not an inertial system due to the rotation of the earth. These aspects are usually neglected, as is the case here as well.

[3] If the coordinates of two vectors with respect to an orthonormal basis are (x_1, y_1, z_1) and (x_2, y_2, z_2), then the vector product can be calculated by $(y_1 z_2 - y_2 z_1, -(x_1 z_2 - x_2 z_1), x_1 y_2 - x_2 y_1)$.

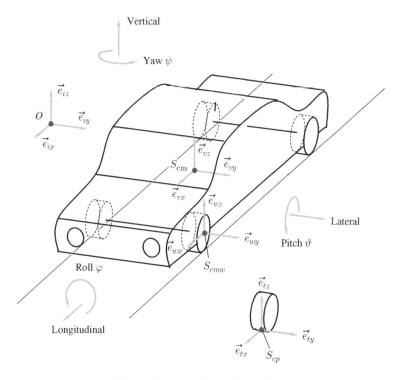

Figure 1.8 Motion of a vehicle

result from the $\vec{e}_{ix}$-axis due to the two rotations with the angles ψ and ϑ. The angle of the third rotation is the roll angle[4] φ.

Some more frame systems are necessary to describe the motion of the vehicle. Figure 1.8 depicts a frame system, $(S_{cmw}, \vec{e}_{wx}, \vec{e}_{wy}, \vec{e}_{wz})$, fixed to the wheel at its centre of mass and an additional system at the contact patch, $(S_{cp}, \vec{e}_{tx}, \vec{e}_{ty}, \vec{e}_{tz})$.

Figure 1.8 shows the frame systems and the angles ψ, ϑ and φ. The angles are not depicted as angles of a sequence of rotations, but as angles of single rotation. This simplification is made in several considerations of this book. For most of them, it is sufficient to look at a single rotation and neglect the interaction of rotation. If the interactions of the rotations are to be investigated, the complexity of the equations will increase significantly. Simple, analytical results are not available for these investigations and the motion of the vehicle should be modelled by MBS.

---

[4] These three angles are called the Tait–Bryan angles in the literature; a characteristic feature is that every axis (with index x, y and z) occurs in the sequence of rotational axes. In German literature, these angles are sometimes called Cardan angles. Another possible definition of the orientation is the use of so-called Euler angles. In this definition, the first axis of rotation is, for example, the $\vec{e}_{ix}$-axis, the second is the $\vec{e}_{iz}''$-axis and the third, again about an x-axis, i.e. the $\vec{e}_{ix}''$-axis. In some MBS software as well as in ISO 8855 2011 we find the name Euler associated with the definition of Tait–Bryan angles. Consequently, you should read the exact definition of the sequence of rotations carefully and you should not simply assume that a particular convention applies.

When the vehicle is driving in a straight line, the $\vec{e}_{vx}$ and $\vec{e}_{ix}$ directions coincide. The first part of this course is limited to the straight-line motion of a vehicle (longitudinal dynamics) and considers resistances, driving performance and braking and acceleration processes. In this aspect of longitudinal dynamics, rotation of the vehicle always occurs about the $\vec{e}_{iy}$-axis. As mentioned above, this rotational motion about the $\vec{e}_{iy}$-axis is called pitch. Hence pitch and straight-line, forward motion are connected: as the centre of mass, S_{cm}, is above the road, every acceleration or braking manoeuvre causes inertia forces to act on S_{cm}, which yields a moment and therefore a pitch motion.

The second class of movements is caused by uneven roads. It is grouped together under the concept of vehicle vibrations. The movements are translations of the vehicle in the $\vec{e}_{iz}$ direction (bounce), rotation about $\vec{e}_{iy}$ direction (pitch) and the $\vec{e}_{ix}$ direction (roll).

In cornering, i.e. a non-constant yaw angle and in general the $\vec{e}_{ix}$ direction do not coincide with the $\vec{e}_{vx}$ direction, the vehicle, in addition to rotating about the $\vec{e}_{iz}$-axis, rotates about the $\vec{e}_{vx}$-axis (roll) and for deceleration or acceleration it rotates about the $\vec{e}_{vy}$-axis (pitch). A lateral motion also occurs. Cornering is investigated in the third part of this book, the lateral dynamics or cornering part.

These short considerations show that nearly always more than one degree of freedom is involved in the motion of the vehicle.

1.3 Questions and Exercises

Remembering

1. What kinds of models often describe the dynamics of vehicles?
2. How many degrees of freedom does the body of a vehicle have?
3. What are the names of the six degrees of freedom of the body associated with movements?
4. The dynamics of motor vehicles is usually divided into three main forms of movement. What are they?
5. What form of movement plays an important role in longitudinal dynamics?
6. What form of movement plays an important role in vertical dynamics?
7. What form of movement plays an important role in lateral dynamics?

Understanding

1. Which degrees of freedom of a vehicle are involved when passing a speed bump (same height for left and right wheels)?
2. Which degrees of freedom of a vehicle are involved when passing a pothole (at one side of the vehicle only)?

Applying

1. Which effects are the same when comparing a handcart with nearly stiff wheels which are fixed to the body without suspension and a vehicle with a centre of mass at the height of the road during acceleration or braking? Consider inertia forces and resulting moments.
2. Which effects are the same when comparing a handcart with nearly stiff wheels which are fixed to the body without suspension and a vehicle with a centre of mass at the height of the road during cornering? Consider centrifugal forces.

2

The Wheel

The wheels are the link between the road and the vehicle structure. For this reason, they play a central role in the dynamics of a vehicle. This chapter derives the equations of motion for the wheels. Furthermore, it also explains wheel resistances (rolling resistance, rolling resistance on a wet road, bearing resistance, toe resistance and curve resistance).

2.1 Equations of Motion of the Wheel

The equations of motion of a wheel which are derived here are only for the motion in the $\vec{e}_{wx}$–$\vec{e}_{wz}$ plane. The translational motion of the wheel is described by the motion of its centre of mass, S_{cmw}, by the coordinates x_w and z_w (see Figure 2.1), the rotation about the $\vec{e}_{wy}$-axis is described by φ_w. The distance between the wheel's centre of mass, S_{cmw}, and the road is given by r_{wst}. It is important that the radius r_{wst} should not be the radius of the undeformed and unloaded wheel. We call this radius, r_{wst}, the static radius of the wheel. The wheel rolls on an inclined plane, the angle of inclination is α_g (index 'g' for grading). In order to establish the equations of motion, the wheel is cut free from both the inclined plane and the wheel hub (or the bearing). The tyre does not touch the plane only at one point but, instead, at a contact surface (contact patch).

Contact patch: This is the contact area where the tyre and the road are in contact. The size of the contact patch depends on the geometry and design of the tyre, the internal pressure and the wheel load. For a passenger car tyre, it has the magnitude of a postcard. (In comparison, the contact area of a railway wheel with the track is the size of a thumb nail).

If the wheel does not move, the normal stress distribution in the contact patch is symmetrical with respect to the $\vec{e}_{wy}$–$\vec{e}_{wz}$ plane and the tangential force distribution

Vehicle Dynamics, First Edition. Martin Meywerk.
© 2015 John Wiley & Sons, Ltd. Published 2015 by John Wiley & Sons, Ltd.
Companion Website: www.wiley.com/go/meywerk/vehicle

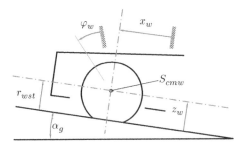

Figure 2.1   Coordinates of the wheel

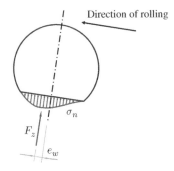

Figure 2.2   Normal stresses in the contact patch of the tyre

(tangential to the flat road) is almost zero. However, when the wheel rolls, the normal force distribution is no longer symmetrical. Figure 2.2 shows the basic trend of the normal stress distribution. The wheel rolls up the plane. It can be seen that the normal stress distribution is asymmetrical and that the line of action of the resultant force, F_z, (which is obtained by integrating the normal stress distribution over the contact area) is shifted by the eccentricity, e_w. Considering the moment due to the force, F_z, and the eccentricity, e_w, we recognize that it counteracts the rolling movements of the wheel.

The occurrence of the asymmetric normal stress distribution can be illustrated with the help of Figure 2.3, which shows a wheel in motion at three different points in time (the wheel rolls from right to left). To illustrate the elastic and damping properties, the wheel may be imagined as a flexible ring which is supported by spring-damper elements against the rim. Only one single spring-damper element is shown in Figure 2.3. This spring-damper element is considered for a sequence of three points in time t_1, t_2 and t_3. The other spring-damper elements have to be imagined as arranged radially around the circumference. The wheel and the (imaginary) unloaded spring-damper element are visible in the left section of Figure 2.3 for $t = t_1$.

In the middle section of Figure 2.3 for $t = t_2$, the tyre and therefore the spring and the damping elements are compressed. The force that is necessary depends firstly on the compression Δs (from the spring) and secondly on the compression rate $\Delta \dot{s}$

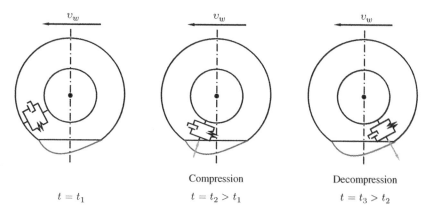

Figure 2.3 Illustration of the asymmetric normal stress distribution in the contact patch

(from the damper). At this particular time, both compression and compression rate are positive: $\Delta s > 0$, $\Delta \dot{s} > 0$. Consequently, the resulting force is

$$F = \underbrace{k\Delta s}_{>0} + \underbrace{b\Delta \dot{s}}_{>0} .$$ (2.1)

In the right-hand section of Figure 2.3 for $t = t_3$, the damping element under consideration is relaxed again. And, of course, there are normal forces acting on the tyre as compressive forces, the resultant external force is the sum of the spring force diminished by the force which is necessary to relax the damper, i.e. $\Delta s > 0$, $\Delta \dot{s} < 0$. This leads to a reduction in normal forces in this area since

$$F = \underbrace{k\Delta s}_{>0} + \underbrace{b\Delta \dot{s}}_{<0} .$$ (2.2)

All things considered, the normal stresses in the front part of the contact patch are larger than in the rear part.

The decelerating torque from the rolling resistance is illustrated below with reference to the equations of motion of the wheel. For this purpose, we cut the wheel free from the structure (section forces X and Z; section torque M_w) and the road (section forces F_x and F_z). The free-body diagram is shown in Figure 2.4. The torque, M_w, represents a driving or braking torque. The normal stress (normal to the road) and tangential stress (tangential to the road) distributions in the contact area are summarized in the free-body diagram by the resultant forces F_z and F_x, respectively. In addition, the moment, M_w, acts at the centre of the wheel hub and the section forces X and Z, too. Moreover, the weight force, $G_w = m_w g$ (g is the acceleration due to gravity) acts on the centre of mass[1] of the wheel. In a further simplification, it is assumed that

---

[1] Strictly speaking, the centre of gravity is the correct point, but centre of mass and centre of gravity differ only very slightly for such small objects as vehicles.

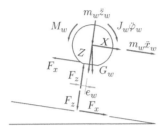

Figure 2.4 Free-body diagram of the wheel

the weight force acts at the wheel centre and also that the wheel centre is equal to the centre of mass. The d'Alembert forces of inertia, $m_w \ddot{x}_w$ and $m_w \ddot{z}_w$, act on the centre of mass. Furthermore, the moment of inertia, $J_w \ddot{\varphi}_w$, completes the free-body diagram. We determine the sum of forces in the $\vec{e}_{wx}$-direction and set this sum to zero. By rearranging, we then obtain the following equation of motion:

$$m_w \ddot{x}_w = F_x - X - G_w \sin \alpha_g. \tag{2.3}$$

Similarly, we obtain the sum of the forces in the $\vec{e}_{wz}$-direction:

$$m_w \ddot{z}_w = F_z - Z - G_w \cos \alpha_g. \tag{2.4}$$

From the condition that the sum of the moments disappears, we obtain the sum with respect to the centre of mass (here by rearranging the equation):

$$J_w \ddot{\varphi}_w = M_w - F_x r_{wst} - F_z e_w \, . \tag{2.5}$$

2.2   Wheel Resistances

Rolling of the wheel creates several wheel resistances (these forces act against the rolling direction). In this section, we consider the rolling resistance in more detail, other resistances such as the resistance from a wet road, from friction in the wheel bearing or from toe-in or toe-out are outlined roughly.

2.2.1   Rolling Resistance

We first give the definition, and the derivation of the formula follows from the definition.

Rolling resistance: If a wheel is rolling on a road, an asymmetric normal stress distribution occurs between road and wheel in the contact patch (Figure 2.2). The line of action of the resultant force, F_z, of the asymmetric normal stress distribution does not intersect the centre of the wheel, but is shifted in the rolling direction. The distance between the wheel centre and the line of action of F_z is the eccentricity, e_w. This results in a moment $M = e_w F_z$. To overcome this moment, a tractive torque, M_w, in the case of a driven wheel or a tractive force, F_r, in the case of a towed wheel is necessary. This force, F_r, is called the rolling resistance. It can be derived by solving the sum of moment $0 = r_{wst} F_r - e_w F_z$ for F_r:

$$F_r = \frac{e_w}{r_{wst}} F_z .$$ (2.6)

In the case of a driven wheel, the rolling resistance is

$$F_r = \frac{M_w}{r_{wst}} .$$ (2.7)

Starting from the equation of motion (2.3) of the wheel in the $\vec{e}_{wx}$ direction, we derive the relationship for the force F_r. We assume that the wheel rolls in a steady state on a non-inclined roadway (hence $\alpha_g = 0$). Steady state means that the velocity, $\dot{x}_w$, is constant. Hence,

$$\ddot{x}_w = 0 .$$ (2.8)

Taking the results from Equation (2.3) and Equation (2.8) we obtain

$$F_x = X .$$ (2.9)

From the moment equation (2.5), we obtain (the wheel is neither driven nor braked; $\ddot{\varphi}_w = 0$ applies here because of the steady rolling):

$$0 = -r_{wst} F_x - e_w F_z .$$ (2.10)

With $F_x = X$ from Equation (2.9) and from Equation (2.10) we get

$$X = -\frac{e_w}{r_{wst}} F_z .$$ (2.11)

We call $F_r = -X$

$$F_r = \frac{e_w}{r_{wst}} F_z$$ (2.12)

the rolling resistance.

Comparing this result with the free-body diagram (Figure 2.4), we see that the tractive force, X, must act in the rolling direction to overcome the rolling resistance. The dimensionless factor e_w / r_{wst} is called the coefficient of rolling resistance, f_r,

$$f_r = \frac{e_w}{r_{wst}} .$$ (2.13)

Coefficient of rolling resistance: The coefficient of rolling resistance f_r is the ratio of rolling resistance, F_r, to the resulting normal force, F_z, in the contact patch:

$$f_r = \frac{F_r}{F_z} .$$

The rolling resistance may be approximated by means of the following empirical formula (depending on the speed) ($v_0 = 100\,\text{km/h}$)

$$f_r(v) = \tilde{f}_{r0} + \tilde{f}_{r1}\frac{v}{v_0} + \tilde{f}_{r4}\left(\frac{v}{v_0}\right)^4 . \tag{2.14}$$

Remark 2.1 It can be seen that a quadratic term is missing from this formula. This member is omitted because it is negligibly small when compared with the air resistance increasing quadratically with speed. Furthermore, the term v^4 exceeds the quadratic member.

Remark 2.2 The values f_r are in the range 0.005–0.015. The coefficients $\tilde{f}_{r0}$, $\tilde{f}_{r1}$ and $\tilde{f}_{r4}$ depend, amongst other factors, on the type of tyre and the inflation pressure. Average values for HR tyres are $\tilde{f}_{r0} = 9.0 \times 10^{-3}$, $\tilde{f}_{r1} = 2.0 \times 10^{-3}$, $\tilde{f}_{r4} = 3.0 \times 10^{-4}$.

Remark 2.3 The rolling resistance defined here is based on the effect of asymmetric normal forces and dissipative effects in the material of the tyre. A further resistance on the wheel is based on slip in the contact patch.

Remark 2.4 The coefficient of rolling resistance falls with an increase in pressure and also with an increase in the wheel load. The dependency of the coefficient f_r on the wheel load means that the rolling resistance, F_r, is not linearly dependent on the wheel load, F_z.

Remark 2.5 The coefficients in the empirical formula (2.14) of the coefficient of rolling resistance as a function of the speed depend, amongst other factors, on the internal pressure. The coefficients $\tilde{f}_{r0}$ and $\tilde{f}_{r4}$ tend to decrease with increasing inflation pressure whereas $\tilde{f}_{r1}$ increases.

Figures 2.5, 2.6 and 2.7 show rolling resistance coefficients for tyres of different dimensions (namely width) for summer tyres, winter tyres and all-season tyres, respectively. The basis for the diagrams was an investigation by TÜV Süd on behalf of the German Federal Environmental Agency in the year 2002 Reithmaier and Salzinger 2002. These studies investigated tyres (number n) of different sizes from several different manufacturers.

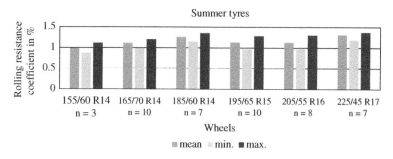

Figure 2.5 Rolling resistance coefficient in percentage for summer tyres (data from Reithmaier and Salzinger 2002)

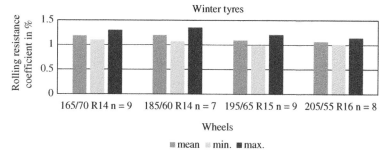

Figure 2.6 Rolling resistance coefficient in percentage for winter tyres (data from Reithmaier and Salzinger 2002)

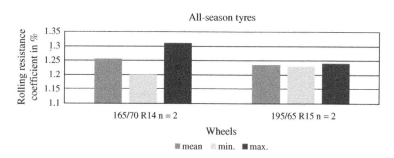

Figure 2.7 Rolling resistance coefficient in percentage for all-season tyres (data from Reithmaier and Salzinger 2002)

2.2.2   Aquaplaning

Another wheel resistance, F_{aq}, is due to water on the roadway. The force F_{aq} depends on the displaced volume of water per unit of time. It is roughly proportional to the tyre width, b, and a power of the velocity

$$F_{aq} \approx b v^{n_{aq}} \ . \tag{2.15}$$

From about 0.5 mm of water height, the exponent is $n_{aq} \approx 1.6$.

At a specific speed, the so-called floating speed or aquaplaning speed, the tyre loses contact with the road, and the lateral and longitudinal forces in the contact patch tend towards zero. This phenomenon is called aquaplaning. The aquaplaning speed depends on several parameters, e.g. width and tread pattern. Figures 2.8 and 2.9 show aquaplaning speeds for different groups of tyres. A tendency can be seen for the aquaplaning speed to decrease with increasing width and for most winter tyres to have higher aquaplaning speeds than summer tyres.

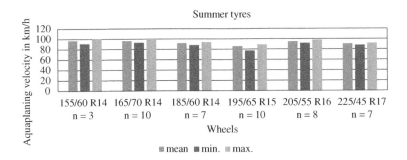

Figure 2.8   Aquaplaning speed for summer tyres (data from Reithmaier and Salzinger 2002)

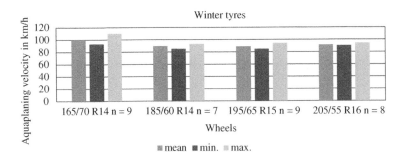

Figure 2.9   Aquaplaning speed for winter tyres (data from Reithmaier and Salzinger 2002)

2.2.3   Bearing Resistance

The friction in the wheel bearing results in frictional forces, which result in a moment, $M_{\rm wb}$:

$$|M_{\rm wb}| - \mu_b r_b \sqrt{X^2 + Z^2} \,. \tag{2.16}$$

Here, $\sqrt{X^2 + Z^2}$ is the resulting normal force from the section forces, X and Z, in the bearing, μ_b is the Coulomb friction coefficient and r_b the radius of the bearing where the friction force occurs. This moment results in a bearing resistance $F_{\rm wb}$:

$$|F_{\rm wb}| = \mu_b \frac{r_b}{r_{wst}} \sqrt{X^2 + Z^2} \,. \tag{2.17}$$

Setting $M_{\rm wb}$ in the sum of moments (2.5) and considering a stationary motion, i.e. $\ddot{\varphi}_w = 0$, yields

$$0 = M_{\rm wb} - F_x r_{wst} - F_z e_w \,. \tag{2.18}$$

Substituting $e_w = f_r r_{wst}$ and $M_{\rm wb} = \mu_b r_b \sqrt{X^2 + Z^2}$ from Equation (2.16) into Equation (2.18), we obtain

$$-F_x = f_r F_z + \mu_b \frac{r_b}{r_{wst}} \sqrt{X^2 + Z^2} \,. \tag{2.19}$$

In addition to the rolling resistance, $f_r F_z$, the bearing resistance, $F_{\rm wb} = \mu_b \frac{r_b}{r_{wst}} \sqrt{X^2 + Z^2}$, also occurs here. The bearing resistance is often negligibly small with respect to the rolling resistance.

2.2.4   Toe-In/Toe-Out Resistance

An additional resistance is caused by the angular position of the wheels (rotation about the $\vec{e}_{wz}$-direction). This inclined position is called toe-in or toe-out (cf. Figure 15.1). The resulting resistance is for small δ_{10}:

$$F_{\rm wtoe} = c_\alpha \delta_{10} \underbrace{\sin(\delta_{10})}_{\approx \delta_{10} \text{ for } \delta_{10} \ll 1}$$

$$\approx c_\alpha \delta_{10}^2 \,. \tag{2.20}$$

Here c_α is the so-called cornering stiffness and δ_{10} is the toe-in or toe-out angle, i.e. the angle by which the wheel is rotated about the $\vec{e}_{wz}$-axis (for more information see Chapter 15).

Remark 2.6 The magnitude of the toe resistance is 1/100th of the rolling resistance.

Another resistance is the curve resistance, F_{wc}, which occurs during cornering due to increased lateral forces. The entire wheel resistance is made up of rolling resistance, F_r, the resistance due to water, F_{aq}, the bearing resistance, F_{wb}, the toe resistance, F_{wtoe} and the curve resistance, F_{wc}.

When the vehicle is driving in a straight line on a dry road, the whole wheel resistance is essentially equal to the rolling resistance (the other resistances may be neglected).

2.3 Tyre Longitudinal Force Coefficient, Slip

We first turn to the concept of slip. Let us consider a wheel with a driving torque, M_w, acting in such a way that $F_x = 0$ (Figure 2.4). This wheel is assumed to roll in a steady state on the roadway. The driving torque has to compensate for the rolling resistance exactly. The angular velocity is ω_{w0} and the velocity is v_{w0}. Since the tangential forces, $F_x = 0$, do not act at the contact area, the parts of the tyre which are in contact with the road adhere to the road. Consequently, a radius R_{w0} can be defined

$$R_{w0} = \frac{v_{w0}}{\omega_{w0}} \ . \tag{2.21}$$

Since no sliding occurs in the contact patch, we call this state a rolling wheel without slip. The radius R_{w0} is defined by v_{w0} and ω_{w0} for a slip-free rolling wheel. This radius, R_{w0}, is called the dynamic rolling radius when the wheel rolls without slip (in this wheel tangential forces do not occur at the contact area)[2].

However, if the wheel is driven such that $F_x \neq 0$, then the relationship between the driving speed, v_v, and the angular velocity, ω, is no longer valid:

$$R_{w0}\omega \neq v_v \ . \tag{2.22}$$

The wheel does not stick to the road but slides or slips. The speed of the individual material particles in the contact area $v_c = R_{w0}\omega$ (the so-called circumferential speed; N.B.: this velocity is defined in the wheel-body fixed coordinate system) is not equal to the driving speed v_v of the wheel (v_w is the velocity of the wheel centre and equal to the velocity of the vehicle $v_w = v_v$) and thus no longer equal to the speed of the roadway seen from the point of view of the centre of the wheel (i.e. from the wheel-body fixed coordinate system). To capture this sliding effect quantitatively, we introduce

---

[2] The dynamic rolling radius R_{w0} is smaller than the radius of the undeformed wheel, r_{w0}, and is larger than the static radius of the wheel, r_{wst}. Other definitions for a rolling radius exist in the literature, with one example being the effective rolling radius, R_e, which is defined for a free rolling wheel, which means that no driving torque, $M_w = 0$, is applied. For this case, the angular velocity, ω_w and the translational velocity of the wheel, v_w, yield the effective rolling radius, $R_e = v_w/\omega_w$. For this wheel, which is not driven but towed (the towing force is the rolling resistance), the tangential force in the contact patch is equal to the rolling resistance, $F_x = F_r \neq 0$. From $F_x \neq 0$ it follows that a non-zero relative velocity occurs between tyre and road. The latter effective rolling radius, R_e, is therefore not as well suited to define relative motion. The relative motion is used for the definition of longitudinal slip. Nevertheless, the latter definition is used in the literature (for more details cf. Pacejka 2002, p. 65). The relation between the slip and the tangential force, F_x, differs slightly for the two definition of slip, using R_{w0} on the one hand or R_e on the other.

the concept of slip, where a distinction in the definition between a driven and braked wheel is necessary:

Slip: For a driven wheel, slip is defined as the difference between the circumferential speed, $v_c = R_{w0}\omega$ and the driving speed, v_v, divided by the circumferential speed, v_c.

$$S = \frac{v_c - v_v}{v_c} \ . \tag{2.23}$$

The slip of a braked wheel is defined as

$$S = \frac{v_v - v_c}{v_v} \ . \tag{2.24}$$

The slip is often given as a percentage value.

Remark 2.7 The percentage value of the slip means that a value of $S = 0.2$, for example, is referred to as 20% slip. Considering the slip in any of the equations, we use in this book, we have to decide whether to use the decimal slip value or the percentage value; usually we prefer the slip as an absolute value and not as a percentage.

Remark 2.8 The definition of the two cases – one for a driven and one for a braked wheel – is necessary for two reasons: the asymmetry of the definitions in the numerator ensures that the slip is always positive. The asymmetry in the denominator prevents division by zero.

Remark 2.9 This representation, in which the tyre slip is a global simplification of local phenomena in the contact patch, is an idealized view. If we examine the contact patch closely, we will find areas in which the material particles slip more than in other areas, where slip is less or even where stick can be observed.

Tyre longitudinal force coefficient: A tangential force, F_x, arises at the driven or the braked wheel, depending on the slip and the normal force F_z:

$$F_x = \mu(S)F_z \ . \tag{2.25}$$

The value μ is referred to as tyre longitudinal force coefficient. This is a function of the slip, S. The functions $\mu_b(S)$ for braking and $\mu_d(S)$ for driving differ slightly: $\mu(S) \approx \mu_b(S) \approx \mu_d(S)$. This is the reason why we do not distinguish between braking and driving with respect to the longitudinal force coefficient: $\mu(S) = \mu_b(S) = \mu_d(S)$.

The function $\mu(S)$ is shown in Figure 2.10. The tyre longitudinal force coefficient rises to the coefficient of adhesion, μ_a, and then falls again. The descending section cannot be driven with a normal motor vehicle in a steady state. When we enter this

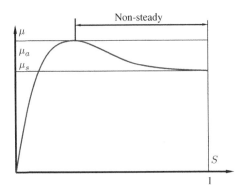

Figure 2.10   Tyre longitudinal force coefficient, μ, as a function of the slip S

region, the slip increases rapidly to 1 and the tyre longitudinal force coefficient falls to the coefficient of pure sliding[3], μ_s.

The coefficient of adhesion, μ_a, is achieved in the range of $S = 0.05$ to $S = 0.2$. The order of magnitude of μ_a ranges from 0.2 (snow) to 1.1 (dry concrete). The coefficient of adhesion μ_a is highly dependent on the weather (rain, snow, ice and temperature). Figures 2.11 and 2.12 show the wet braking capability of different tyres. To illustrate this, a braking manoeuvre on a wet road is investigated. The graph shows the mean deceleration, $a_{\mathrm{mean}} = \Delta v / T$, during braking from 80 km/h to 10 km/h, with reference to gravitational acceleration, g, ($\Delta v = 70$ km/h and T is the time that is necessary to decelerate the car). The sum of the four wheel loads of a car (neglecting aerodynamic lift) is $F_{z\mathrm{tot}} = m_{\mathrm{tot}}g$; the mean longitudinal inertia force during deceleration (neglecting rotational parts of the vehicle) is $F_{xa} = m_{\mathrm{tot}}a_{\mathrm{mean}}$. Consequently, $\mu = F_{xa}/F_{z\mathrm{tot}} = a_{mean}/g$ is a measure of the wet braking capability of the tyre[4].

In the following, we look at two idealized (and in some senses artificial) examples in order to illustrate the interaction between the slip and the rolling resistance. The rolling resistance is due to the asymmetric normal force distribution, which has a resultant moment.

Example 2.1   We consider a wheel rolling on an ideal slippery surface (think of soap or oil on a pane of glass). The wheel moves with a velocity of v_v and rotates with an angular velocity of $\omega = \frac{v_v}{R_{w0}}$ We wish to answer the following question: What forces

---

[3] The expressions *coefficient of adhesion* μ_a and *coefficient of pure sliding* μ_s are short but, strictly speaking, not correct. They are used in the German literature; in ISO 8855 (2011) the expression *maximum longitudinal force coefficient* is used instead of coefficient of adhesion. A detailed look at the friction physics of the tyre shows that, sliding occurs in nearly every situation. Nevertheless, we use and prefer the expressions *coefficient of adhesion*, μ_a, and *coefficient of pure sliding*, μ_s, here for convenience.
[4] It should be emphasized that the different measurements are not comparable to each other, because the road was changed by the braking manoeuvres during the investigations of Reithmaier and Salzinger 2002. This is obvious from the measurements for the summer tyre 205/55 R16, which yield very high values for μ; the reason is that these measurements were taken on a new road with a very high grip level.

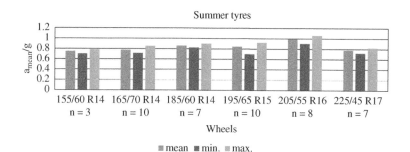

Figure 2.11 Mean deceleration divided by gravitational acceleration during braking from 80 km/h to 10 km/h for summer tyres (data from Reithmaier and Salzinger 2002)

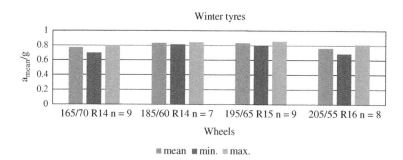

Figure 2.12 Mean deceleration divided by gravitational acceleration during braking from 80 km/h to 10 km/h for winter tyres (data from Reithmaier and Salzinger 2002)

and moments must act on the centre of mass, S_{cmw}, so that this motion is converted to a stationary state?

Thus we are looking for the force, X, and the moment, M_d, which are necessary to permit the motion described above (see Figure 2.4).

To calculate these quantities, we consider the equations of motion ($\alpha_g = 0$):

$$0 = F_x - X \,, \tag{2.26}$$

$$0 = F_z - Z - G_w \,, \tag{2.27}$$

$$0 = M_d - F_x r_{wst} - F_z e_w \,. \tag{2.28}$$

Since the road surface is perfectly slippery, no tangential forces can be transmitted, so $F_x = 0$. Consequently, from Equation (2.26) we obtain: $X = 0$. From the free-body diagram of the body (mass of the body m_b, Figure 2.4) we get

$$Z = \frac{m_b}{4} g \,, \tag{2.29}$$

and we assume that the centre of mass lies in the middle of the vehicle (this means that all wheel loads are equal). Eliminating F_z from the above equations, we obtain

$$M_d = F_z e_w$$
$$= \left(\frac{m_b}{4} + m_w\right) g e_w \; . \tag{2.30}$$

It can be seen from this example:

1. that the rolling resistance is not necessarily associated with tangential forces $F_x \neq 0$;
2. that rolling of the wheel (i.e. a movement without the occurrence of sliding in the contact patch) is possible when a moment acts on the wheel; a driving force $X \neq 0$ is not required.

Example 2.2 We consider a wheel rolling on an ideal rough surface (i.e. no sliding at the contact area is possible; mathematically speaking $\mu(S) \to \infty$ for all $S \neq 0$). Due to the ideal rough surface considered on the wheel, we have

$$\omega = \frac{v_v}{R_{w0}} \; . \tag{2.31}$$

The load is the driving torque $M_d = \left(\frac{m_b}{4} + m_w\right) g e_w$ and the weight $Z = \frac{m_b}{4} g$ of the quarter body mass.

We want to calculate the tangential force, F_x, at the contact patch. From Equation (2.27), we obtain

$$F_z = \left(\frac{m_b}{4} + m_w\right) g \; . \tag{2.32}$$

Substituting Equation (2.32) into Equation (2.28) for the sum of the moments, we obtain

$$F_x = 0 \; . \tag{2.33}$$

It can be seen in both examples that the resulting tangential force, F_x, can be zero independent of the sliding or adhesion conditions in the contact patch.

In reality, when pure rolling of the wheel occurs, there are also sliding zones in the contact area, which lead to a slight increase in the rolling resistance. The literature therefore sometimes divides the rolling resistance into two parts, which results from deformation of the tyre and that from friction.

2.4   Questions and Exercises

Remembering

1. What is the contact patch?
2. What does the normal stress distribution in the contact patch of a wheel look like?
3. What is the name of the value by which the resulting normal force of the normal stresses is shifted in the contact area?

4. How great is the magnitude of rolling resistance due to the asymmetric distribution of normal force?
5. What is the name of the ratio of rolling resistance, F_r, to the resultant normal force, F_z?
6. What is the highest exponent in the relation between the coefficient of rolling resistance, f_r, and the speed?
7. On what do the coefficients $\tilde{f}_{r0}$, $\tilde{f}_{r1}$ and $\tilde{f}_{r4}$ in the following formula depend?

$$f_r(v) = \tilde{f}_{r0} + \tilde{f}_{r1}\frac{v}{v_0} + \tilde{f}_{r4}\left(\frac{v}{v_0}\right)^4 \tag{2.34}$$

8. How does toe-in affect the forces on the wheel?
9. Which of the resistances acting on the wheel has the greatest influence on the wheel resistance when driving in a straight line?
10. How is the slip, S, defined?
11. Draw a free-body diagram of a rolling wheel.
12. What is the name of the ratio of the tangential force, F_x, to the vertical force, F_z, in the contact area?
13. How does this ratio depend on the slip?

Understanding

1. What causes the eccentricity, e_w?
2. Explain the necessity of the asymmetric definition in the formula for the slip, S, using the two manoeuvres *racing start* without ASR and *full braking* without ABS!
3. Explain what happens with respect to the longitudinal force, F_x, when the slip, S, increases from a starting value of $S = 0$ (the wheel load, F_z, is constant)!

Applying

1. In ABS (anti-lock braking system) one criterion for sensing the locking of a wheel is that the angular deceleration exceeds a limit value. Estimate this limit for the parameters given below (the mass of the body is equally distributed across all four wheels; you will find more parameters than you need for the calculation in the parameter list)! $\mu_a = 1.1$, $\mu_s = 0.9$, $f_r = 0.011$ $m_w = 20\,\text{kg}$, $m_b = 1200\,\text{kg}$, $J_w = 0.1\,\text{kg m}^2$ and $r_{wst} = 0.3\,\text{m}$.
2. The total mass of a vehicle is $m_{tot} = 1500$ kg. What is the magnitude of the rolling resistance of the whole vehicle?

Analysing

1. In ABS (anti-lock braking system), one criterion for sensing the locking of a wheel could be that the slip, S, of this wheel exceeds a limit value. The difficulty

is that the driving velocity is not known; the driving velocity can be estimated by
using the angular velocities of the wheel.

Analyse the following situation: Your ABS estimates the driving velocity of
the vehicle by calculating the mean of the circumferential velocities, v_{ci}, of the
four wheels, $(i = 1, \ldots, 4)$. During a braking process, the angular velocities
of all four wheels follow the same time dependency $\omega_{wi} = \Omega_0(1 - t/T_0)$, $i =$
$1, \ldots, 4$. The braking process starts at $t = 0$ s and ends at $t = T_0$. Is the ABS
control algorithm able to detect that the slip exceeds the slip limit?

2. Analyse the following situation: The road for the left-hand side of your vehicle differs from the road on the right-hand side with respect to μ_a and μ_s (this
type of road is called a split-μ road). Now you should consider a braking process
assuming that the slip values for all four wheels are equal. What happens to the
vehicle?

3. Calculate the coefficient of rolling resistance for a velocity of $v_v = 30$ m/s ($\tilde{f}_{r0} =$
9.0×10^{-3}, $\tilde{f}_{r1} = 2.0 \times 10^{-3}$, $\tilde{f}_{r4} = 3.0 \times 10^{-4}$).

4. The power P necessary for overcoming the rolling resistance can be calculated by
$P = F_r v_v$, where F_r is the rolling resistance and v_v is the velocity of the vehicle.
Calculate the power necessary for a car ($m = 1500$ kg; all wheel loads equal;
$g = 9.81$ m/s^2) with the above mentioned coefficients of rolling resistance!

5. Consider a wheel (static radius $r_{wst} = 0.3$ m, excentricity $e_w = 3$ mm, wheel
load $F_z = 2500$ N). A driving torque $M = 382.5$ Nm is acting on the wheel. The
wheel is moving with the velocity $v_v = 30$ m/s. Assume for this exercise that the
function μ can be approximated linearly

$$\mu(S) = 10S . \tag{2.35}$$

Calculate the angular velocity of the wheel! As this problem may be challenging, please follow the solution procedure:

(a) Calculate the tangential force F_x in the contact patch! (Remember that you
need one portion of the torque M to overcome the rolling resistance.)

(b) Calculate with the linearized function (2.35) the longitudinal slip S. (Remember that $\mu(S) = F_x/F_z$.)

(c) Calculate with the longitudinal slip S the angular velocity $\dot{\varphi}_w$.

3

Driving Resistances, Power Requirement

In the following chapter, we turn to other resistances (apart from the rolling resistance) which have no effect on the wheel, but are mainly attributable to the whole vehicle. In Section 3.1, we concentrate on the air forces that lead to aerodynamic drag. Section 3.2 deals with the gradient resistance that occurs when driving on an inclined road. In Section 3.3, we discuss d'Alembert's inertial forces resulting in the so-called acceleration resistance. In Section 3.4, we use the driving resistances (total running resistance) to set up the equations of motion for the entire vehicle. These form the basis for consideration of the power in Section 3.5, in which the concept of supply and demand is explained in the characteristic graphs.

3.1 Aerodynamic Drag

The air flow around the vehicle causes turbulence losses to occur in some areas, which are reflected in aerodynamic drag. The largest contribution comes from the vortex behind the car (cf. Figure 3.1). Small vortices at wheels, mirrors, the engine compartment and at the A-column (cf. Figures 3.2 and 3.3) contribute to the aerodynamic drag, too. The force exerted by the formation of these vortices on the vehicle is

$$F_a = c_d A \frac{\rho_a}{2} v_r^2 \ . \tag{3.1}$$

Here c_d is the aerodynamic drag coefficient, A is projected area of the vehicle in the longitudinal direction, ρ_a the density of air and v_r the resulting velocity of the air, which comes from the driving velocity of the vehicle, v_v, and wind velocity, v_a, which have to be added with reference to their directions.

Vehicle Dynamics, First Edition. Martin Meywerk.
© 2015 John Wiley & Sons, Ltd. Published 2015 by John Wiley & Sons, Ltd.
Companion Website: www.wiley.com/go/meywerk/vehicle

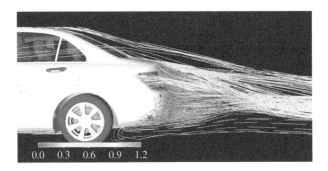

Figure 3.1   Vortex behind a car (reproduced with permissions of Daimler AG)

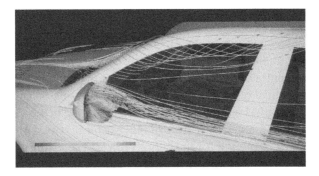

Figure 3.2   Vortex at A-column (reproduced with permissions of Daimler AG)

Figure 3.3   Vortex at engine compartment and at A-column (reproduced with permissions of Daimler AG)

The drag coefficient depends on the flow direction. In order to achieve a better comparison of vehicles, it is common to use a simplified approach in which the wind speed is not considered.

Aerodynamic drag force: On a vehicle with the projected frontal area, A, travelling at a speed v_v in the longitudinal direction, a longitudinal force, F_a, the so-called aerodynamic drag force, acts as follows (wind velocity $v_a = 0$):

$$F_a = c_d A \frac{\rho_a}{2} v_v^2 \ . \tag{3.2}$$

Here c_d is the coefficient of aerodynamic drag. The value of c_d for modern passenger cars is about 0.2–0.3. A typical size for the frontal area A is $2\,\text{m}^2$.

3.2 Gradient Resistance

Gradient resistance: The gradient resistance (or climbing resistance), F_g, is the portion of the weight of the vehicle which acts parallel to the road:

$$F_g = m_\text{tot} g \sin \alpha_g \ . \tag{3.3}$$

The slope of roads is determined by the gradient, p, which is the rise of the road divided by the (horizontal) run. Thus the gradient, p, is equal to the tangent of the inclination angle:

$$p = \tan \alpha_g \ . \tag{3.4}$$

The gradient may also be indicated in percentage. For small angles ($\alpha_g \leq 17°$), we can replace $\sin \alpha_g$ in the formula (3.3) for the gradient resistance by $\tan \alpha_g$ (the error is less than 5%). Performing this substitution yields a simplified formula

$$F_g = m_\text{tot} g \sin \alpha_g$$

$$\approx m_\text{tot} g \tan \alpha_g$$

$$= m_\text{tot} g p \ . \tag{3.5}$$

The error resulting from $\tan \alpha_g \approx \sin \alpha_g$ can be calculated (for $\alpha_g \leq 17°$):

$$\frac{|m_\text{tot} g \sin \alpha_g - m_\text{tot} g \tan \alpha_g|}{|m_\text{tot} g \sin \alpha_g|} = \frac{|\sin \alpha_g - \tan \alpha_g|}{|\sin \alpha_g|}$$

$$\leq 0.045 \ldots \ . \tag{3.6}$$

For small values of the angle α_g (3.3), a second approximate formula can be derived by substituting $\sin \alpha_g \approx \alpha_g$:

$$F_g \approx m_\text{tot} g \alpha_g \ . \tag{3.7}$$

In this formula (3.7), it is essential to use radians as the units for the angle α_g. The specification of p can be found on traffic signs.

3.3 Acceleration Resistance

Acceleration resistance: Another resistance is due to d'Alembert's inertial forces. These inertial forces (from translational and rotational motions) are combined and are referred to as the acceleration resistance (or inertial resistance), F_i (see Figure 3.4). The acceleration resistance not only takes into account the forces due to the translatory acceleration, but also the forces in the longitudinal direction, which arise due to the angular acceleration of the rotating masses.

We proceed by assuming that the excitation of the pitching motion is negligible. Similarly, we neglect the inertial forces of the wheels and axles in the first step. These are considered at the end and added to the acceleration resistance. The rotational inertia of the wheels and the axles here are initially neglected, but can also be considered analogous to the other rotational inertias. Figure 3.5 is a sketch of a vehicle from the top. It shows the front axle (angular velocity $\dot\varphi_{a1}$ and moment of inertia J_{a1}) and the engine-driven rear axle (angular velocity $\dot\varphi_{a2}$, moment of inertia J_{a2}).

The vehicle is driven by the engine (angular velocity $\dot\varphi_e$ and moment of inertia J_e with clutch and portion of the transmission) which is connected through the clutch and transmission (transmission i_g) and the Cardan shaft (moment of inertia J_c with portion of the transmission) and the final drive or differential (transmission ratio i_d) to the rear axle.

The question to be answered in the following concerns how large a force F_i^* acting on the centre of mass S_{cm} has to be in order to accelerate the vehicle[1].

We assume that the slip on the front wheels, S_1, and the rear wheels, S_2, is constant. In order to calculate the force, F_i^*, we derive the kinetic energy of the entire vehicle,

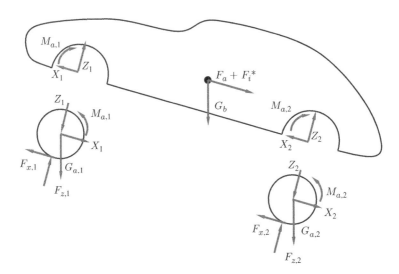

Figure 3.4 Free-body diagram of the entire vehicle (without inertia forces at the axles)

---

[1] In this first step we denote the force, F_i^*, which is necessary to accelerate the car with an asterisk to distinguish it from the complete acceleration resistance, F_i. It is important that this force, F_i^*, should not act as d'Alembert's inertial force in the centre of mass, but, nevertheless, we depict it in this position for convenience; later, when we derive the axle loads with the equation of the moment of equilibrium in Chapter 6, only d'Alembert's force, $m_b\ddot{x}_v$, acts in the centre of mass.

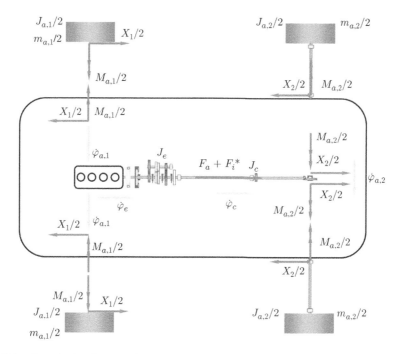

Figure 3.5 Free-body diagram for determining the acceleration resistance (without inertia forces at the axles)

which we specify in terms of the velocity, $v_v = \dot{x}_v$, of the vehicle. The acceleration resistance can then be derived by Lagrange equations[2].

We will proceed with the calculation of F_i^*:

$$F_i^* = \frac{\mathrm{d}}{\mathrm{d}t}\left(\frac{\partial E_{\mathrm{kin}}}{\partial v_v}\right). \tag{3.8}$$

The expression E_{kin} is the kinetic energy of the vehicle. The potential energy term, V, and the derivative with respect to the spatial coordinate, x_v, yield the gradient resistance and are therefore not considered here.

The kinetic energy of the system (without the axles and wheels) is

$$E_{\mathrm{kin}} = \frac{1}{2}m_b\dot{x}_v^2 + \frac{1}{2}J_c\left(\frac{i_d\dot{x}_v}{(1-S_2)R_{w0}}\right)^2 + \frac{1}{2}J_e\left(\frac{i_d i_g\dot{x}_v}{(1-S_2)R_{w0}}\right)^2. \tag{3.9}$$

---

[2] The Lagrange function, $L = T - V$, is the difference between the kinetic, T, and the potential energy, V. The variables in L are velocities and displacements, e.g. x_i and $v_i = \dot{x}_i$ ($i = 1, \ldots, n$) for translational motions. The equations of motion can be established by simply calculating derivatives:

$$\frac{\mathrm{d}}{\mathrm{d}t}\left(\frac{\partial L}{\partial v_i}\right) - \frac{\partial L}{\partial x_i} = 0, i = 1, \ldots, n.$$

We assume that the transmission ratios i_d and i_g are not dependent on time or velocity[3] and obtain

$$F_i^* = \left(m_b + J_c \left(\frac{i_d}{(1 - S_2) R_{w0}} \right)^2 + J_e \left(\frac{i_d i_g}{(1 - S_2) R_{w0}} \right)^2 \right) \ddot{x}_v . \tag{3.10}$$

3.4 Equation of Motion for the Entire Vehicle

In the following section, we present the equations of motion for a vehicle moving in the longitudinal direction. We proceed from Figure 3.4. Equations of motion for the front and rear axles are ($j = 1, 2$):

$$m_{aj} \ddot{x}_{aj} = -X_j + F_{xj} - G_{aj} \sin \alpha_g , \tag{3.11}$$

$$J_{aj} \ddot{\varphi}_{aj} = M_{aj} - F_{xj} r_{wstj} - F_{zj} e_{wj} . \tag{3.12}$$

The moment of inertia, J_{aj}, includes the wheels, brakes and hubs and, in the case of a driven axle, the drive shafts. Assuming ideal bearings for the front wheels results in $M_{a1} = 0$. The acceleration of the axis, $\ddot{x}_{aj}$, is equal to the acceleration, $\ddot{x}_v$, of the vehicle. We resolve the equations for the sum of the moments (3.12) for F_{xj} and insert this into Equation (3.11) for the sum of forces in the longitudinal direction and obtain ($\ddot{x}_{aj} = \ddot{x}_v$):

$$m_{aj} \ddot{x}_v = -X_j - G_{aj} \sin \alpha_g + \frac{1}{r_{wstj}} (-J_{aj} \ddot{\varphi}_{aj} + M_{aj} - F_{zj} e_{wj}) . \tag{3.13}$$

The equation of motion for the vehicle is obtained from the free-body diagram in Figure 3.4:

$$X_1 + X_2 = F_a + F_i^* + F_g . \tag{3.14}$$

Here F_a is the aerodynamic drag force, F_i^* is the acceleration resistance (without axles, wheels, etc.) and F_g the gradient resistance. As in the previous sections, we do not consider the pitching motion of the vehicle. We express the angular velocities of the axles, $\dot{\varphi}_{aj}$, by the velocity, $\dot{x}_v$, of the vehicle. For the driven rear axle, this means

$$\dot{\varphi}_{a2} = \frac{\dot{x}_v}{(1 - S_2) R_{w02}} \tag{3.15}$$

and for the front axle

$$\dot{\varphi}_{a1} = \frac{(1 - S_1) \dot{x}_v}{R_{w01}} . \tag{3.16}$$

At this point, the asymmetry returns as the definition of the slip.

---

[3] For continuous variable transmissions (CVT), the transmission ratio of this CVT may be dependent on time-or velocity. Consequently, the derivative with respect to time and/or with respect to $\dot{x}_v$ has to be taken into account. The formula then becomes more complicated.

If we replace the angular accelerations, $\ddot{\varphi}_{aj}$, in the equations of motion for the axles (3.13) by the body acceleration $\ddot{x}_v$ according to the relationship in Equations (3.15) and (3.16), then resolve Equation (3.13) for X_1 and X_2, and substitute X_1 and X_2 in (3.14) we obtain (wind velocity $v_a = 0$)[4]

$$\frac{M_{a1}}{r_{wst1}} + \frac{M_{a2}}{r_{wst2}} = c_d A \frac{\rho_a}{2} \dot{x}_v^2 \tag{3.17}$$

$$+ \left(m_b + J_c \left(\frac{i_d}{(1 - S_2)\, R_{w02}} \right)^2 + J_e \left(\frac{i_d i_g}{(1 - S_2)\, R_{w02}} \right)^2 \right) \ddot{x}_v$$

$$+ \left(m_{a1} + m_{a2} + J_{a1} \frac{(1 - S_1)^2}{R_{w01}^2} + J_{a2} \frac{1}{(1 - S_2)^2 R_{w02}^2} \right) \ddot{x}_v$$

$$+ G \sin \alpha_g$$

$$+ f_{a1} F_{z1} + f_{a2} F_{z2} \; .$$

Here, f_{aj} are the rolling resistance coefficients for the front and the rear axle. Adding the masses m_b, m_{a1} and m_{a2} to give the total mass m_{tot} of the vehicle yields the so-called rotational mass factor, λ,

$$\lambda = 1 + \frac{1}{m_{tot}} \left(J_c \left(\frac{i_d}{(1 - S_2)\, R_{w02}} \right)^2 + J_e \left(\frac{i_d i_g}{(1 - S_2)\, R_{w02}} \right)^2 \right.$$

$$\left. + J_{a1} \frac{(1 - S_1)^2}{R_{w01}^2} + J_{a2} \frac{1}{(1 - S_2)^2 R_{w02}^2} \right) \; . \tag{3.18}$$

Now, Equation (3.17) can be rewritten as

$$\frac{1}{r_{wst1}} M_{a1} + \frac{1}{r_{wst2}} M_{a2} = \underbrace{c_d A \frac{\rho_a}{2} \dot{x}_v^2 + F_i + F_g + f_{a1} F_{z1} + f_{a2} F_{z2}}_{=F_a} \; . \tag{3.19}$$

Here $F_i = \lambda m_{tot} \ddot{x}_v$ is the acceleration resistance of the entire vehicle; thus including the wheels, axles, drive shafts and brakes. This equation yields the moments at the wheels required to overcome the resistances. It is one of the fundamental equations in longitudinal vehicle dynamics.

Remark 3.10 Equation (3.19) considers the wheels of one axle considered together. We can also consider the wheels separately and then proceed in the same way to a similar equation.

---

[4] Here $M_{a1} < 0$ is a small braking torque from the bearing. If the front axle is driven, too, the torque M_{a1} is a portion of the torque of the engine. In this case, we have an all-wheel-drive vehicle and there is an additional differential between front and rear axles. The equation must then be modified, and this modification determines how the angular velocities between front and rear axles depend on each other.

Remark 3.11 These equations include many parameters that may be time dependent or may depend on other variables, such as $f_{aj}, F_{zj}, \lambda, r_{wstj}, R_{w0j}, S_j$. Not substituting these parameters, which are not constant, with time-dependent variables enables the equation of motion to be represented in this simple form.

3.5   Performance

The power which is necessary at the axles to move the vehicle is

$$P_{\text{tot}} = M_{a1}\dot{\varphi}_{a1} + M_{a2}\dot{\varphi}_{a2} \, . \tag{3.20}$$

Substituting the angular velocities $\dot{\varphi}_{aj}$ by the velocity $\dot{x}_v$ of the vehicle, we obtain

$$P_{\text{tot}} = M_{a1}\underbrace{\frac{(1 - S_1)\dot{x}_v}{R_{w01}}}_{\geq 0} + M_{a2}\underbrace{\frac{\dot{x}_v}{(1 - S_2)R_{w02}}}_{\geq 0} \, . \tag{3.21}$$

Assuming that only one axle is driven (in this example, the rear axle with $S = S_2$) and that we are interested only in the power of this driven axle, we obtain from Equation (3.19) by neglecting the difference between R_{w0} and r_{wst}

$$P_{\text{tot}} = \frac{1}{(1 - S)}\dot{x}_v\underbrace{(F_a + F_i + F_g + F_r)}_{=F_{\text{tot}}} \, , \tag{3.22}$$

where $F_r = f_r G$ was simplified assuming that the coefficients of rolling resistances are equal at rear and front axle. Further neglecting slip ($S \approx 0$) and making use of the approximation $\sin\alpha_g \approx \tan\alpha_g = p$ for the gradient resistance, we obtain the power necessary at the wheels of one axle ($v = \dot{x}_v$; velocity of wind $v_a = 0$):

$$P_{\text{tot}} = \left(f_r + p + \lambda\frac{\ddot{x}_v}{g}\right)Gv + c_d A\frac{\rho_a}{2}v^3 \, . \tag{3.23}$$

If there is wind in the negative $\vec{x}_v$-direction with a speed v_a, Equation (3.23) reads as follows:

$$P_{\text{tot}} = \left(f_r + p + \lambda\frac{\ddot{x}_v}{g}\right)Gv + c_d A\frac{\rho_a}{2}(v + v_a)^2 v \, . \tag{3.24}$$

The graph in Figure 3.6 shows different components of the power required by the various resistances. The parameters were chosen as follows:
$\tilde{f}_{r0} = 0.0087, \tilde{f}_{r1} = 0.0022, \tilde{f}_{r4} = 5.7258 \times 10^{-4}, m_{\text{tot}} = 1350 \, \text{kg}, c_d = 0.32, A = 2.2 \, \text{m}^2, \rho_a = 1.226 \, \text{kg/m}^3, v_a = 0$.
Included in the graph are the power at the axle: $P_r = F_r v$ from the rolling resistance, $P_a = F_a v$ from the aerodynamic drag force, and the sum of both $P_{\text{basic}} = P_r + P_a$.

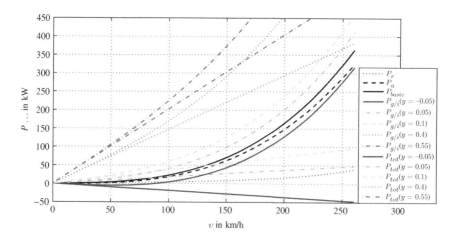

Figure 3.6   Power demand of several resistances

The other curves $P_{g/i}(y) = (F_g(p) + F_i(\lambda \ddot{x}_v/g))v$ give the power needed to over-come the gradient and acceleration resistances for the value

$$y = p + \lambda \frac{\ddot{x}_v}{g} \,. \tag{3.25}$$

The power, $P_{g/i}$, could be the result of a pure gradient resistance with $p = y$, but it could also be the power of a pure acceleration $\ddot{x}_v = gy/\lambda$ or a combination of both in accordance with Equation (3.25). The total power

$$P_{\text{tot}} = \underbrace{P_r + P_a}_{=P_{\text{basic}}} + \underbrace{P_g + P_i}_{P_{g/i}} \tag{3.26}$$

is also shown.

The basic demand for power increases with the third power of the vehicle velocity, v (velocity of the wind neglected) from F_a and the fifth power from F_r. Consequently, there is a significant amount of power for high velocities (for example at 200 km/h, approximately 165 kW is needed).

If we look at the total amount of power needed to climb a hill with a gradient of 5% ($p = 0.05$), for example, it is obvious that for high velocities a large amount of power is needed to overcome the gradient resistance.

The graph in Figure 3.7 shows the tractive forces (the resistances and the sums of resistances) for the same circumferences as in Figure 3.6.

In the following, we assume that the maximum and minimum inclination, and the maximum accelerations to be attained by a vehicle should lie between $y = p + \lambda \frac{\ddot{x}_v}{g} = 0$ and $y = p + \lambda \frac{\ddot{x}_v}{g} = 0.55$

This means that the drive power has to cover a particular region at a certain speed. This area is marked on the graph in Figure 3.8. This power area corresponds to an area

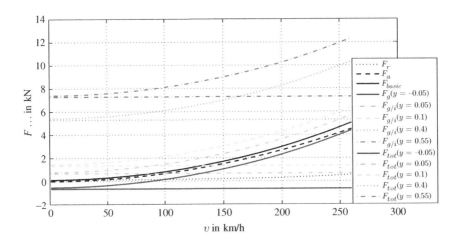

Figure 3.7   Tractive forces demand of several resistances

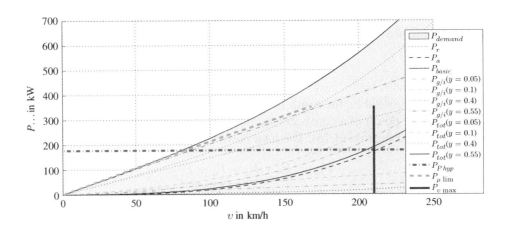

Figure 3.8   Power demand area

for the tractive force (see Figure 3.9) which is also marked. This area represents all the required power-vehicle speed or tractive forces-vehicle speed points which would be necessary to have the capability to climb a hill with $p = 0.55$ at every possible velocity.

We call this area the vehicle characteristic demand map. This demand map has to be compared with the delivery map of the power train.

Before we start with this comparison, we will take a closer look at some boundary conditions. These boundary conditions define the ideal engine delivery map.

We first assume that an ideal power train can deliver the maximum engine power over the entire speed range. In Figure 3.8, this is shown by the horizontal line $P_{P\,hyp}$. In the tractive force graph in Figure 3.9 this horizontal line becomes a hyperbola,

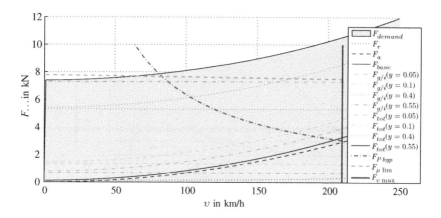

Figure 3.9   Forces demand map

the so-called ideal traction hyperbola, because the constant power yields the tractive forces by $F = P_e/v$. The traction hyperbola $F_{P\ \mathrm{hyp}}$ is depicted in Figure 3.9.

A further constraint to the ideal engine delivery map is determined by the maximum speed of the vehicle (identified by the vertical lines $P_{v\ \mathrm{max}}$ and $F_{v\ \mathrm{max}}$ in the graphs in Figure 3.8 and Figure 3.9, respectively).

The third limitation is due to the maximal tyre longitudinal force coefficient, the coefficient of adhesion, μ_a. This limit depends on the axle load. Figures 3.8 and 3.9 depict the limit for a vehicle with a driven front axle. This means that the limit is $F_{z1}\mu_a$, where F_{z1} is the section force between the wheels of the front axle and the road. Neither graph takes into account the load transfer from rear to front axle during acceleration nor other effects on axle loads (for example air lift forces), and, furthermore, the coefficient of adhesion, μ_a, decreases slightly with increasing velocity (in these graphs the value of μ_a decreases linearly from 1.1 for the velocity $v = 0$ to a value of 1.05 at the velocity $v \approx 72$ m/s). This limit is marked by the lines $P_{\mu\ \mathrm{lim}}$ and $F_{\mu\ \mathrm{lim}}$.

These three boundaries result in an ideal engine delivery map. The ideal engine delivery map is highlighted in Figures 3.10 and 3.11.

Contrasting with this is the characteristic map of real engines. Figure 3.12 shows the characteristic map of a diesel engine as an example. We also call this characteristic map a real engine delivery map.

The parameters shown are the power ratio P_{100}/P_{max} and the moment ratio $M_{100}/M(P_{\mathrm{max}})$ over the related rotational speed $n/n(P_{\mathrm{max}})$. Here $P_{100} = P(100\%)$ is the maximum power which can be delivered from the engine at a particular rotational speed while $M_{100} = M(100\%)$ is the full load torque, P_{max} the maximum power of the engine, $M(P_{\mathrm{max}})$ and $n(P_{\mathrm{max}})$ are, respectively, the moment and rotational speed where the power of the engine reaches a maximum.

We compare the real engine delivery map (as shown in Figure 3.12) with the ideal delivery characteristic map (Figure 3.10 or 3.11). Three differences are obvious:

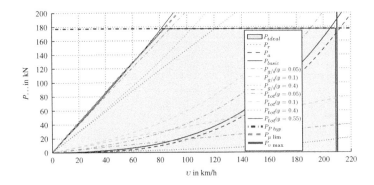

Figure 3.10   Ideal engine delivery map for the power

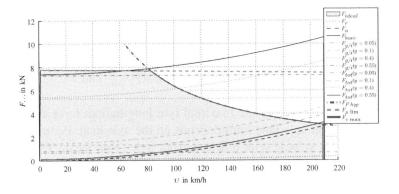

Figure 3.11   Ideal engine delivery map for the tractive force

Remark 3.12 In the ideal delivery characteristic map, there should be a large region in which the power is constant. In the case of a real characteristic map, however, this is only possible in a very small region at the end of the speed range.

Remark 3.13 In the ideal characteristic graph, there is a short drop due to the traction limit. This area is not covered in the real characteristic map.

Remark 3.14 The ideal characteristic map covers all speeds, the real characteristic map of the engine is not able to deliver power below a certain speed. (We call this gap in the characteristic map the speed, torque or power gap.)

One notable feature in the comparison of the ideal delivery characteristic map for the torque is that the torque varies only a little over the range of rotational speed in the real map, whereas the variations in the real map are large.

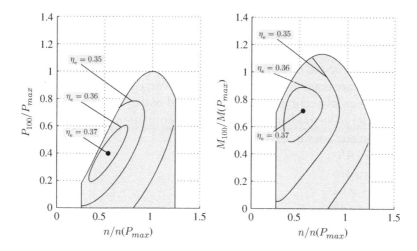

Figure 3.12   Real engine delivery map for the tractive force

Furthermore, the efficiency, η_e, of the engine is plotted in Figure 3.12. The maximum efficiency of $\eta_e = 0.37$ is attained at one point in the characteristic graph. This point is not on the full load characteristics (boundary line in Figure 3.12).

3.6   Questions and Exercises

Remembering

1. What is the gradient resistance?
2. What is the acceleration resistance?
3. How does the aerodynamic drag depend on the velocity?
4. What is the order of magnitude of the drag coefficient and the aerodynamic area?
5. The translatory inertias and the rotating masses are usually combined. Which factor plays a role here?
6. What order of magnitude does the rotational mass factor have?
7. How do we calculate the performance?
8. How do we obtain the required characteristic values for the traction force and performance?
9. What limits the required characteristic value?

Understanding

1. Describe the ideal delivery characteristics for traction and performance.
2. Describe the real delivery characteristic map of an internal combustion engine.
3. What are the implications from comparing a real and an ideal characteristic map?
4. Why $\lambda_i > \lambda_{i+1}$ holds for the rotational correction factor?

5. Why $\lambda_i > 1$ holds for the rotational correction factor?
6. Is it possible to build a vehicle with $\lambda_i = 1$?
7. Which of the following equations hold for the basic demand of power (no wind)?
   f_r is the coefficient of rolling resistance, it is assumed that the coefficients of all four wheel are equal; G is the weight of the car; c_d the aerodynamic drag coefficient; A the cross-sectional area; v_v the speed of the vehicle; a_v the acceleration of the vehicle; p the gradient; and λ the mass correction factor.
   More than one answer may be correct.
   (a) $Gf_r + c_d A \frac{\rho_a}{2} v_v^2$
   (b) $Gf_r + c_d A \frac{\rho_a}{2} v_v^3$
   (c) $Gf_r v_v + c_d A \frac{\rho_a}{2} v_v^3$
   (d) $(Gf_r + c_d A \frac{\rho_a}{2} v_v^2) v_v$
   (e) $(Gp + c_d A \frac{\rho_a}{2} v_v^2) v_v$
   (f) $(Gy + c_d A \frac{\rho_a}{2} v_v^2) v_v$ where $y = p + \lambda a_v / g$
8. The following statements have to be assigned to the points in the diagram of power demand curves (cf. Figure 3.13).
   (a) The total power demand for $v = 150$ km/h for a grading of $p = 0.05$ without acceleration.
   (b) The total power demand for $v = 150$ km/h for a grading of $p = 0.00$ without acceleration.
   (c) The total power demand for $v = 200$ km/h for a grading of $p = 0.1$ without acceleration.

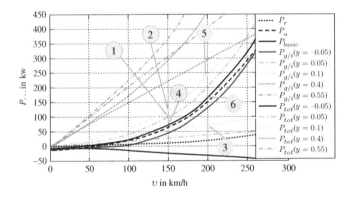

Figure 3.13   Demand curve for the power

9. The following limits have to be assigned to the curve in the diagram of the ideal characteristic map in Figure 3.14.
   (a) Limit of adhesion μ_a.
   (b) Limit of maximum power.
   (c) Limit of maximum velocity of the vehicle.

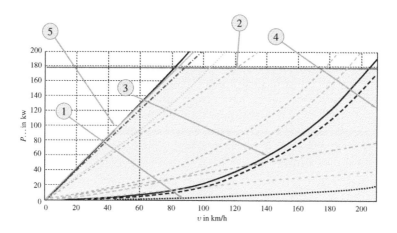

Figure 3.14 Ideal characteristic map for the power

Applying

1. Consider a wheel (static radius $r_{wst} = 0.3$ m, excentricity $e_w = 3$ mm, wheel load $F_z = 2500$ N). A driving torque $M = 382.5$ Nm is acting on the wheel. The wheel is moving at a velocity $v_v = 30$ m/s. Assume for this exercise that the function μ can be approximated linearly

$$\mu(S) = 10S . \tag{3.27}$$

   Calculate the angular velocity of the wheel!

2. Calculate for the following parameters the aerodynamic resistance and the power necessary to overcome this resistance at driving speeds of $v_v = 10$ and $v_v = 60$ m/s: aerodynamic drag force coefficient $c_d = 0.3$, cross-sectional area $A = 2$ m^2, mass density of air $\rho_a = 1.226$ kg/m^3. The speed of wind is zero. To calculate the power P_a, please use $P_a = F_a v_v$.

3. The mass density of air depends on the temperature, the pressure and the humidity[5].

Parameter set	Temperature $T(\mathrm{K})$	Pressure $p(\mathrm{Pa})$	Mass density ρ_a (kg/m^3)	
1	223.15 ($-50\,^\circ$C)	1.1×10^5	1.717	(3.28)
2	323.15 ($50\,^\circ$C)	6.24×10^4	0.648	

   Using the vehicle parameters from the second task of application calculate the power of the aerodynamic resistance for the two densities in the table ($v_v = 60$ m/s).

---

[5] For a calculation of the density see for example `http://wind-data.ch/tools/luftdichte.php?lng=en`.

4

Converters

Chapter 3 presented discrepancies between the ideal and the real delivery characteristic map. These discrepancies require a conversion of the real characteristic map supplied by an internal combustion engine with the aim of

1. closing the speed gap and
2. approximating the ideal delivery characteristic map by the real delivery characteristic map.

Other aspects such as environmental protection play an important role in the development of motor vehicles. Consequently, further environmental requirements have to be fulfilled, such as the reduction in fuel consumption and emissions of pollutants. Here we will look primarily at the two objectives listed above, i.e. closing the speed gap and attaining an approximation to the ideal delivery characteristic map.

Figure 4.1 shows the real and the ideal characteristic maps for the torque and the power together. The abscissa represents the speed, while the power or torque is plotted on the ordinate. For simplicity, only one rotational speed axis has been drawn in the diagram here. However, there should in fact be two axes: one for the speed of the wheels and one for the speed of the motor. However, we assume that the rotational speed of the motor has been translated in such a way that the speed limit of the real characteristic diagram coincides with the boundary of the ideal characteristic field (see the following example).

Similarly, the moment and the power of the engine have to be converted to the corresponding magnitudes at the wheels.

Example 4.1 The maximum speed of a motor is 6000 rpm (revolutions per minute). This is achieved in the highest gear at a speed of $v_{\max} = 50$ m/s. The radius of the wheel is $r_{wst} = 0.32$ m. From the radius r_{wst} and the maximum velocity $v_{\max}$, we

Vehicle Dynamics, First Edition. Martin Meywerk.
© 2015 John Wiley & Sons, Ltd. Published 2015 by John Wiley & Sons, Ltd.
Companion Website: www.wiley.com/go/meywerk/vehicle

obtain the maximum revolutions per minute of the wheel $n_{w\ max}$ (here the last term (s rpm = s rev/min) is necessary to convert the units):

$$n_{w\ max} = 60\frac{v_{max}}{2\pi r}\text{s}\frac{\text{rev}}{\text{min}}$$

$$\approx 1492\frac{\text{rev}}{\text{min}}\ .$$

This means that, between the engine speed, n_e, and the wheel speed, n_w, there should be a transmission ratio (from the differential) of $i_d = 4$:

$$i_d = \frac{n_e}{n_w}\ . \tag{4.1}$$

The index d here indicates the ratio of the differential.

This transmission ratio from the example is already considered in Figure 4.1. We see that it is necessary to have a basic transmission ratio, which is required to downsize the speed of the engine. The differential is necessary in addition to this basic gear ratio, which is why this basic transmission ratio is often implemented in the differential. One of the higher gears of the transmission can then be designed with a transmission ratio of 1, without a gear pair, in order to maximize the efficiency (cf. Figure 4.5).

The imperfections can be seen in Figure 4.1: the engine merely delivers torque and power above a certain speed, and yet large parts of the ideal characteristic map are not covered by the real characteristic map.

Converters are necessary to overcome these gaps.

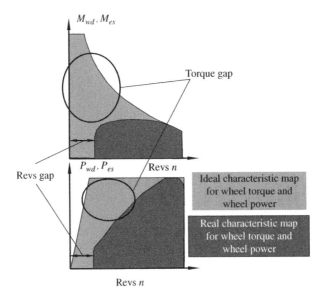

Figure 4.1   Real and ideal characteristic maps

4.1   Clutch, Rotational Speed Converter

To overcome the speed gap, we need a converter that ideally converts the torque and the power from high speeds to low speeds. This is not possible without power losses in the converter (as we see in the following), but it is possible for the torque. Hence, the input and the output torques are equal, but the speeds are not. These two characteristics can be used to define the speed converter with the following two equations:

$$M_o = M_i \, , \tag{4.2}$$

$$n_o \neq n_i \, . \tag{4.3}$$

Here, M_i is the torque at the input of the speed converter, M_o the torque at the output and n_i and n_o are the speeds of the input and output, respectively (here with the units rev/s). If we look at the following brief calculation, it is immediately evident that the input power P_i and the output power P_o have to be different

$$P_o \neq P_i \, , \tag{4.4}$$

where

$$P_o = \underbrace{2\pi n_o}_{\omega_o} M_o \, , \tag{4.5}$$

$$P_i = \underbrace{2\pi n_i}_{\omega_i} M_i \, . \tag{4.6}$$

The principle of speed converters becomes clear when they are illustrated using Figure 4.2. Two discs are fastened at the ends of two shafts. The left-hand shaft is joined to the engine, and the right-hand shaft is connected to the transmission. The engine shaft rotates at the angular velocity $\omega_i = 2\pi n_i$, and the torque is M_i. The analogue parameters of the transmission shaft are $\omega_o = 2\pi n_o$ and M_o.

The two discs are now pressed by a coupling force, F, against each other. This means that the normal force acts on the friction surface and, if $\omega_i \neq \omega_o$ holds true, then a frictional force occurs. From Figure 4.2, we obtain the sum of the moments about the rotational axis (for the stationary condition, i.e. $\dot{\omega}_i = 0$ and $\dot{\omega}_o = 0$ which results in vanishing torques due to inertia effects):

$$M_o = M_i \, . \tag{4.7}$$

This condition for a speed converter is thus fulfilled. It can be seen in the free-body diagram of Figure 4.2 that there are no inertial forces. Consequently, the relation $M_o = M_i$ holds true only during stationary operation. In non-stationary operation, the error would be small if the moments of inertia were small (which is generally not the case for a speed converter coupled with a reciprocating internal combustion engine).

With the help of Figure 4.3, we show an example of how a speed converter operates in a vehicle starting on a hill with a slope of 10%. The diagram shows the moments

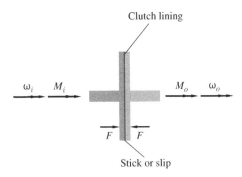

Figure 4.2   Principal mode of operation of a speed converter

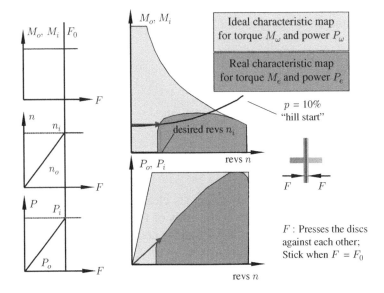

Figure 4.3   Engagement of clutch when starting on a hill

M_o and M_i and the power P_o and P_i in each case, i.e. the output and the input of the speed converter. The inputs M_i and P_i correspond to the motor torque $M_e = M_i$ and the motor power $P_e = P_i$ (here the index e stands for engine). The ideal characteristic map and the real characteristic map (for the fourth gear) are shown in both diagrams as well as the torque demand curve for $p = 10\%$ is highlighted in the diagram for the torque. The intersection of this torque demand curve with the real characteristic map indicates the necessary rotational engine speed, n_e, at which the engine must rotate to provide the required torque. We consider the process in which the contact force, F (see Figure 4.2), is slowly increased until F attains the value F_0. At this value, the two clutch plates stick to each other. The motor speed is kept constant during this process with the clutch in. If we increase the contact force, then the output moment, M_o, of the speed converter rises immediately, which indicates: $M_o = M_i$ (see the small diagram at the top left of Figure 4.3). The output speed, n_o, also increases with the contact force, F. When the latter reaches F_0, the discs stick and the speeds $n_o = n_i$ are the same (if the output torque is slightly higher than the torque for driving on the slope). This excess torque is necessary so that the vehicle can also be accelerated and n_o can increase. From $M_o = M_i$ and with Equations (4.5) and (4.6) we obtain

$$\frac{P_o}{2\pi n_o} = \frac{P_i}{2\pi n_i} , \tag{4.8}$$

which can be rearranged as

$$P_o = P_i \frac{n_o}{n_i} . \tag{4.9}$$

The efficiency $\eta = \frac{n_o}{n_i}$ of the clutch increases to 1 with increasing force, F. If the real characteristic map is attained, the speed converter is no longer needed. At this point it should be noted that, as illustrated here, the vehicle generally does not move forward from standstill in higher gears but in first gear. The characteristic diagram for the fourth gear was chosen here purely for the convenience and ease of presentation.

Figure 4.4 shows more details of a friction clutch. The shaft of the motor is fixed to a flywheel. The flywheel is rotatably mounted on the transmission shaft. In addition to the flywheel, the clutch cover contains the diaphragm spring, the pressure plate, the clutch lining and the throwout bearing (or release bearing). To disengage the clutch, a force presses against the throwout bearing. In the engaged state, the diaphragm spring presses the pressure plate with the clutch lining against the flywheel. The clutch lining is firmly connected by the clutch disc to the transmission with the shaft. To disengage

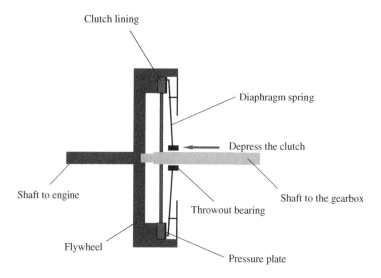

Figure 4.4 Significant functional components of a friction clutch

the clutch, a force, F, acts in the direction of the arrow, the pressure of the pressure plates then decreases and the friction lining starts to slip on the flywheel.

4.2 Transmission, Torque Converter

We now turn to the second part, completing the coverage of the ideal delivery characteristic map (Figure 4.1) with the help of the transmission (or torque converter). Here we will only consider the stepped transmission. This raises the question of how the ratios of the individual stages of the transmission should be chosen. We use the transmission ratio to identify a specific stage of the transmission.

Transmission ratio: The transmission ratio, i_z, is the ratio (the quotient) of the input speed n_{iz} to the output speed n_{oz} of a transmission:

$$i_z = \frac{n_{iz}}{n_{oz}} \quad z = 1, \ldots, N_{z\,\text{max}} . \tag{4.10}$$

The index z indicates the stage of transmission with $N_{z\,\text{max}}$ gears. The transmission ratio, i_z, is independent of the speed. The name gear ratio is used, too.

Progression ratio: The progression ratio, α_{gz}, denotes the ratio (quotient) of the transmission ratios for two adjacent gears:

$$\alpha_{gz} = \frac{i_{z-1}}{i_z} \quad z = 2, \ldots, N_{z\,\text{max}} . \tag{4.11}$$

We consider the layout design of a five-speed transmission ($N_{z\,\text{max}} = 5$). The fifth gear of a step-variable transmission can be designed so that the maximum speed of the vehicle can be achieved at the highest speed of the engine. This is only possible for vanishing resistances, which means that the gradient resistance for negative gradients, $p < 0$, compensates for the sum of the rolling resistance and the aerodynamic drag force. Furthermore, this maximum speed is not the maximum speed that can be attained by the vehicle on a non-inclined road. In Figure 4.1, the real characteristic diagrams have been drawn so that this is the case. The transmission of the final drive or differential is used to adjust the maximum speed of the engine to the maximum speed of the vehicle. The fifth or one of the higher gears of the transmission is often selected as a direct gear with $i_5 = 1$ by choosing an interlocking connection between the input and output shafts of the transmission[1], to minimize mechanical losses. In higher gears, fuel consumption plays an important role. This is the reason for minimizing losses and maximizing efficiency in the highest gear. The transmission ratio of the first gear is selected so that the requirement for the maximum torque is met. The transmission ratios have to be chosen such that a good approximation to the ideal

---

[1] This locking interconnection is not possible for all transmission designs.

characteristic map is achieved. Figure 4.7 shows the ideal characteristic map and the real characteristic map on the left. The real characteristic diagram is already converted by the differential (transmission ratio i_d and efficiency η_d) on the right boundary of the ideal characteristic map. Apart from the gear ratio, $i_z = n_{iz}/n_{oz}$, converting the torque characteristic map also requires the efficiency, η_z:

$$P_o = \eta_z P_i \ . \tag{4.12}$$

With $P_i = 2\pi n_{iz} M_i$ and $P_o = 2\pi n_{oz} M_o$ (n_{iz} and n_{oz} in rev/s), we also obtain from this equation:

$$M_o = M_i \eta_z \frac{n_{iz}}{n_{oz}} \ . \tag{4.13}$$

The efficiencies η_z for manual transmissions are close to 1.

Figure 4.5 shows examples of transmission ratios[2].

In some of the designs, the transmission ratio of 1 can be recognized, but not all of the designs are geometric, which is obvious from Figure 4.6. In the last three cars, the same automatic transmission is used; this is the reason why no differences can be seen.

The design of gears 2, 3 and 4 can be prepared by two formal methods: the geometric design and the progressive design. For the geometric design, we obtain

$$\alpha_{g5} = \alpha_{g4} = \alpha_{g3} = \alpha_{g2} \ . \tag{4.14}$$

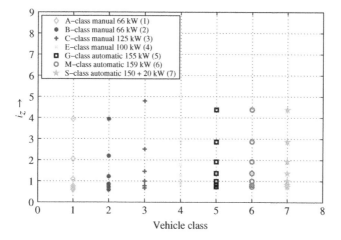

Figure 4.5 Gear ratios for several vehicles from Mercedes

[2] Source: http://www.mercedes-benz.de/content/germany/mpc/mpc_germany_website/de.../home_mpc/passengercars /home/new_cars/models/a-class/w176/facts_/technicaldata/models.html substitute a-class/w176 by: b-class/w246, c-class/w205,e-class_ w212, g-class/w463_crosscountry, m-class_ w166, s-class/w222, Sept. 18th 2014.

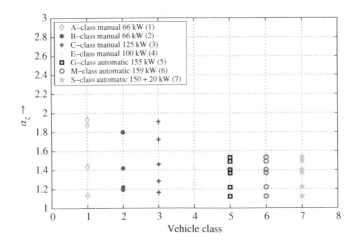

Figure 4.6   Progression ratios for several vehicles from Mercedes

The increments are therefore constant. Using the definition of the progression ratio from Equation (4.11), Equation (4.14) is equivalent to

$$\frac{i_4}{i_5} = \frac{i_3}{i_4} = \frac{i_2}{i_3} = \frac{i_1}{i_2} . \tag{4.15}$$

From Equation (4.15), we obtain $i_1 i_5 = i_2 i_4 = i_3^2$, which yields

$$i_3 = \sqrt{i_1 i_5} \, , i_2 = \sqrt{i_1 i_3} \text{ and } i_4 = \sqrt{i_3 i_5} . \tag{4.16}$$

In a progressive design, the increments are chosen according to the following equation:

$$\alpha_{gz} = \alpha_{p1} \alpha_{p2}^{5-z} . \tag{4.17}$$

Defining the transmission ratios allows the characteristic maps for the power and for the torque to be converted.

Figure 4.7 shows an example.

For both torque and power maps, we still detect differences between ideal and real maps; in particular, gaps can be identified in comparison of the maps. In the geometric design the gaps are approximately of the same size. The geometric design is used mainly in utility vehicles such as lorries. The progressive design, in which the gaps become smaller with increasing speed, is usually found in private or passenger cars (cf. Figure 4.7). Due to the low driving resistance for private cars at lower speeds, these gaps do not drop as much in weight. For higher velocities of 60 m/s or higher, the significant increase in aerodynamic drag force requires smaller gaps. These small gaps enable the driver to use the power limit (or the level slightly below this) to accelerate the car without causing a collapse in torque and hence a collapse in the acceleration.

Figures 4.9 and 4.10 compare geometric and progressive transmission ratios (P_{wsi} and F_{wsi} denote the supplied power and forces, respectively, at the wheels). A

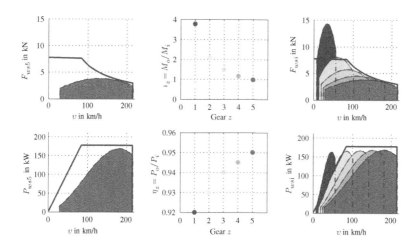

Figure 4.7 Basic operating principle of the manual transmission

reduction in the gaps can be clearly seen in the progressive grading gaps, while the gaps in the geometrical grading remain almost constant. In the shift chart (in the left graphs of Figures 4.9 and 4.10) it can be seen, that the speed decreases when shifting is constant for the geometric design, whereas in progressive grading it decreases with increasing gear.

One aim of the transmission is to cover the ideal characteristic map. Another aim is that, for a broad area in the tractive force–velocity plane, the powertrain should be able to deliver good efficiencies of the engine (the same applies to the power–velocity plane). Achieving this latter goal calls for a smooth design of the transmission. An absolutely smooth design can be achieved with a continuously variable transmission, which means that every transmission ratio within a certain range can be attained. This continuous design can be realized approximately by means of a transmission with a large number of discrete gears. This is one reason why the number of gears in automatic transmissions has risen over the course of time.

Figure 4.8 shows a nine-speed automatic gear unit. The spread of the transmission ratio is $i_1/i_9 = 9.15$, the efficiency of the Trilok converter is $\eta = 0.92$. The efficiency is increased by a lock-up clutch in non-converting mode of the Trilok converter. A twin torsional damper and a centrifugal pendulum absorber reduce the torsional oscillations. The transmission has four planetary wheel sets and six shifting elements for braking single wheels of wheel sets (brakes) or for joining wheels or wheel sets (clutches).

For both the speed converter and torque converter, there are other operating principles and constructive design reactions, such as a continuously translating transmission or a speed converter with fluid operation (cf. Chapter 17).

At this point, we have derived the demand of power and forces at the wheel resulting from driving resistances, on the one hand, while, on the other, we have deduced the

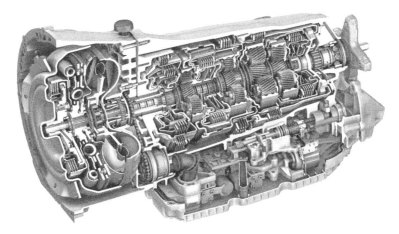

Figure 4.8 Nine-speed Mercedes automatic gear unit (reproduced with permissions of Daimler AG)

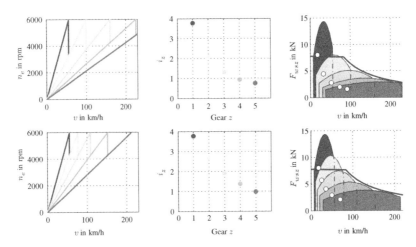

Figure 4.9 Geometric (second row diagrams) and progressive transmission (first row diagrams) ratios represented for the forces

supply of forces and power from a powertrain (internal combustion engine with a transmission).

These quantities are summarized in Figures 4.11 and 4.12 for the forces and the power, respectively. For the sake of completeness, the ideal characteristic maps are depicted, too. In the next steps (in the following chapter), we will compare the supply of forces and power from the powertrain with the demands of a specific driving situation, e.g. the demand when climbing a hill with $p = 0.1$.

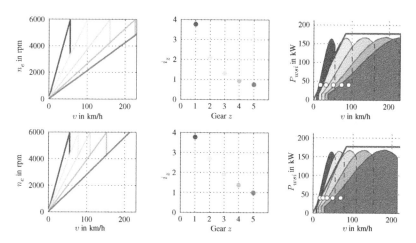

Figure 4.10 Geometric (second row diagrams) and progressive (first row diagrams) transmission ratios represented for the power

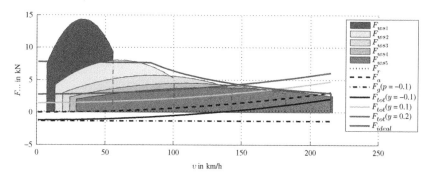

Figure 4.11 Driving performance diagram (forces)

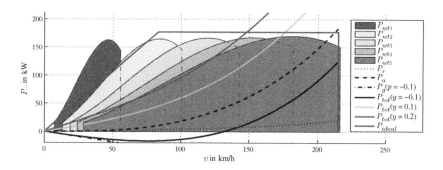

Figure 4.12 Driving performance diagram (power)

4.3   Questions and Exercises

Remembering

1. What is the aim of a speed converter?
2. What is the aim of a torque converter?
3. What is the definition of the transmission ratio?
4. What is the definition of the progression ratio?
5. Name two ways of designing a manual transmission.
6. What conditions apply to the input and output sides of torque and speed of the speed converter?
7. What is the efficiency of a speed converter?
8. What is the ratio of the input-to-output speed of a torque converter?

Understanding

1. In Figure 4.13 you see a transmission.

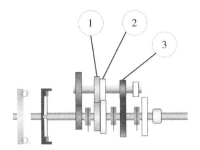

Figure 4.13   Transmission

   Which number of the gears shown belongs to the third gear?
2. Is it possible to design a combined speed and torque converter with $P_o > P_i$ for steady-state condition?
3. Describe non-steady-state situations, where $P_i < 0$ and $P_o < 0$ for a clutch or a transmissions.

Applying

1. The full-load characteristic for the torque of an engine is approximated by a parabola. This parabola interpolates the points in the table (M: moment; n: speed; for convenience indices are omitted):

$$
\begin{array}{c|c}
n \text{ in rpm} & M \text{ in Nm} \\
\hline
n_1 = 1000 & M_1 = 100 \\
n_2 = 3000 & M_2 = 200 \\
n_3 = 5000 & M_3 = 100
\end{array}
\tag{4.18}
$$

Calculate the interpolation parabola:

$$M(n) = a_2 n^2 + a_1 n + a_0 .\tag{4.19}$$

The maximum of the parabola (which is the maximum torque of the engine) is: $M_{\max} = M_2$ at $n = n(M_{\max})$. The minimum speed is $n_{\min} = n_1$, the maximum speed $n_{\max} = n_3$, the corresponding torques are $M(n_{\min})$ and $M(n_{\max})$, respectively.

Hint: You can use Lagrange's interpolation formula:

$$M(n) = \sum_{i=1}^{3} M_i \prod_{\substack{j=1\\j\neq i}}^{3} \frac{n - n_j}{n_i - n_j} .\tag{4.20}$$

2. Calculate the maximum power of the engine and the revolutions per minute, where the power becomes maximum.

   Hint: You obtain the power from $P = M\omega$, and thus the revolutions per minute, where the power becomes maximum, by

   $$\frac{\partial P}{\partial n} = 0 .\tag{4.21}$$

3. The aim of the remaining parts of this application (3 to 6) is to design a six-gear progressive transmission. The fourth gear should be a gear with a direct joint between the input and output shafts; thus, the gear ratio is $i_4 = 1$.

   The radii of the wheels are $r_{wst} = \frac{10.5}{10\pi}$ m.

   Calculate the gear ratio i_d of the differential under the condition that the vehicle speed of 180 km/h in the fourth gear is attained for the maximum engine speed of n_3.

4. We have the parameters for the aerodynamic drag of the car: $c_d = 0.3$, $\rho_a = 1.2$ kg/m^3, $A = 2$ m^2.

   Calculate the gear ratio i_6 of the sixth gear assuming that the maximum velocity of the vehicle in the sixth gear becomes maximum, i.e. for other gear ratios $\tilde{i}_6 \neq i_6$ the maximum velocity is always less than that which you should calculate for the gear ratio i_6. Please follow the hints for the simplification of this calculation; otherwise, the calculations become too complicated.

   Hints:
   - You obtain this gear ratio i_6, if you calculate the section between the basic demand of power (without the small rolling resistance F_r) of the vehicle and the full load characteristic curve of the sixth gear assuming that this section point is at the maximum power (cf. task 2 of application) of the engine.
   - You are recommended to neglect the rolling resistance; otherwise, you will have to calculate the roots of a fourth-order polynomial.
   - You should assume that the efficiency in the sixth gear is $\eta_6 = 1$.

5. The mass of the car is 1200 kg, the maximum payload 800 kg and a towed trailer 500 kg (for calculation of the gradient resistance use $10\,\text{m/s}^2$ and the simplified formula $F_g = Gp$).

   Calculate the gear ratio of the first gear i_1. The car with payload and trailer should be able to drive on an inclined road with $p = \pi/10$ at the maximum available engine torque of 200 Nm.

6. The progression ratio for a progressive design is

$$\alpha_{gz} = \alpha_{p1}\alpha_{p2}^{6-z} = \frac{i_{z-1}}{i_z} \ . \tag{4.22}$$

Calculate α_{p1} and α_{p2}.

Hint:
This or other similar derived formulas will help you:

$$\alpha_{gz}\alpha_{g(z-1)} = \frac{i_{z-1}}{i_z}\frac{i_{z-2}}{i_{z-1}}$$

$$= \frac{i_{z-2}}{i_z}$$

$$= \alpha_{p1}^2\alpha_{p2}^{12-(2z-1)} \tag{4.23}$$

5

Driving Performance Diagrams, Fuel Consumption

In this chapter, we will discuss driving performance diagrams.

Driving performance diagrams: A driving performance diagram comprises
1. the (real) supply characteristic maps of the engine converted to forces and power at the wheels as a function of the driving speed and in the same diagram
2. the required tractive effort (the driving resistances) or the effort for the power.

With the help of these diagrams, we can, for example, determine the maximum speed without gradient, the climbing ability in any gear and the acceleration capability.

Figure 5.1 shows characteristic supply maps for each of the five gears (force and power) and the demands for several driving situations for the standard set of parameters. These demand curves are based on the driving resistance and consist of the essential components of the gradient resistance F_g, acceleration resistance (inertial resistance) F_i, rolling resistance F_r and aerodynamic drag F_a.

The basic difference to Figures 4.11 and 4.12 is that Figure 5.1 shows an all-wheel drive with respect to the adhesion limit. It is obvious that the limit in Figure 5.1 is nearly twice as high as the limit in the other diagrams of Figures 4.11 and 4.12.

The supply characteristic diagram of the engine must be converted (by the transmission ratios i_z and the efficiency η_z of the transmission and the transmission ratios i_d and the efficiency η_d of the differential) to be comparable with the demand curves. Hence, the power and the torque of the engine are converted to tractive force, and the power at the driven wheels and the angular velocity (or speed) of the engine must be converted to the rotational speed of the wheels, n_w, and then to the driving speed, v, of the vehicle:

$$v = r_{wst} 2\pi \underbrace{\frac{n_e}{i_z i_d}}_{n_w} . \tag{5.1}$$

Vehicle Dynamics, First Edition. Martin Meywerk.
© 2015 John Wiley & Sons, Ltd. Published 2015 by John Wiley & Sons, Ltd.
Companion Website: www.wiley.com/go/meywerk/vehicle

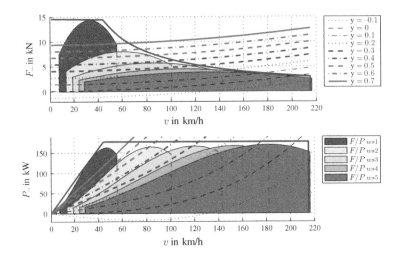

Figure 5.1   Driving performance diagrams; $y = p + \lambda \ddot{x}/g$

First the torque, M_e, from the engine must be converted using the total transmission ratio of the torque and speed converter, i_t, to the torque at the wheel. The total transmission ratio $i_t = i_z i_d$ consists of the transmission ratio of the differential i_d and that of the transmission, i_z. The torque at the wheel is, however, reduced due to torque losses. The torque losses yields the efficiency of the torque and speed converter $\eta_t = \eta_z \eta_d$ (η_t is the efficiency of the transmission in zth gear and the differential), so the supply torque, M_{ws}, at the wheels is

$$M_{ws} = \eta_t i_t M_e \qquad (5.2)$$

for driving, and

$$M_{ws} = \frac{1}{\eta_t} i_t M_e \qquad (5.3)$$

for braking.

The distinction between driving and braking is necessary, since the torque losses are always braking torques, thus reducing the driving torque of the engine or increasing the braking or drag torque of the engine. The equation for braking (5.3) holds approximately for efficiency η_t close to unity:

$$\eta_t = 1 - \zeta_t , \qquad (5.4)$$

with $\zeta_t \ll 1$.

The torque loss, M_l, of the speed and torque converter is

$$M_l = \zeta_t i_t M_e . \qquad (5.5)$$

The braking torque at the wheels is composed of the drag torque of the engine, $i_t M_e$, and the torque loss, $M_l = \zeta_t i_t M_e$:

$$i_t M_e + \zeta_t i_t M_e = (1 + \zeta_t) i_t M_e \,. \tag{5.6}$$

If we consider the following series expansion:

$$\frac{1}{\eta_t} = \frac{1}{1 - \zeta_t}$$

$$= \left(1 + \zeta_t - \frac{1}{2} \eta_t^2 + \cdots \right) \tag{5.7}$$

we recognize the following relationship and hence the approximate validity of Equation (5.3):

$$M_{ws} = (1 + \zeta_t) i_t M_e$$

$$\approx \frac{1}{1 - \zeta_t} i_t M_e \,. \tag{5.8}$$

Similarly for the power

$$P_{ws} = \eta_t P_e, \tag{5.9}$$

$$P_{ws} = \frac{1}{\eta_t} P_e, \tag{5.10}$$

where in (5.9) the power for the driven axle is positive $P_e > 0$ and in Equation (5.10) the power for the braked axle is negative $P_e < 0$.

We obtain Equation (5.9) as follows:

$$P_{ws} = \omega_w M_{ws}$$

$$= \omega_w \eta_t i_t M_e$$

$$= \eta_t \overbrace{\omega_w i_t M_e}^{=P_e}$$
$$\qquad \underbrace{}_{\omega_e}$$

$$= \eta_t P_e \,. \tag{5.11}$$

The derivation of Equation (5.10) is similar.

Remark 5.1 In these calculations, we are simplifying the assumption of a constant, speed-independent efficiency, η_t. However, the efficiency is a function of the angular velocity, the transmission ratio and the torque itself.

Figure 5.1 shows the power and the tractive forces supply characteristic maps for the different gears, including the loss. The demand curves are shown for different values $y = p + \lambda \ddot{x}_v / g$. In the following sections, we will have a closer look at these diagrams in order to obtain characteristics of a vehicle.

5.1 Maximum Speed without Gradient

To determine the maximum speed of a vehicle without any gradient of the road, we determine the intersection of the demand power curve with the full load characteristic of the fifth gear (see the upper diagram in Figure 5.2). The case illustrated yields the speed v_1. In the upper part of Figure 5.2, we see that the demand curve for $y = 0$ intersects the full load curve at its maximum. In this setting, the highest value for the maximum speed can be attained in the fifth gear, $z = 5$, P_{ws5}. In the fourth gear, $z = 4$, P_{ws4}, the vehicle reaches a maximum speed $v_2(z = 4) < v_1(z = 5)$.

If we consider the maximum speed for a gradient of $p = 0.12$, it is obvious, i.e. v_3 for the fourth gear and v_4 for the fifth gear that the reverse holds: $v_4(z = 5) < v_3(z = 4)$.

There are also possible settings for the fifth gear, other than the design described above, in which the demand curve intersects the full load curve at the maximum power. The lower part of Figure 5.2 indicates two other possibilities. For a characteristic map, P_{ws1}, the transmission, $\hat{i}_5$, is larger than that for P_{ws2}: $\hat{i}_5 > i_5$. For the third characteristic map, P_{ws3}, the following holds: $\tilde{i}_5 < i_5$. It can be seen that for both P_{ws1} and P_{ws3}, the maximum velocities v_3 and v_2 are, respectively, smaller than v_1: $v_2 < v_1$ and $v_3 < v_1$. Comparing the maximum velocities for a gradient with $p = 0.12$, we observe that the maximum velocity v_5 for P_{ws1} is higher than v_6 for P_{ws2}: $v_5 > v_6$.

The decrease in the maximum velocity reaches a maximum for P_{ws3} and a minimum for P_{ws1}:

$$v_2 - v_4 > v_1 - v_6 > v_3 - v_5 . \tag{5.12}$$

Here the variant P_{ws3} of the engine rotates at a low speed, i.e. in a range with a higher efficiency. The lower speed and higher efficiency (see Figure 3.12) mean that less fuel is consumed. This gear is therefore called an economic drive or overdrive.

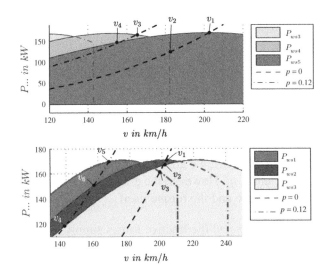

Figure 5.2 Determination of the maximum speed without gradient

In the layout P_{ws1}, the engine rotates at a higher speed for the same vehicle speed. The two variants P_{ws1} and P_{ws3} differ significantly in power reserve, which can be recognized by the drop in maximum velocities for the gradient (cf. Equation (5.12)).

5.2 Gradeability

The gradeability is the ability of a vehicle to drive on a road with a certain inclination of p. If the velocity v lies below the maximum speed, the excess of power (the difference between the demand for driving with no gradient, $p = 0$, and the full load power of the powertrain) can be used for accelerating or driving on an inclined road. The gradient resistance, F_g, can be written as a function of the remaining resistances F_r and F_a as,

$$F_g = Z(z) - (F_r + F_a) , \qquad (5.13)$$

where $Z(z)$ is the maximum tractive force in gear z (on the full load characteristic curve). The acceleration resistance should not be considered since no acceleration $\ddot{x}_v = 0$ occurs when the vehicle is driving on the slope. By replacing $F_g = pG$ and $F_r = f_r G$, it follows that

$$p = \frac{1}{G}(Z(z) - F_a) - f_r . \qquad (5.14)$$

The full load characteristic curve $Z(z)$ can be replaced by the supplied power from the powertrain in gear z: P_{wsz}/v.

5.3 Acceleration Capability

The acceleration capability can be considered as similar to the climbing ability. The demand characteristic maps in Figure 5.1 are shown for the parameter $y = p + \lambda \frac{\ddot{x}_v}{g}$. To obtain the acceleration capability, we have to replace p by $\lambda \frac{\ddot{x}_v}{g}$ in Equation (5.14). We obtain

$$\ddot{x}_v = \frac{g}{\lambda_z G}((Z(z) - F_a) - G f_r) . \qquad (5.15)$$

The main difference between the gradeability and acceleration ability is the rotating mass factor, λ_z, which depends on the gear engaged.

In order to make statements about the acceleration ability of a vehicle, we often use average data such as the time required for a vehicle to change from a speed of $v_1 = 0$ km/h to a speed of $v_2 = 100$ km/h. To calculate this, the equation

$$\ddot{x}_v = \frac{dv_v}{dt} \qquad (5.16)$$

is rearranged (by reading dv_v and dt as a differential in the mathematical sense)

$$dt = \frac{1}{\ddot{x}_v} \, dv_v . \qquad (5.17)$$

Integration of Equation (5.17) yields

$$\Delta t = \int_{v_1}^{v_2} \frac{1}{\ddot{x}_v} \mathrm{d}v_v \ . \tag{5.18}$$

The average acceleration, $\overline{\ddot{x}}_v$, for this case would be

$$\overline{\ddot{x}}_v = \frac{v_2 - v_1}{\Delta t} \ . \tag{5.19}$$

The travel distance during the acceleration process can be calculated by using the following relationship:

$$\begin{aligned} \ddot{x}_v &= \frac{\mathrm{d}v_v}{\mathrm{d}t} \\ &= \frac{\mathrm{d}v_v}{\mathrm{d}x_v} \frac{\mathrm{d}x_v}{\mathrm{d}t} \\ &= \frac{\mathrm{d}v_v}{\mathrm{d}x_v} v_v \end{aligned} \tag{5.20}$$

which, after solving for $\mathrm{d}x_v$, yields

$$\Delta x_v = \int_{v_1}^{v_2} \frac{v_v}{\ddot{x}_v} \ \mathrm{d}v_v \ . \tag{5.21}$$

Due to the different rotational mass factors in different gears, the speed, v_{optg}, for which the gear needs to be changed to achieve a maximum climbing ability differs when compared with the speed, v_{opti}, at which maximum acceleration ability is attained. To illustrate this, we consider the upshift point from the first to the second gear (cf. Figure 5.3). When we consider the free tractive force $Z_f(z)$ ($Z(z)$ is the maximum tractive force on the full load characteristic curve for the zth gear)

$$Z_f(z) = Z(z) - F_r - F_a \tag{5.22}$$

the intersection point of $Z(z = 1)$ and $Z(z = 2)$ yields the optimum velocity, v_{optg}, for upshifting from the first to second gear in order to achieve optimum climbing ability or in order to achieve maximum towed load for an inclined road. However, when we look for this speed, v_{optg}, in the acceleration capability, we obtain the following for these two gears ($\hat{Z}_f = Z_f(z = 1) = Z_f(z = 2)$):

$$\ddot{x}_{1.\mathrm{gear}} = \frac{g}{\lambda_1 G} \hat{Z}_f \ , \tag{5.23}$$

$$\ddot{x}_{2.\mathrm{gear}} = \frac{g}{\lambda_2 G} \hat{Z}_f \ . \tag{5.24}$$

Since $\lambda_1 > \lambda_2$, the maximum acceleration at this speed v_{optg} for the first gear is significantly below the maximum acceleration for the second gear. The speed for the optimum upshift point $v = v_{\mathrm{opti}}$ for optimum acceleration capability is obtained if the ratio

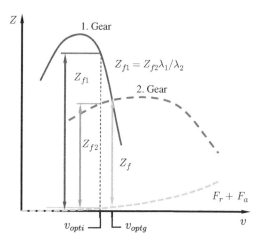

Figure 5.3 Free traction, Z_f, at the intersection of traction full load characteristics for the first and second gears

of free traction for the first gear, $Z_{f1} = Z_f(z = 1)$, to the second gear, $Z_{f2} = Z_f(z = 2)$, corresponds to the ratio of the rotational mass factors:

$$\frac{Z_{f1}}{Z_{f2}} = \frac{\lambda_1}{\lambda_2} . \tag{5.25}$$

The optimum upshift velocity, v_{opti}, thus lies below the optimum point, v_{optg}, for optimum climbing ability: $v_{\mathrm{opti}} < v_{\mathrm{optg}}$.

5.4 Fuel Consumption

First, we will assume constant efficiency of the engine and transmission for the fuel consumption, which involves a simplification because the efficiency of the engine depends on the torque and the speed, while the efficiency of the powertrain depends (slightly) on these quantities. From Figure 3.12, we know that the efficiency of the engine depends on the torque (or power) and the rotational speed. Furthermore, the efficiency of the entire speed and torque converter (clutch, transmission and differential) depends on the rotational speed and torque, too. In the first approach to fuel consumption, the efficiency of the motor, $\bar{\eta}_e$, and the speed and torque converter including the differential, $\bar{\eta}_t$, is assumed to be constant. To calculate the fuel consumption, we need the work (or energy) W_w required for moving the vehicle. This work is calculated from the power, P_w at the wheels:

$$W_w = \int_0^T P_w \, \mathrm{d}t , \tag{5.26}$$

where the power is approximately equal to the product $P_w \approx F_w v_v$ (the slip is neglected; F_w is the tractive force at the driven wheels). If B is the amount of fuel (e.g. in $\ell = (\mathrm{dm})^3$) and H_l the lower heating value (e.g. in J/ℓ), we obtain from the amount of fuel, B, an energy which is equal to the work W_w:

$$\int_0^T F_w v_v \ \mathrm{d}t = W_w = \overline{\eta}_e \overline{\eta}_t B H_l \ . \tag{5.27}$$

The lower heating value, H_l, is the amount of heat per unit volume (or mass unit) of a fuel that is released during complete combustion, along with water that is produced in the gaseous form. The lower heating value is relevant for internal combustion engines. It differs from the upper heating value, H_u, by the heat of vaporization of water Q_v: $H_l = H_u - Q_v$. Solving Equation (5.27) for B, we obtain the fuel consumption in relation to the distance travelled, L:

$$\frac{B}{L} = \frac{1}{\overline{\eta}_e \overline{\eta}_t H_l} \frac{1}{L} \int_0^T F_w v_v \ \mathrm{d}t \ . \tag{5.28}$$

When driving with a constant speed v_0 (i.e. $F_i = 0$ and $F_w = F_r + F_a + F_g = $ const.), we obtain,

$$\frac{B}{L} = \frac{F_r + F_a + F_g}{\overline{\eta}_e \overline{\eta}_t H_l}. \tag{5.29}$$

Here $v_0 = L/T$ is substituted. The fuel consumption is therefore low if the efficiency of the motor and of the speed and torque converter are large and when the driving resistances, F_r, F_a and F_g, are small.

The case of varying velocity $v_v(t) = \overline{v} + \Delta v(t)$ (Figure 5.4) will be considered in more detail below. Here $\overline{v}$ is the average velocity (with respect to time, not with respect to the distance travelled)

$$\overline{v} = \frac{1}{T} \int_0^T v_v(t) \ \mathrm{d}t \tag{5.30}$$

and Δv denotes the deviation

$$\Delta v(t) = v_v(t) - \overline{v} \ . \tag{5.31}$$

Here we restrict ourselves to non-inclined roads because the gradient resistance provides a component independent of the velocity. With reference to the driving distance,

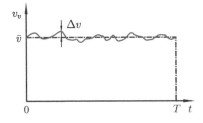

Figure 5.4   Example of varying velocity

the fuel consumption is given by

$$\frac{B}{L} = \frac{1}{\bar\eta_e \bar\eta_t H_l L} \int_0^T (F_r + F_i + F_a) v_v \ \mathrm{d}t \ . \tag{5.32}$$

If we set $F_r = f_r G$, $F_i = \lambda \frac{G}{g} \ddot{x}_v$ and $F_a = \frac{\rho_a}{2} c_d A v_v^2$, we obtain the following for the integral:

$$\int_0^T (F_r + F_i + F_a) v_v \ \mathrm{d}t = f_r G \int_0^T v_v \mathrm{d}t + \lambda \frac{G}{g} \int_0^T \dot{v}_v v_v \ \mathrm{d}t + \frac{\rho_a}{2} c_d A \int_0^T v_v^3 \mathrm{d}t \ . \tag{5.33}$$

Since the rolling resistance, F_r, plays a minor role, we neglect the velocity dependence of f_r and assume a constant coefficient of the rolling resistance f_r. The first term is then

$$f_r G \underbrace{\int_0^T v_v \ \mathrm{d}t}_{\bar v} = f_r G \bar v T \ , \tag{5.34}$$

where $\bar v$ is the average velocity. The second term can be substituted as follows $\left(\frac{1}{2} \frac{\mathrm{d}}{\mathrm{d}t} \left(v_v^2 \right) = \dot{v}_v v_v \right)$:

$$\int_0^T \dot{v}_v v_v \mathrm{d}t = \int_0^T \frac{1}{2} \frac{\mathrm{d}}{\mathrm{d}t} (v_v^2) \mathrm{d}t$$

$$= \frac{1}{2} (v_v^2(t = T) - v_v^2(t = 0)) \ . \tag{5.35}$$

When we use the integral to calculate the power in this way, we assume that the energy of the braking process (usually the thermal energy in brake discs) can be completely recovered. In a conventional motor vehicle with an internal combustion engine, this is not the case but would at least be approximately possible in a hybrid or electric vehicle with a very high efficiency. We assume in the vehicle under consideration that the initial velocity $v_v(t = 0)$ and the final velocity $v_v(t = T)$ are the same, so that

$$v_v(t = 0) = v_v(t = T) \ . \tag{5.36}$$

This is not an essential restriction because we consider a long period of time, T (e.g. a trip on a highway of 20 or 30 min or even longer). It follows that

$$\int_0^T \dot{v}_v v_v \ \mathrm{d}t = 0 \ . \tag{5.37}$$

For the deviation, $\Delta v(t)$, the following two equations hold:

$$\int_0^T \Delta v(t) \ \mathrm{d}t = 0 \ , \tag{5.38}$$

$$\int_0^T (\Delta v)^3(t) \ \mathrm{d}t \ll \bar v \int_0^T (\Delta v)^2(t) \mathrm{d}t \ . \tag{5.39}$$

Equation (5.38) results directly from the definition of Δv

$$\int_0^T \Delta v(t)dt = \int_0^T \left(v(t) - \bar{v} \right) \mathrm{d}t$$

$$= \underbrace{\int_0^T v(t)\ \mathrm{d}t}_{\bar{v}T} - \underbrace{\int_0^T \bar{v}\ \mathrm{d}t}_{\bar{v}T} \tag{5.40}$$

$$= 0\ . \tag{5.41}$$

The second relation (5.39) holds because firstly the velocity deviation Δv is small compared to the average velocity $\bar{v}$ and secondly because in the integral of $(\Delta v)^3$ positive and negative parts cancel each other out to some extent, whereas no compensation occurs in the other integral of $(\Delta v)^2$.

The integral of the third term can then be written as follows:

$$\int_0^T v_v^3 \mathrm{d}t = \int_0^T \bar{v}^3 + 3\bar{v}^2 \Delta v + 3\bar{v}(\Delta v)^2 + (\Delta v)^3\ \mathrm{d}t$$

$$= \underbrace{\int_0^T \bar{v}^3\ \mathrm{d}t}_{=T\bar{v}^3} + 3\bar{v}^2 \underbrace{\int_0^T \Delta v\ \mathrm{d}t}_{=0} + 3\bar{v} \underbrace{\int_0^T (\Delta v)^2\ \mathrm{d}t}_{=\sigma_v^2 T} + \underbrace{\int_0^T (\Delta v)^3\ \mathrm{d}t}_{\ll 3\bar{v}\sigma_v^2 T}$$

$$\approx T(\bar{v}^3 + 3\bar{v}\sigma_v^2)\ , \tag{5.42}$$

where σ_v is the standard deviation of v:

$$\sigma_v = \sqrt{\hat{T} \int_0^T (\Delta v)^2 \mathrm{d}t} \tag{5.43}$$

Overall, for consumption the result is

$$\frac{B}{L} = \frac{1}{\eta_e \eta_t H_l} \left[f_r G + \frac{\rho_a}{2} c_d A \bar{v}^2 \left(1 + \frac{3\sigma_v^2}{\bar{v}^2} \right) \right]. \tag{5.44}$$

This case uses the relation between the distance, L, the velocity, $\bar{v}$, and the time, T:

$$L = \int_0^T v(t)\ \mathrm{d}t$$

$$= \int_0^T \left(\bar{v} + \Delta v(t) \right)\ \mathrm{d}t$$

$$= \underbrace{\int_0^T \bar{v}\ \mathrm{d}t}_{\bar{v}T} + \underbrace{\int_0^T \Delta v(t)\ \mathrm{d}t}_{=0}\ . \tag{5.45}$$

On one hand, we notice the constant consumption due to the rolling resistance. On the other hand, we can see the consumption change as a result of the speed and aerodynamic drag.

In spite of the assumed complete recovery of energy when braking, the term $3\frac{\rho_a}{2}c_d A\sigma_v^2$ due to velocity changes remains in the equation. This term is a consequence of the quadratic dependence of the aerodynamic drag on the velocity.

Finally, we go to the consumption for speed-dependent efficiency η_e of the engine. Figure 5.5 shows the characteristic diagram of the diesel from Figure 3.12, where it had already been converted to the vehicle speed v_v. Shown in the left of the diagram is the fourth gear. The graph on the left shows three points of the greatest efficiency, too, which can be joint by the curve of the greatest efficiency. We can obtain this curve of the greatest efficiency by drawing a horizontal line for each value of the power (the dashed horizontal line on the left-hand graph of Figure 5.5) and seek the constant efficiency curve (these are the dot and dash lines) for which the horizontal, constant power line is a tangent at the maximum of the constant efficiency curve. In the graph, this is the example curve for $\eta_e = 0.36$.

The greatest efficiency curve is also the curve of the lowest specific fuel consumption. The right-hand graph compares the lowest specific fuel consumption curves (for two points) for the fourth gear and the fifth gear with each other.

If we compare a driving speed ($v_v \approx 25$ m/s, this is indicated in the graph on the right by the vertical dashed line) for the two gears at a low power, P_1, (lower horizontal line), the efficiency in the fifth gear is then $\eta_e(z = 5, P_1) = 0.36$ and in the fourth gear it lies below $\eta_e(z = 4, P_1) < 0.36$. If we consider the power P_2 (upper horizontal line), the efficiency in fourth gear is $\eta_e(z = 4, P_2) = 0.35$, in the fifth gear it lies slightly below $\eta_e(z = 5, P_2) < 0.35$. This means that there are driving speed performance regions in which the fourth gear is more efficient and those in which

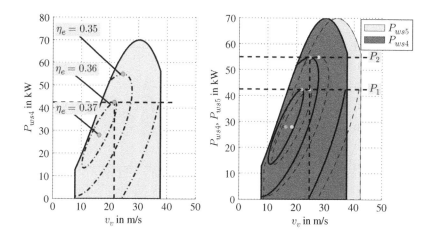

Figure 5.5 Points of best efficiency

the fifth gear is more favourable. However, this last case is the exception. In general, choosing a higher gear is more efficient for fuel consumption at a given speed and a given power than choosing a lower gear.

5.5 Fuel Consumption Test Procedures

A lot of test cycles exist for measuring fuel consumption and exhaust emissions and for comparing different cars. Examples include the NEDC (New European Driving Cycle), EPA (Environmental Protection Agency, United States) FTP-75 (Federal Test Procedure) or the SFTP (Supplemental FTP US06, SC06, Cold Cycle) or the 10 mode or 10–15 mode from Japan. The Worldwide harmonized Light vehicles Test Procedures (WLTP) is a test cycle developed by experts from the European Union, Japan, and India to harmonize the various test cycles used in different countries. These tests are usually executed on a chassis dynamometer. The road loads and the parameters of the vehicle have to transferred to the dynamometer. In these cycles, driving speed is defined as a function of time, with periods of standstill or stopping of the engine. Figure 5.6 shows the velocity for the NEDC. In the acceleration and deceleration periods, the velocity depends linearly on the time, which means that in these periods the acceleration is constant (in the other periods the acceleration is zero). The accelerations and the decelerations are low, with the maximum value for the acceleration in the city part of the cycle being: $(3.75 \text{ km/h})/(1\text{s}) \approx 1.042 \text{ m/s}^2$.

Periods of cold starts can also form part of the cycles, whereas gradients are not included. This means that energy recovery during negative inclination of a road is not included in the tests. Several boundary conditions are defined, such as temperatures, tyres and their inflation pressure, payload. In some older cycles, e.g. the NEDC, shift points for manual transmissions are defined, too. As modern vehicles use manual transmissions with up to six gears, fixed gear shift velocities are difficult, as these shift points influence the results. One aim of the cycles is to compare different cars

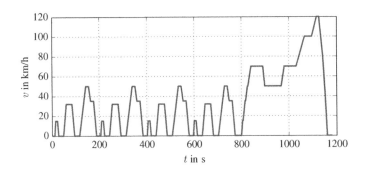

Figure 5.6 The velocity–time dependence of the NEDC

Table 5.1 WLTP vehicle classes

Class	Description	Power to mass ratio P/m_{tot}
1	Low-power vehicles	$P/m_{tot} \leq 22\,\text{kW/t}$
2	Medium-power vehicles	$22\,\text{kW/t} < P/m_{tot} \leq 34\,\text{kW/t}$
3	High-power vehicles	$P/m_{tot} > 34\,\text{kW/t}$

Power, P, in kW, mass m_{tot} in t

with respect to fuel consumption or exhaust emissions (e.g. CO or NO_x). The test procedures may differ according to the power and weight of the vehicles. In WLTP, for example, there are three classes (cf. Table 5.1) of vehicles distinguished by the power to weight ratio (in kW/t).

Different WLTC (test cycles) are defined for these three classes. The main differences are the velocity ranges, which for class 3, for example, cover a total of four parts: low, medium, high and extra-high velocities.

In order to assess a vehicle with respect to a test cycle, the demand of power or tractive force of this vehicle with respect to the test cycles can be plotted in the driving performance diagram of a vehicle. The driving performance diagram for this purpose should be extended by efficiency lines. Then the different demands can be assessed with respect to efficiency. An overall assessment with respect to emissions is not possible with simple diagrams, because the engine is operated in a non-steady-state mode.

Figure 5.7 depicts the power and the tractive forces at the wheels from a powertrain with an internal combustion engine and a five-speed transmission. It also shows the points of maximum efficiency in the five gears, and an elliptical approximation of one

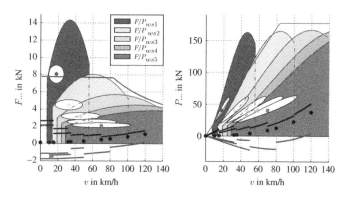

Figure 5.7 Power and tractive forces supplied by the powertrain with an internal combustion engine and the demand for power from NEDC

isoline of constant efficiency. In addition to the information depicted in the preceding diagrams, this figure also shows the demand for tractive force and power for the NEDC. Apart from the demand for power and tractive forces, the negative portions are shown, too. These negative parts can be used by hybrid or electric vehicles to recover parts of the kinetic energy. The parabolic shape of all pieces of the force curves is the result of the basic demand, and mainly the aerodynamic drag forces.

It is obvious that the demand for power for the NEDC is very low, which is the result of the low value of acceleration, and that the ranges for good efficiencies are not attained by these demand curves.

5.6   Questions and Exercises

Remembering

1. What is a driving performance diagram?
2. What are the essential components needed to determine the driving resistance?
3. On which variables do these components depend?
4. What data are needed for a driving performance diagram?
5. How should the engine characteristics be converted for the driving performance diagram?

Understanding

1. How do we determine the switching point for the optimum acceleration capability?
2. How do you determine the fuel consumption?
3. Are higher or lower gears more fuel efficient? Explain the relationships.
4. Why do the shift points of the optimum climbing ability not coincide with those of optimum acceleration?

Applying

1. The full load characteristic for the moment M of an engine is approximated by a parabola:

$$M(n) = a_2 n^2 + a_1 n + a_0 ,\qquad (5.46)$$

where $a_2 = -0.00005\,\text{Nm/(rpm)}^2$, $a_1 = 0.3\,\text{Nm/(rpm)}$, $a_0 = -50\,\text{Nm}$

Calculate the full load curve for the tractive force as a function of the vehicle velocity v

$$F(v) = b_2 v^2 + b_1 v + b_0 ,\qquad (5.47)$$

in the fourth gear $i_4 = 1$; the gear ratio of the differential is $i_d = 3$, the radii of the driven wheels are $r_{wst} = 0.3\,\text{m}$; the slip should be neglected.

Hint: You have to transform the revolutions of the engine **and** the moment.

2. Use the parameters from task 1 of 'Applying'.

   Estimate the climbing ability in the first gear $i_1 = 3$. For this, you should neglect the basic demand $F_r + F_a$ (otherwise the calculations become complicated).

   Please, use $g = 10 \, \text{m/s}^2$ and the simplified formula for the gradient resistance $F'_g = Gp$ (the value p is too high for this formula, but the calculation is very easy using with this value). The total mass of the vehicle is 1200 kg.

6

Driving Limits

Chapter 5 examined the derivation of maximum speed, climbing ability and acceleration capability for vehicles. The main focus there was on the power or tractive forces. In the case of $\mu > \mu_a$, it is not possible to apply the longitudinal force to the road. This relationship was captured by the third limit of the real characteristic map (page 37 in Chapter 3). We assume for this third limit in Chapters 3 and 4[1] that

1. only one axle of the vehicle is driven;
2. the centre of mass is in the middle of the car, which means (G is the total vehicle weight):

$$F_z = G/2;\tag{6.1}$$

3. that there is no transfer of axle load from the rear axle to the front axle and vice versa;
4. there are no other effects which influence the wheel or axle load; and
5. vertical motion and pitching motion may be neglected.

   In this chapter, we therefore turn in Section 6.1 to the vertical forces at the axles depending on different factors. These vertical forces are essential for the maximum transferable longitudinal forces when braking and accelerating the vehicle. Furthermore, the vertical forces affect the maximum possible tangential forces during cornering. Section 6.2 is dedicated to the braking process. Section 6.3 examines the distribution of braking forces.

   We restrict our considerations to motions without oscillations in the vertical direction and without pitching oscillations.

---

[1] In Figure 5.1, we assume an all-wheel-drive vehicle; load transfer from the rear to front axle and vice versa are neglected and so the other effects on axle loads.

Vehicle Dynamics, First Edition. Martin Meywerk.
© 2015 John Wiley & Sons, Ltd. Published 2015 by John Wiley & Sons, Ltd.
Companion Website: www.wiley.com/go/meywerk/vehicle

6.1 Equations of Motion

In the derivation of the equations of motion, we assume the free-body diagram in Figure 6.1. In addition to the forces previously considered, a lifting force by the air, F_{az}, and a moment, M_a, due to the aerodynamic forces are also included. The equations of motion for the body and for the front and rear axles are given directly in the free-body diagram.

The moment of inertia of the engine, J_e, is not taken into account, and this would be dependent on the orientation of the rotational axis of the engine: if the axis is longitudinal, the moment of inertia does not appear, if the axis is lateral the moment of inertia of the engine appears. In the latter case, the inertial moment of the engine is $M_{ie} = J_e \ddot{\varphi}_e$, where J_e is the moment of inertia and φ_e is the angle of rotation of the engine; the sign of M_{ie} depends on the direction of the rotation. For vehicles with a front engine and driven rear axles, the inertia of the drive train (clutch, transmission, Cardan shaft without drive shafts), which rotates about the longitudinal axis, leads to changes in wheel load between the left- and right-hand sides but not between the front and the rear axles. Consequently, these inertia terms do not appear here, either.

For the pitch and vertical oscillations, we consider only steady states, so that $\ddot{z}_v = 0$ and $\ddot{\varphi}_v = 0$ hold.

In the following, we will examine the axle loads F_{z1} and F_{z2} between the road and the front and rear axle respectively. For simplicity, it is assumed here that the radii of the deformed wheels and the eccentricities are the same for all wheels: $r_{wst} = r_{wst1} = r_{wst2}$ and $e_w = e_{w1} = e_{w2}$.

The sum of the forces in the z-direction gives

$$0 = F_{z1} + F_{z2} - (G_{a1} + G_{a2})\cos\alpha + F_{az} - G_b\cos\alpha . \tag{6.2}$$

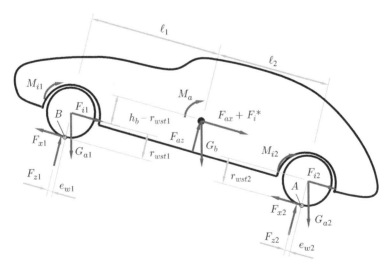

Figure 6.1 Free-body diagram

The equilibrium conditions for the forces in the x-direction are (the force $F_i^* = m_b \ddot{x}_v$ is d'Alembert's inertial force and not the acceleration resistance of the body from (3.8)):

$$F_i^* + F_{i1} + F_{i2} = F_{x1} + F_{x2} - (G_{a1} + G_{a2} + G_b)\sin\alpha - F_{ax} . \qquad (6.3)$$

The sum of the moments with respect to the point A on the rear tyres (see Figure 6.1) ($\ell = \ell_1 + \ell_2$) is

$$\begin{aligned}
0 = {} & F_{z1}(e_w + \ell) + F_{z2}e_w \\
& + (F_{i1} + F_{i2})r_{wst} + (G_{a1} + G_{a2})r_{wst}\sin\alpha \\
& - G_{a1}\ell\cos\alpha + M_{i1} + M_{i2} - G_b\ell_2\cos\alpha \\
& + (F_{ax} + F_i^*)h_b + F_{az}\ell_2 + M_a + G_b h_b \sin\alpha + M_{ie} .
\end{aligned} \qquad (6.4)$$

The inertial forces are rewritten with the help of acceleration $\ddot{x}_v$ and the angular accelerations $\ddot{\varphi}_{a1}$ and $\ddot{\varphi}_{a2}$, where we choose the weight divided by the acceleration due to gravity instead of the masses as a parameter. The overall result is

$$\begin{aligned}
F_{z1}\ell = {} & (G_{a1}\ell + G_b\ell_2)\cos\alpha - ((G_{a1} + G_{a2})r_{wst} + G_b h_b)\sin\alpha \\
& - F_{ax}h_b - F_{az}\ell_2 - M_a - ((G_{a1} + G_{a2})r_{wst} + G_b h_b)\frac{\ddot{x}_v}{g} \\
& - (J_{a1}\ddot{\varphi}_{a1} + J_{a2}\ddot{\varphi}_{a2}) - M_{ie} \\
& - F_r r_{wst} .
\end{aligned} \qquad (6.5)$$

Here the rotating inertia terms for the wheels have been replaced by the corresponding expressions $J_{a1}\ddot{\varphi}_{a1}$ and $J_{a2}\ddot{\varphi}_{a2}$

If we form the sum of the moments about point B (Figure 6.1), we obtain the axle load for the rear axle:

$$\begin{aligned}
F_{z2}\ell = {} & (G_{a2}\ell + G_b\ell_1)\cos\alpha + ((G_{a1} + G_{a2})r_{wst} + G_b h_b)\sin\alpha \\
& + F_{ax}h_b - F_{az}\ell_1 + M_a + M_{ie} \\
& + ((G_{a1} + G_{a2})r_{wst} + G_b h_b)\frac{\ddot{x}_v}{g} + (J_{a1}\ddot{\varphi}_{a1} + J_{a2}\ddot{\varphi}_{a2}) \\
& + F_r r_{wst} .
\end{aligned} \qquad (6.6)$$

The individual summands can be divided into four groups, in which we do not consider the negligibly small rolling resistance, $F_r r_{wst}$, in detail.

1. **Static parts:** The major portion of static parts is due to the weight, G_b, of the body. For $\alpha = 0$, the distribution of the total weight, $G = G_b + G_{a1} + G_{a2}$, depends on the position of the centre of mass. For the front axle load, F_{z1}, at $\alpha = 0$ we obtain

$$F_{z1\,\text{stat}} = G_{a1} + \frac{\ell_2}{\ell}G_b , \qquad (6.7)$$

and for the rear axle load, F_{z2},

$$F_{z2\,\text{stat}} = G_{a2} + \frac{\ell_1}{\ell}G_b \, . \tag{6.8}$$

When the vehicle is on an inclined road $\alpha_g > 0$, the front axle load decreases and the rear axle load increases. In the following, we estimate how large the angle α_g must be, so that the front axle load F_{z1} is just not equal to zero. To this end, we neglect the axle weights of G_{a1} and G_{a2} and obtain from

$$0 = G_b\ell_2\cos\alpha_g - G_bh_b\sin\alpha_g \, . \tag{6.9}$$

the tangent of the limit angle:

$$\tan\alpha_g = \frac{\ell_2}{h_b} \, . \tag{6.10}$$

In this limiting case, the line of action of the weight, G_b, runs straight through the point A. It can be seen that in practice this limiting case is of no importance, since in general $l_2 > h_b$ applies, from which $\alpha > 45°$ would follow. However, the reduction of the front axle load plays a role when we look at the tractive forces at the front wheels. If the front wheels are driven, the reduction of the front axle load results in a reduction of transmittable tangential forces. With a driven rear axle, the transmittable tangential force increases because of the higher normal forces.

2. **Air forces:** The overall aerodynamic lift forces, F_{az}, and the aerodynamic torque, M_a, can be represented via two forces, F_{az1} and F_{az2}, acting at the front and rear axle respectively. The axle loads due to these forces can then be written as:

$$F_{z1\,\text{aero}} = -F_{ax}\frac{h_{pp}}{\ell} - F_{az1} \, , \tag{6.11}$$

$$F_{z2\,\text{aero}} = F_{ax}\frac{h_{pp}}{\ell} - F_{az2} \, . \tag{6.12}$$

Here, h_{pp} is the distance between the centre of pressure S_{pp} and the road.

The lift forces, F_{az1} and F_{az2}, can be calculated in a similar way to the aerodynamic drag forces by using the lift coefficients c_{l1} and c_{l2}:

$$F_{az1} = c_{l1}\frac{\rho}{2}Av_v^2 \, , \tag{6.13}$$

$$F_{az2} = c_{l2}\frac{\rho}{2}Av_v^2 \, . \tag{6.14}$$

Sample values given for the lift coefficients are based on the historical development of the BMW 3 Series in Figure 6.2 and for the Porsche 911 in Figure 6.3.

A large lift coefficient at the rear axle has a destabilizing effect on the driving behaviour. The aerodynamic drag forces decrease the front axle load and increase

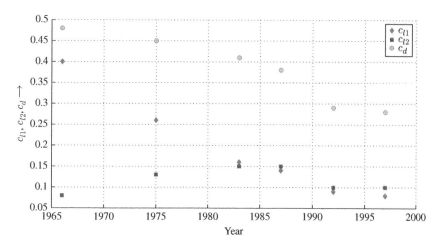

Figure 6.2 Lift coefficients of BMW 3 Series (data from Braess 1998)

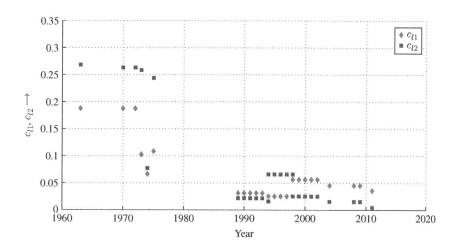

Figure 6.3 Lift coefficients of Porsche 911 (data from Harrer et al. 2013)

the rear axle load, whereas the aerodynamic lift reduces the load on both axles in conventional passenger cars (however, this is not true for racing cars, for example).

3. **Dynamic parts:** We summarize the dynamic components and obtain

$$Gh_{\mathrm{cm}} = (G_{a1} + G_{a2})r_{wst} + G_b h_b , \qquad (6.15)$$

where G is the total weight of the entire vehicle, and h_{cm} is the distance of the centre of mass for the entire vehicle from the roadway. Neglecting the slip and setting $R_{w0} = r_{wst}$ in the definition of the slip, we obtain

$$((G_{a1} + G_{a2})r_{wst} + G_b h_b)\frac{\ddot{x}_v}{g}$$

$$+ (J_{a1}\ddot{\varphi}_{a1} + J_{a2}\ddot{\varphi}_{a2}) = \left(\frac{Gh_{cm}}{g} + \frac{J_{a1}}{r_{wst}} + \frac{J_{a2}}{r_{wst}}\right)\ddot{x}_v$$

$$= G(h_{cm} + (\lambda^* - 1)r_{wst})\frac{\ddot{x}_v}{g}, \qquad (6.16)$$

where

$$\lambda^* = 1 + \frac{1}{mr_{wst}^2}(J_{a1} + J_{a2}). \qquad (6.17)$$

However, this relationship is valid for an engine with an axis of rotation in the longitudinal direction of the vehicle. If this rotational axis is in the lateral direction, the term must be extended. Assuming that the direction of rotation of the motor is equal to the direction of rotation of the wheels, we obtain

$$\lambda^* = 1 + \frac{1}{mr_{wst}^2}(J_{a1} + J_{a2} + i_d^2 i_g^2 J_e). \qquad (6.18)$$

The sign of the last term must be negative if the engine rotates in the opposite direction.

During acceleration, the inertial forces diminish the vertical forces on the front axle and increase them on the rear axle, whereas the opposite effect occurs during braking. This means that in a front-wheel drive vehicle the maximum tractive force decreases as acceleration increases because the tractive force is limited by the coefficient of adhesion, μ_a.

All together, the following holds for the dynamic portion of the axle loads:

$$F_{z1\ dyn} = -G(h_{cm} + (\lambda^* - 1)r_{wst})\frac{\ddot{x}_v}{g}, \qquad (6.19)$$

$$F_{z2\ dyn} = G(h_{cm} + (\lambda^* - 1)r_{wst})\frac{\ddot{x}_v}{g}. \qquad (6.20)$$

Aside from the rolling resistance forces, it may be noted that the front axle loads reduce when the vehicle is driving on a gradient, during acceleration and due to air forces. Figure 6.4 (the middle diagram) depicts the three principal components of the axle loads: the static, the aerodynamic and the dynamic components. To calculate the dynamic components, we assume that the vehicle is accelerated at maximum, i.e. the vehicle is accelerated at acceleration capability.

The maximum acceleration of a vehicle is not only determined by the power, but also the coefficient of adhesion, μ_a, and the axle load of the driven axle. Since the axle loads vary during acceleration, not every acceleration that could be theoretically achieved at the acceleration capability limit is transferable to the vehicle.

The upper graph in Figure 6.4 shows the dynamic axle loads. We can see how the axle loads for front and rear axles approach each other with increasing speed because of the associated decline in acceleration. The total wheel load decreases due to the

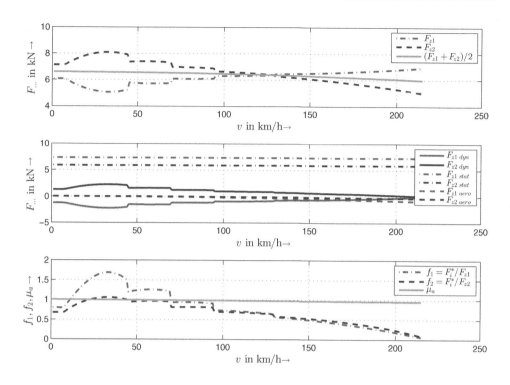

Figure 6.4 Components of axle loads

aerodynamic lift forces. In this example, the forces arising from the available acceleration, i.e. d'Alembert's inertial force $F_i^* = m_b \ddot{x}_v$, shows that the adhesion limit in first and second gear prohibits the huge theoretically attainable accelerations. This can be seen when comparing the longitudinal force coefficient, $f_1 = F_i^*/F_{z1}$, with the adhesion limit, μ_a. In third gear, the longitudinal force coefficient, $f_1 = F_i^*/F_{z1}$, for the front driven axle is close to the adhesion limit, μ_a, whereas $f_2 = F_i^*/F_{z2}$ for the rear driven axle is well below the adhesion limit.

6.2 Braking Process

In the following, we turn to the braking process. This is divided into different time segments as follows.

Reaction time: The time from the first appearance of an obstacle to the start of build-up of the braking force at the pedal (force at the foot, F_{foot}), this period is called reaction time, t_r. This phase includes perception (*There is something on the road.*), recognition (*It is a child.*) and the decision time (*It is better to brake than to steer to the right.*) as well as the time required to move the foot from the accelerator to the brake pedal. In Figure 6.5, the length of this phase is $t_r = 0.9$ s.

Foot pressure build-up time: After time t_r, the foot force, F_{foot}, rises. The time
to build the maximum braking force is called the pressure build-up time, t_{fb}. In
Figure 6.5, the length of this phase is $t_{\text{fb}} = 0.8\,\text{s}$.
   The deceleration of the vehicle, however, starts after the expiry of the time $t_r + t_t$
(t_t: transmission time)

Transmission time: The transmission time, t_t, is the time in which the tolerances in
the joints and bearings must be overcome. In Figure 6.5, the length of this phase is
$t_t = 0.2\,\text{s}$.

Rise time until maximum pressure: The time elapsed from the start of deceleration
to the maximum deceleration is called the rise time until the maximum pressure,
t_b, or pressure build-up time. This is greater than t_{fb}. In Figure 6.5, the length of
the phase is $t_b = 0.8\,\text{s}$.

Remark 6.1 A typical value for $t_r + t_t$ is 0.6 s when the obstacle shows up in front of
the driver and 0.9 s when the driver has to turn his head to detect the obstacle. The time
that passes until the vehicle comes to a standstill is called the total stopping time, t_s.
The total stopping distance, s_{tot}, consists of three parts, namely the reaction distance
(or thinking distance), s_1, the brake engagement distance, s_2, and the physical braking
distance, s_3 (see Figure 6.5, lower graph).

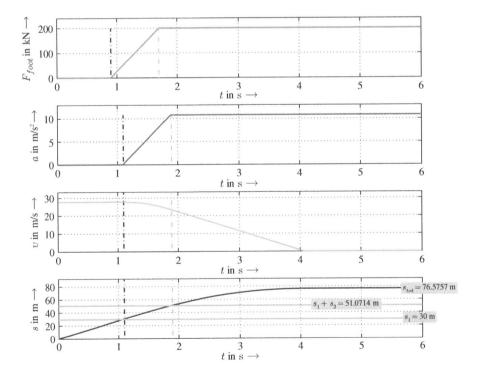

Figure 6.5   Foot force, acceleration, velocity and distance during the braking process

Remark 6.2 In modern brake systems, some of these times can be reduced to a certain extent, with examples being the foot pressure build-up time, t_{fb}, or the transmission time, t_t. Brake-assist systems are able to anticipate the driver's wish for full braking and can therefore reduce the stopping distance. These systems detect the sudden change of foot from the accelerator to the brake pedal and calculate the desired braking application, then build-up the maximum brake pressure independently of the pedal force. Furthermore, the reaction time can be positively influenced by warning signals in a head-up display. As a result, the actuation pressure build-up time and the initial response time are shortened.

In the following, we derive an equation for the total stopping distance, s_{tot}.

The deceleration in the time interval $t_r + t_t$ is zero, so the velocity is constant. We immediately obtain the distance s_1 (where v_i is the initial velocity):

$$s_1 = v_i(t_r + t_t) \, . \tag{6.21}$$

We assume a linear relationship for the deceleration up to the maximum value of $\ddot{x}_f$ during the pressure build-up time, t_b:

$$\ddot{x} = \frac{\ddot{x}_f}{t_b} t \, . \tag{6.22}$$

From this we then obtain the velocity during the pressure build-up period

$$v(t) = v_i + \int_0^t \frac{\ddot{x}_f}{t_b} t \, dt$$

$$= v_i + \frac{\ddot{x}_f}{2t_b} t^2 \, . \tag{6.23}$$

Further integration yields the distance

$$s_2 = \int_0^{t_b} v(t) \, dt$$

$$= v_i t_b + \frac{\ddot{x}_f}{6} t_b^2 \, . \tag{6.24}$$

Note the negative value of $\ddot{x}_f$. At this point, we assume that the vehicle does not stop during the pressure build-up period (otherwise the equation will have to be modified, because the upper limit of integration is less than t_b). This assumption means that the velocity at the end of the pressure build-up period is greater than zero. From Equation (6.23), we derive

$$0 < v_i + \frac{\ddot{x}_f}{2} t_b \, . \tag{6.25}$$

Dividing this equation by $\ddot{x}_f < 0$ yields

$$0 > \frac{v_i}{\ddot{x}_f} + \frac{t_b}{2} \, . \tag{6.26}$$

The speed and time dependence during the full braking time is given by

$$v = v_2 + \ddot{x}_f \int_0^t \mathrm{d}t$$

$$= v_2 + \ddot{x}_f t . \tag{6.27}$$

Here v_2 is the velocity at the beginning of the full braking phase:

$$v_2 = v_i + \frac{\ddot{x}_f}{2} t_b . \tag{6.28}$$

Putting these together, we obtain the speed after the full braking phase which has to be zero:

$$v = v_i + \frac{\ddot{x}_f}{2} t_b + \ddot{x}_f t_f$$

$$= 0 . \tag{6.29}$$

Solving Equation (6.29) for the time t_f yields

$$t_f = -\frac{v_i}{\ddot{x}_f} - \frac{t_b}{2} . \tag{6.30}$$

As the condition (6.26) holds, the time is positive: $t_f > 0$. The distance s_3 is

$$s_3 = \int_0^{t_f} v \, \mathrm{d}t$$

$$= \int_0^{t_f} (v_2 + \ddot{x}_f t) \, \mathrm{d}t$$

$$= v_2 t_f + \frac{\ddot{x}_f}{2} t_f^2$$

$$= -\frac{v_i^2}{2\ddot{x}_f} - \frac{v_i t_b}{2} - \frac{\ddot{x}_f t_b^2}{8} . \tag{6.31}$$

The total braking distance (or stopping distance), s_{tot}, is given by the sum of Equations (6.21), (6.24) and (6.31),

$$s_{\text{tot}} = s_1 + s_2 + s_3$$

$$= v_i \left(t_r + t_t + \frac{t_b}{2} \right) - \frac{v_i^2}{2\ddot{x}_f} + \frac{\ddot{x}_f}{24} t_b^2 . \tag{6.32}$$

This equation shows factors that influence the stopping distance. Some systems in vehicles are developed in order to reduce s_{tot}. Some can be seen in Figure 6.6.

- To reduce the reaction time, systems can help to enhance the driver's perception, recognition and decision making. Some such systems are shown in Figure 6.6:

infrared systems or cornering lights, for example, help the driver to recognize obstacles earlier. Head-up displays can direct the attention of the driver towards certain, critical situations and hence reduce the time for perception, recognition or even decision-making.

- The time for transmission and full pressure build-up can be reduced by electrical systems such as the electrohydraulic brake (EHB) introduced by Daimler or the electronic wedge brake announced by Siemens VDO (but not yet put into practice). The electronic systems can on the one hand reduce the above-mentioned times while, on the other, they can be used to amplify the driver's braking input. The latter functionality is part of brake-assist systems.
- In the last group of systems, the maximum deceleration, $\ddot{x}_f$, can be influenced. In brake-assist systems, a sudden change of the driver's foot from accelerator pedal to brake pedal is taken as an indicator that the situation requires emergency braking. In this case the brake-assist system amplifies the driver's input and increases the brake pressure to maximum, until the anti-lock braking system (ABS) limits the pressure in the brake cylinders. These systems have been introduced because drivers who are short on experience do not fully exploit the capabilities of braking systems (including ABS onset). Hence, a significant amount of braking capability is not used. In this respect, the brake-assist system helps to reduce the braking distance.

The second means of enhancing deceleration capability is to improve the coefficient of adhesion. This can be achieved by improving tyre characteristics. If we consider the mean deceleration capabilities of tyres in Figures 2.11 and 2.12, it is obvious that there are differences in the magnitude of approximately 10%.

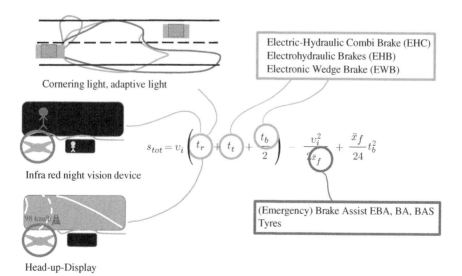

Figure 6.6   Factors influencing total braking distance

6.3 Braking Rate

When considering the deceleration of the vehicle, we first refer to Figure 6.1. When the vehicle is on a road without any inclination, by neglecting the rolling resistance, air resistance and the rotational inertia, we obtain the equilibrium of forces in the longitudinal direction of the vehicle:

$$F_{x1} + F_{x2} = m_{\text{tot}} \ddot{x}_v \ . \tag{6.33}$$

Remark 6.3 Neglecting the rotational inertia does not cause any major errors, as a significant proportion of the braking force is required for deceleration of the translational inertia when the clutch is not engaged. The air forces support the braking process. The rolling resistances are neglected because they are small compared to the braking forces.

The longitudinal forces, F_{xi}, and the acceleration, $a = \ddot{x}_v$, are negative during braking. In order to avoid the negative signs, positive braking forces $B_1 = -F_{x1}$, $B_2 = -F_{x2}$ and a positive deceleration or braking ratio $\mathcal{Z} = -a/g$ are introduced. With these, we obtain ($G = m_{\text{tot}} g$):

$$B_1 + B_2 = G\mathcal{Z} \ . \tag{6.34}$$

The quotients B_1/F_{z1} and B_2/F_{z2} yield the longitudinal force coefficient, μ (see the diagram in Figure 6.7).

The maximum tangential force is achieved for the coefficient of adhesion μ_a. If the quotients B_1/F_{z1} and B_2/F_{z2} simultaneously become μ_a, the maximum braking ratio $\mathcal{Z}_{\text{max}}$ of the vehicle ($G = F_{z1} + F_{z2}$) can be derived from

$$\mu_a \underbrace{(F_{z1} + F_{z2})}_{=G} = B_1 + B_2$$

$$= G\mathcal{Z}_{\text{max}} \ . \tag{6.35}$$

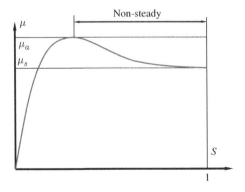

Figure 6.7 Tyre longitudinal force coefficient, μ, as a function of the slip S

Consequently, the maximum braking ratio is:

$$Z_{max} = \mu_a \, . \tag{6.36}$$

The maximum braking ratio Z_{max} is equal to the value of the coefficient of adhesion, μ_a. In most cases, it will be below this value, as we shall see in the following. To obtain the ratio of the braking force, B_j, to the wheel load F_{zj} ($j = 1$ or $j = 2$), we look at the longitudinal force coefficient, which we denote by f_1 and f_2:

$$\frac{B_1}{F_{z1}} = f_1 \leq \mu_a \, , \tag{6.37}$$

$$\frac{B_2}{F_{z2}} = f_2 \leq \mu_a \, . \tag{6.38}$$

The demanded maximum braking ratio, Z_{max}, and therefore the shortest braking distance can be achieved if neither the wheels at the front axle nor at the rear axle are blocking (except in the case when both simultaneously achieve μ_a). This leads to the condition in which the longitudinal force coefficients f_1 and f_2 at the front and rear axles, respectively, must be equal, and in the case of the maximum braking ratio Z_{max} equal to μ_a:

$$f_1 = f_2 (= \mu_a \text{ for } Z_{max}) \tag{6.39}$$

$$\Rightarrow \frac{B_1}{F_{z1}} = \frac{B_2}{F_{z2}} \tag{6.40}$$

$$\Rightarrow \frac{B_1}{B_2} = \frac{F_{z1}}{F_{z2}} \, . \tag{6.41}$$

This ratio of the braking forces will be referred to as the ideal braking force distribution.

The ratio of the braking forces is of importance because in a vehicle only one brake pedal is usually available, but an ideal braking force distribution requires individual braking forces for the front and rear axles. The desired braking force is transmitted using the pedal to the appropriate transmission paths (usually this is a hydraulic system) of the braking system at the front and rear axles. Here, on the one hand, the distribution between the front and rear axles is of crucial importance in order to achieve good braking performance and, on the other hand, to prevent the wheels from locking. If there is too much braking force on the front axle and the wheels slip, this means that no more cornering forces can be transmitted and the vehicle can no longer be steered; it therefore travels in a straight line. Locking of the rear wheels results in a loss of cornering forces, which leads to instability: with a small lateral disturbance, the vehicle turns from the longitudinal direction. In the ideal braking force distribution, there is no premature axle locking (or all four wheels lock simultaneously) so the maximum possible deceleration is achieved.

Due to the hydraulic transmission from the foot power, there are few possibilities of influencing the distribution. The hydraulic pressure can be divided into a constant ratio for the rear and front brakes or a pressure limiter can be present in the system, which prevents a further rise in braking pressure for the control of the rear axle. In the ideal braking force distribution, the ratio of braking forces, B_1/B_2, is not constant, since the ratio of wheel loads is again dependent on the deceleration. Now we derive the ideal braking force distribution; to do this, we use only simplified equations for the wheel loads. From Figure 6.8, it follows that (we omit the index cm here, in Figure 6.8 and in the following, thus $h_{cm} = h$)

$$F_{z1} = \frac{\ell_2 G}{\ell} + \frac{G \mathcal{Z} h}{\ell}$$

$$= \frac{G}{\ell}(\ell_2 + \mathcal{Z}h) , \tag{6.42}$$

$$F_{z2} = \frac{\ell_1 G}{\ell} - \frac{G \mathcal{Z} h}{\ell}$$

$$= \frac{G}{\ell}(\ell_1 - \mathcal{Z}h) . \tag{6.43}$$

This results in the following ratio of the braking forces:

$$\frac{B_1}{B_2} = \frac{\ell_2 + \mathcal{Z}h}{\ell_1 - \mathcal{Z}h} . \tag{6.44}$$

Together with

$$B_1 + B_2 = G\mathcal{Z} \tag{6.45}$$

we have two Equations (6.44) and (6.45), which can be solved for the two variables B_1 and B_2:

$$B_1 = \frac{\ell_2 + \mathcal{Z}h}{\ell} G\mathcal{Z} , \tag{6.46}$$

$$B_2 = \frac{\ell_1 - \mathcal{Z}h}{\ell} G\mathcal{Z} . \tag{6.47}$$

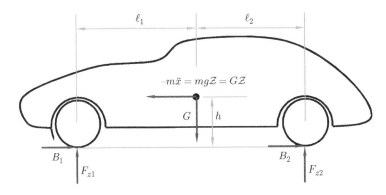

Figure 6.8   Dynamic wheel load under braking

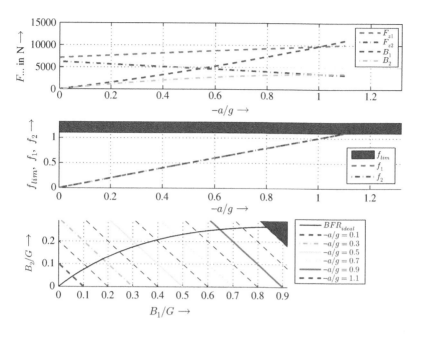

Figure 6.9   Ideal braking force distribution

It can be seen that the ratio is not constant but depends on the deceleration. In the graph at the top of Figure 6.9, the wheel loads, F_{zi}, and the ideal braking forces, B_i ($i = 1, 2$), are plotted against the braking ratio, $\mathcal{Z} = -a/g$. It is very good to see the steady growth of the braking force gradient on the front axle and the decreasing gradient of the braking force on the rear axle. With stronger deceleration, the braking force on the rear axle drops because of the dynamic reduction of the wheel loads.

The middle graph is a plot of the longitudinal force coefficients, f_1 and f_2, for both axles against the braking ratio, $\mathcal{Z} = -a/g$. It additionally shows a maximum longitudinal force coefficient $f_{\lim} = \mu_a = 1.1$. Here the ideal trends are obvious, since the maximum braking ratio, $\mathcal{Z}$, is achieved for any given longitudinal force coefficients. The graph at the bottom shows the data of ratio B_1/G and B_2/G. This representation is common in the literature. The black line is the ideal braking force ratio.

For implementation in a vehicle, a fixed ratio of the braking forces is derived from the above design constraints. In the following, we shall consider various fixed braking force ratios along with their advantages and disadvantages.

First, a design based on the static wheel loads can be implemented at a deceleration of $\mathcal{Z} = 0$. Then we have

$$\frac{B_1}{B_2} = \frac{F_{z1\text{stat}}}{F_{z2\text{stat}}}. \tag{6.48}$$

Solving Equation (6.48) for B_2 and setting this in $G = B_1 + B_2$, it follows

$$B_1 = F_{z1\mathrm{stat}}\, \mathcal{Z} \ . \tag{6.49}$$

Using Equation (6.48), we obtain

$$B_2 = F_{z2\mathrm{stat}}\, \mathcal{Z} \ . \tag{6.50}$$

In this braking force distribution, the rear axle is strongly over-braked, i.e. the rear wheels lock prematurely, thus limiting the achievable deceleration. This relationship can be seen clearly in the middle graph of Figure 6.10. With a longitudinal force coefficient of $f_2 = 1$, only a braking ratio $\mathcal{Z} \approx 0.7$ can be attained. On the other hand, the front axle is clearly below its potential. If the braking force of an axle is below the ideal possible force in the diagram of braking force versus braking ratio $\mathcal{Z}$, we call this axle under-braked, if it is above the ideal force, we call it over-braked. Hence, in the static design of the braking force distribution, the front axle is under-braked, the rear axle over-braked.

For $\mu_a = 1.1 = f_{\lim}$ the (ideal) deceleration of $\mathcal{Z} = 1.1$ could be achieved. In the present case, however, only $\mathcal{Z} \approx 0.7$ is possible. It will therefore not use much braking potential. One especially bad feature of this design is the locking of the rear axle. If the wheels lock prematurely, which by definition they should not, then that should happen on the front axle. This will only eliminate the ability to steer the vehicle, but the vehicle will remain stable.

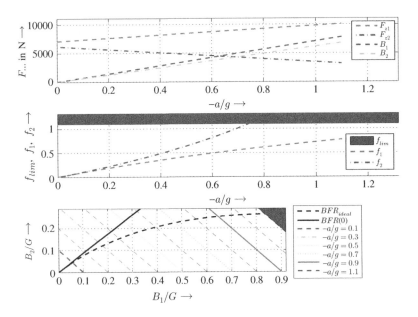

Figure 6.10   Braking force distribution on the basis of the static wheel loads

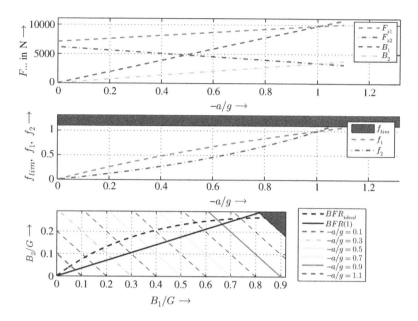

Figure 6.11 Braking force distribution on the basis of the dynamic wheel loads at $\mathcal{Z} = 1$

The design based on the dynamic wheel loads at $\mathcal{Z} = 1$ gives us a better braking force distribution (see Figure 6.11). This shows the entire range up to $\mathcal{Z} = 1$ $f_2 < f_1$. One disadvantage is the strong convexity of the longitudinal force coefficient, f_2, on the rear axle at medium decelerations, which again results in much loss of braking potential. An improvement can be achieved if the dimensioning is oriented towards dynamic wheel loads at lower decelerations, e.g. $\mathcal{Z} = 0.8$ (see Figure 6.12). Up to a deceleration of $\mathcal{Z} = 0.8$, $f_2 < f_1$ holds. The f_i-curves are in the vicinity of the ideal design $f_1 = f_2 = \mathcal{Z}$, and therefore up to a value of $\mathcal{Z} = 0.8$ not as much brake potential is wasted as in the design with $\mathcal{Z} = 1.0$. A disadvantage in the design $\mathcal{Z} = 0.8$ in comparison to $\mathcal{Z} = 1.0$ is that $\mathcal{Z} = 0.8$ leads to a locking of the rear axle at lower braking ratios than for a design with $\mathcal{Z} = 1.0$.

Using a braking force limiter for the rear axle makes it possible to avoid the above-mentioned locking of the rear axle and therefore further improve the distribution. First, there is a layout as usual but this is only for very low decelerations, usually $\mathcal{Z} = 0.6$. Up to this point, over-braking of the front axle and under-braking of the rear axle take place, but both differ only slightly from the ideal path. From this point, for example $\mathcal{Z} = 0.6$, the braking force B_2 at the rear axle is constant, the rear axle is under-braked. (see Figure 6.13). In the diagram, the improvement over a continuous linear distribution is clearly seen. Both longitudinal force coefficients are close to the ideal distribution. Over the entire range of braking, the front axle is over-braked, while the rear axle is simultaneously under-braked, causing the vehicle to remain stable.

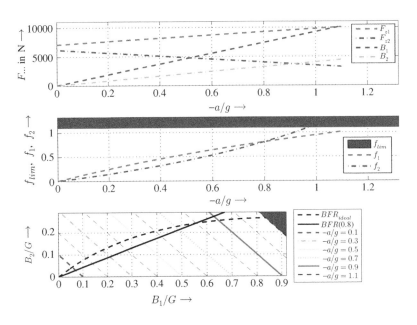

Figure 6.12   Braking force distribution on the basis of the dynamic wheel loads at $\mathcal{Z} = 0.8$

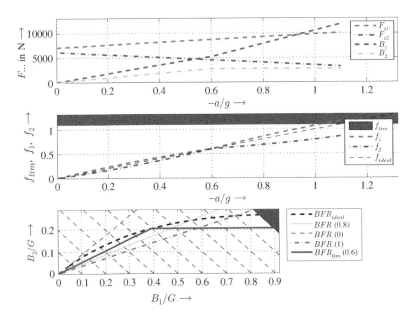

Figure 6.13   Braking force distribution on the basis of the dynamic wheel loads at $\mathcal{Z} = 0.6$ and rear braking force delimiter

The graph at the bottom of Figure 6.13 allows a comparison of the designs explained above. The design with the force delimiter is close to the ideal design.

6.4 Questions and Exercises

Remembering

1. What are the essential parameters that constitute the wheel loads?
2. Explain the terms reaction time, t_r, transmission time, t_t and rise time until maximum pressure, t_b.
3. What are typical magnitudes for reaction time, t_r, transmission time, t_t and rise time until maximum pressure, t_b.

Understanding

1. Which parameters affect the essential proportions of wheel loads?
2. What does the wheel load distribution depends on in the front and rear wheels and in the left and the right wheels?
3. Explain the effect of air forces on wheel loads.
4. Explain the effect of inertial forces on the wheel loads.
5. What role does the mounting direction of the motor play on the wheel loads?
6. Why are static braking force distributions unfavourable?
7. Explain a dynamic braking force distribution.
8. Explain the braking force distribution with a force delimiter.

Applying

1. Calculate the aerodynamic lift forces ($v = 60$ m/s, $A = 1.8$ m^2) for a vehicle built between 1960 and 1970 on rear and front axles and the sum of those lift forces.
2. The following parameters are given: $t_r = 0.9$ s, $t_t = 0.2$ s, $t_b = 0.8$ s, $v_i = 30$ m/s, $\ddot{x}_f = 8$ m/s^2. Calculate the total braking distance (stopping distance) and the effect of doubling t_r, t_t or t_b.
3. The height of the centre of mass S_{cm} of a car is $h_{cm} = h = 0.8$ m, the mass $m = 1200$ kg, the distance of S_{cm} to the front and rear axles is $\ell_1 = 2.0$ m, $\ell_2 = 2.5$ m, resp.
   Calculate the braking forces B_1 and B_2 for an ideal braking force distribution for an acceleration of $\ddot{x} = -5$ m/s^2 (please, use $g = 10$ m/s^2).
4. The height of the center of mass S_{cm} of a car is $h_{cm} = h = 0.8$ m, the mass $m = 1200$ kg, the distance of S_{cm} to front and rear axle is $\ell_1 = 2.0$ m, $\ell_2 = 2.5$ m, resp.
   Calculate the braking forces B_1 and B_2 for a braking force distribution that is ideal for $Z = 0.8$, for an acceleration of $\ddot{x} = -5$ m/s^2 (please, use $g = 10$ m/s^2).

Analysing

1. Explain the effect on taking slip at the driven axle into account for the wheel load.
2. Explain the effect of increasing the pressure build-up time, t_b, on the full braking duration, t_f?

7

Hybrid Powertrains

Some of the first (automotive) vehicles were electrically driven, but for more than 100 years now vehicles with internal combustion engines have dominated most areas of automotive engineering. Since the last decade, more and more hybrid or purely electric driven passenger cars have entered the market, with several reasons being responsible for this. This chapter explains some basic concepts of hybrid powertrains. As the new developments are continuously changing the situation, the description here is restricted to the basics.

7.1 Principal Functionalities

This section outlines the principal functionalities of hybrid powertrains. The idea of hybrid powertrains involves combining an electric motor and an internal combustion engine in order to combine the advantages of both and in order to avoid the disadvantages.

Two disadvantages of internal combustion engines that seem to be the main reason for the rise in the numbers of hybrid vehicles are the limited resources of fossil fuel available and the air pollution they cause. In order to reduce dependency on fossil fuel and to reduce air pollution, internal combustion engines are being combined with electric motors in hybrid powertrains. The following describes some modes of operation using a so-called parallel hybrid powertrain. In parallel hybrid powertrains (details of different kinds of powertrains are explained in Section 7.2) an internal combustion engine and the electric motor are mechanically joined by a shaft (the joint may be interruptible by a clutch). An example of a vehicle with a parallel hybrid powertrain is shown in Figure 7.1.

With this combination of the internal combustion engine and the electric motor, it is possible in some situations to meet the demand for power (or tractive force) exclusively by the electric part of the powertrain, or, alternatively, it is possible to

Vehicle Dynamics, First Edition. Martin Meywerk.
© 2015 John Wiley & Sons, Ltd. Published 2015 by John Wiley & Sons, Ltd.
Companion Website: www.wiley.com/go/meywerk/vehicle

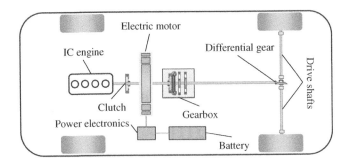

Figure 7.1   Vehicle with a parallel hybrid powertrain

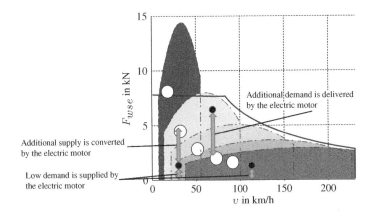

Figure 7.2   The hybrid idea

reduce the amount of power provided by the combustion engine because of the power delivered by the electric motor. An example is depicted in Figure 7.2. It shows the tractive force supplied by a conventional powertrain with an internal combustion engine and a five-speed manual transmission. The centres of the white circles are the points of optimum efficiency for the combustion engine. Principal modes of operation are able to reduce emissions, and here we explain the mode to reduce CO_2 (or reduce fuel consumption) and a mode to reduce NO_x emissions. If the demand for tractive forces in a specific situation is greater than the tractive force at optimum efficiency, the electric motor can close the gap and deliver the additional amount of tractive force. If the demand in the situation is below the best efficiency curve, the excess energy can be converted by the electric engine into electrical power, which can be stored as chemical energy in the battery, and then used later to drive the electric motor. Consequently, one advantage of a hybrid powertrain is a mode of operation for the combustion engine at nearly optimum efficiency in order to reduce CO_2 emissions. To achieve low demands for power, another mode of operation is important to reduce NO_x, CO and HC emissions. For low power demand, the combustion engine runs in lean mode with higher emissions. Two possible hybrid modes can be applied to reduce these emissions: with pure electric driving the emissions can be avoided, with an increase in power (the additional power is converted to chemical energy in the battery) the combustion engine can be operated in regions with lower emissions.

After explaining one advantage of hybrid powertrains, we will look at different modes of hybrid powertrains.

The first, explained in the preceding passage, is illustrated in Figure 7.3. One portion of the power is necessary for the driven wheels to overcome the driving resistances, the rest is converted by the electric motor into electric power, and then stored as chemical energy in the battery. The amount of storable energy depends on the characteristics of the electric motor, on the capability of the power electronics converter, and, lastly, on the capacity of the battery.

In hybrid powertrains, the electric motor can be integrated in the powertrain, with a small amount of additional space being necessary. Figure 7.4 shows an example in which the electric motor is positioned immediately after the clutch; the transmission, which would be the next component of the powertrain, is not shown in the figure. This is, of course, a small electric motor, which cannot convert high amounts of power. We call this mode the generator mode.

There is a secondary generator mode called regenerative braking. In this situation, the demand for tractive force or for power is negative, because the driver wants to decelerate the car and the driving resistances from the air, the tyres and any gradient are too small to deliver the braking moment that the driver wants to act on the car. In this situation, the driver applies the brakes. The electronic control unit recognizes the driver's request and switches the electric motor to generation mode in order to convert a portion of the kinetic energy into electrical energy (cf. Figure 7.5). To increase the efficiency, it is advantageous to disengage the clutch between the combustion engine and the electric motor in order to avoid drag torques from the combustion engine (which would be converted to heat energy in the combustion engine).

In the so-called boost mode, the torque of the combustion engine is raised by the torque of the electric motor. This can be performed, as described at the beginning of this section, to operate the combustion engine in an optimum or fairly high efficiency

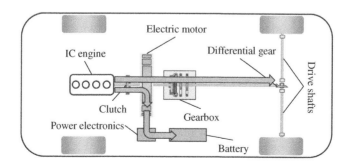

Figure 7.3 Split of the power from the internal combustion engine

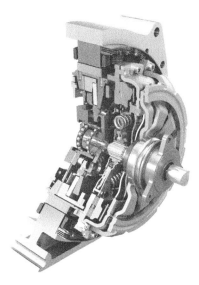

Figure 7.4 Electric engine integrated in the powertrain (reproduced with permissions of Schaeffler)

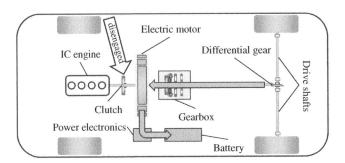

Figure 7.5 Regenerative braking

range, or it can be performed to increase the maximum torque available from the combustion engine, as in the case of a sports vehicle, for example (cf. Figure 7.6).

The last mode of operation is a purely electric mode, in which all the power is delivered by the electric motor. The duration of this mode depends on the capacity of the battery, and whether this mode makes sense within the entire speed range depends on the maximum power of the electric motor and on the driving resistances. For highway coasting, this means driving at a moderate and constant velocity on a highway, where the need for power results mainly from aerodynamic drag, rolling resistance and gradient resistance (cf. Figure 7.7).

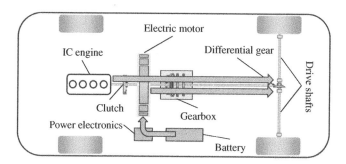

Figure 7.6   Boost mode

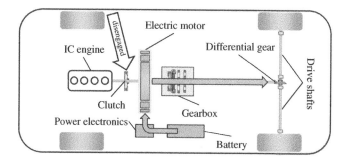

Figure 7.7   Purely electric mode

Another advantage of the hybrid powertrain is that start–stop operations are possible, which means that the combustion engine can be stopped, for example, at a traffic light, and the electric motor can start the engine again. This eliminates the need for an extra starter motor.

There are different hybrid levels, some characteristics of which are summarized in Figure 7.8. The lowest level is the so-called mild hybrid, with a small electric motor, usually in parallel mode. This type permits a start–stop functionality, and the electric motor can deliver an additional torque to support the combustion engine. The magnitude of electric power is about 20 kW, which is used for starting the engine, for delivering an additional torque at low velocities or for boosting the vehicle. Regenerative braking, within limits of course, is also possible. The power necessary at a velocity of 30 m/s, for example, to decelerate a vehicle of 1200 kg at 5 m/s^2 is 180 kW, which is significantly greater than the electric power in a mild hybrid vehicle. To compare these values, it is reasonable to look at half of the desired braking power if only one axle is driven and therefore only one axle can brake regeneratively. A mild hybrid requires an electric power supply system with a higher voltage than the usual 14 V system.

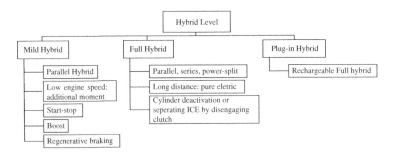

Figure 7.8 Hybrid levels

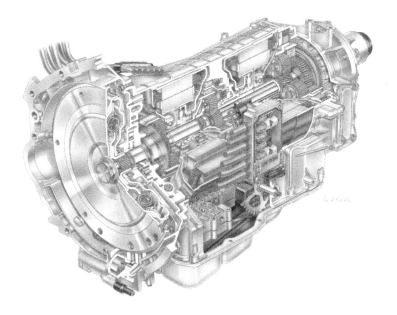

Figure 7.9 Hybrid powerpack from GM, Mercedes and BMW (reproduced with permissions of Daimler AG)

The second hybrid level is the full hybrid, which exists in different topologies (serial, parallel or a combination of the two). The purely electric mode is possible for longer distances and the combustion engine can be separated by a clutch from the electric engine or the drag torque can be avoided by cylinder deactivation. A powerful electric engine is necessary for this kind of powertrain. An example of an electric motor integrated in the automatic transmission is shown in Figure 7.9.

The highest hybrid level is the plug-in hybrid, which is a full hybrid vehicle with a battery that can be recharged by an external power supply system.

The next level in this series will be a purely electric vehicle.

7.2   Topologies of Hybrid Powertrains

Several topologies can be used in hybrid powertrains. The simplest of these is a parallel hybrid in which an electric engine is mounted directly onto the combustion engine (cf. Figure 7.10). A first approach to implementing this powertrain involves using only one clutch (in Figure 7.10, a clutch with a trilok converter is integrated in the gearbox), no clutch between the combustion engine and the electric engine is provided for in this concept. This means that the combustion engine is firmly mounted with the electric engine. This can be accomplished with high package densities (cf. Figure 7.4). This picture shows a clutch with a flywheel and an electric motor. The parallel hybrid is suited for both mild hybrid and full hybrid mode; although the latter needs an electric engine with more power than those shown in Figure 7.4. The configuration with one clutch is suitable for start–stop operation, for boosting and shifting operation point of the combustion engine and for regenerative braking. However, the latter mode is not as efficient in a parallel hybrid with one clutch as in a parallel hybrid with two clutches. In the one-clutch configuration, the braking torque during deceleration is split, with one portion being necessary for the drag torque of the combustion engine (this portion is lost because it is converted to heat in the combustion engine), and only the other portion being capable of conversion into electrical energy by the electric motor. This disadvantage is avoided by a configuration with two clutches. Here, the combustion engine is separated by disengaging the clutch, and the entire power output from braking can be converted into electric energy provided that the electric motor, the power electronics and the battery are able to handle the power.

An essentially different design for a hybrid powertrain is a serial configuration as shown in Figure 7.11. In this case, the combustion engine drives a generator (EM1), which charges the battery. A second electric machine (EM2) is used for driving the vehicle or for regenerative braking operations. In this configuration, the operation mode of the combustion engine is independent of any actual demand for power or tractive force of the vehicle. This means that the combustion engine can be operated at certain points with high efficiency or with low emissions of NO_x, CO and CH_4. As the driving torque must be delivered by the second electric machine, this machine must have enough power to drive the car. The first electric machine must also have

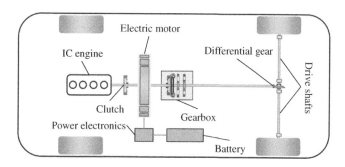

Figure 7.10   Parallel hybrid vehicle

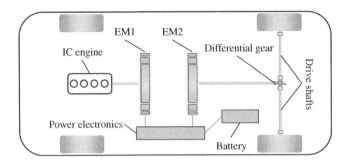

Figure 7.11   Serial hybrid vehicle

enough power to convert the power from the combustion engine. One advantage is that no clutch and no transmission are necessary in the driving powertrain (EM2, Cardan shaft, differential, drive shaft), so that this part of the serial hybrid is significantly simpler than the analogue part of the parallel hybrid. Nevertheless, there are some disadvantages with this type of hybrid powertrain. The first is the efficiency. As the power from the combustion engine has to be converted from mechanical to electrical, from electrical to chemical, from chemical to electrical and, finally, from electrical to mechanical energy, the efficiency decreases as a result of the energy conversion processes. Further drawbacks are evident in the number of components that incur high costs and add a high weight. As both electric machines have to manage the full power (the first machine, EM1, the whole power of the combustion engine, and the second, EM2, the whole power necessary to drive the vehicle), these electric machines have a high weight and are expensive. One advantage in this connection is the high potential of EM2 to regenerate energy during braking.

The disadvantage of the high weight and the capability of the electric machines can be reduced by using a serial–parallel hybrid powertrain (cf. Figure 7.12), in which the two electric machines can be connected by engaging a clutch. If the clutch is engaged, the powertrain is similar to a conventional parallel hybrid powertrain in which the power of the combustion engine can be directly transmitted to the driven axle and hence the operating mode is the same as that of a parallel hybrid. Direct connection of

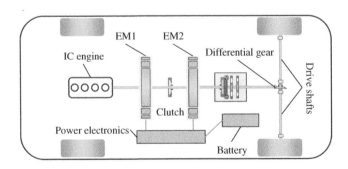

Figure 7.12   Serial–parallel hybrid vehicle

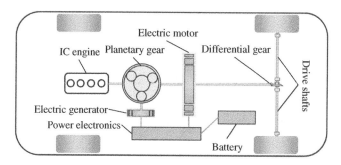

Figure 7.13   Power–split hybrid vehicle

the combustion engine to the differential requires a transmission. Consequently, more components have to be attached to the powertrain, but smaller electric machines can be selected because the combustion engine can deliver power for high demands.

The last hybrid powertrain described here is the power split system (cf. Figure 7.13). The central part of this system is a planetary gear. The combustion engine, a small generator and the Cardan shaft are joined to the planetary gear: the combustion engine to the planet carrier, the ring gear to the Cardan shaft and the sun gear to the generator. The power from the combustion engine can be used directly to drive the wheels, or a portion of this power can be converted by the generator into electrical energy and then this electrical energy can be used to drive the electric motor (in terms of efficiency, it does not make sense to charge the battery). This means that the combustion engine can be operated at a favourable point with high efficiency. The electric motor can be used for boosting or for regeneration of energy during braking. In the case of regeneration, the torque of the generator can be set to zero. The other torques in the planetary gear are then zero, too, and there is no drag torque from the combustion engine to decrease the efficiency of energy regeneration. One advantage of the power–split hybrid system is that the planetary transmission ratio is continuously variable, which is made possible by the fact that the torque of the generator is continuously variable, too. This results in a free choice of the engine operating point, which means that this type of hybrid powertrain is similar to a continuously variable transmission (CVT), with respect to the revolutions of the combustion engine.

Other possibilities also exist such as a combustion engine on one axle and an electric motor on the other axle, giving a dual clutch hybrid. The various components for the different types of hybrid powertrains are summarized in Figure 7.14. It is evident that the number of components differ.

7.3   Regenerative Braking and Charging

Considering the fundamental equation of the longitudinal dynamics, we recognize that the existing driving resistance from the air resistance and the rolling resistance needs

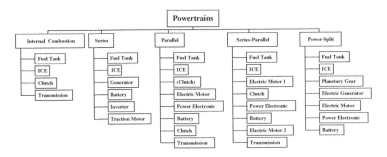

Figure 7.14   Components of hybrid powertrains

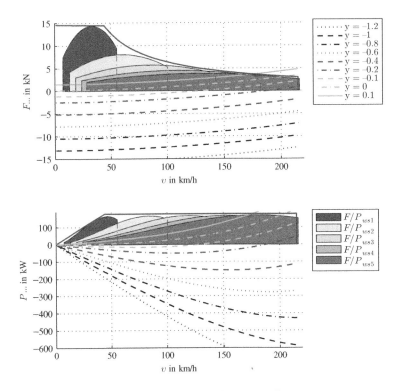

Figure 7.15   Driving performance diagrams

a portion of the available energy (kinetic energy and potential energy), so that only the remaining power can be recovered during braking. Figure 7.15 shows the tractive forces (which are mainly due here to negative values of y braking forces) and the power for different negative values of $y = p + \lambda \ddot{x}_v / g$. As there are absolute values up to 1.2, it is obvious that the high values are the result of a braking process, but not of a negative inclination of the road. Looking at the curve for $y = -0.1$, we recognize that, up to a velocity of $v = 131$ km/h, the power (or tractive force) is negative. This means that between 0 and 131 km/h, it is possible to recover energy from kinetic energy. For higher values of $y \approx -0.3$ the power and the tractive forces are both negative, and therefore it is possible to recover energy during regenerative braking over nearly the whole velocity range.

The point of intersection with the abscissa defines the speed range in which energy recovery is possible, and the shape of the curve, especially the distance of the curve to the zero line, determines the portion of power that can be recovered. The portion of recoverable energy due to the potential energy is generally small because the slope does not attain very large values. The proportion of the energy recovery that can be theoretically achieved due to a braking operation achieves significantly higher values. The following considerations are used to judge these theoretically attainable values.

We will examine four scenarios below, for which we can determine the traction, the moment and the recoverable power. The starting point is a vehicle with the following data: $c_d = 0.3$, $A = 2\,\mathrm{m}^2$, $\rho_a = 1.2\,\mathrm{kg/m}^3$, $f_r = 0.01$, $m_{\mathrm{tot}} = 1500\,\mathrm{kg}$, $g = 10\,\mathrm{m/s}^2$, $r_{wst} = 0.3\,\mathrm{m}$, $i_t = i_d i_g$ is the total transmission ratio which is the product of the transmission ratio of the differential i_d and the gearbox i_g.

Stop and go: We start with the first scenario, which corresponds to a stop-and-go situation. It is assumed that the vehicle is decelerated from a speed of $v_0 = 5\,\mathrm{m/s}$ to a speed of $0\,\mathrm{m/s}$. The acceleration used as the basis here is $a_0 = 2.5\,\mathrm{m/s}^2$. This, for example, is a reasonable value for an ACC in city traffic or when continuously stopping and starting in a traffic jam. We assume a constant negative acceleration; this means that there is a linear decrease in the velocity:

$$v(t) = v_0 \left(1 - \frac{t a_0}{v_0} \right) . \tag{7.1}$$

The duration of the braking is $T_0 = v_0/a_0 = 2\,\mathrm{s}$. The mass correction factor is $\lambda = 1.5$. Assuming these values, we obtain a relatively small aerodynamic drag force $F_a = 9\,\mathrm{N}$ at $v_0 = 5\,\mathrm{m/s}$, a rolling resistance of $F_r = 150\,\mathrm{N}$ and an acceleration resistance of $F_i = -5625\,\mathrm{N}$. Hence the force, $F_{\mathrm{max\ rec}}$ that can be used for energy recovery is minimal in the velocity interval considered at $v_0 = 5\,\mathrm{m/s}$, and this force is $F_{\mathrm{max\ rec}} = -5466\,\mathrm{N}$. The maximum force, $F_{\mathrm{max\ rec}}$, results in a maximum power of $P_{\mathrm{max\ rec}} = 27.33\,\mathrm{kW}$. Since we assume a constant acceleration, the velocity decreases linearly, so that the following relationship for the recovery power holds:

$$P_{\mathrm{rec}} = P_{\mathrm{max\ rec}}(1 - t/T_0) . \tag{7.2}$$

In order to determine whether this maximum power of 27.33 kW from an electric motor is recoverable, we have to convert the power into a torque at the engine using the radius of the wheels and the velocity of the vehicle and the total transmission ratio of $i_t = 12$. In this configuration, we assume that we have a parallel hybrid vehicle in which the electric motor is connected to the driven wheels via the gear unit. The result is an engine torque of

$$M_{\mathrm{max\ rec}} = F_{\mathrm{max\ rec}} r_{wst}/i_t = 136.65\,\mathrm{Nm} \tag{7.3}$$

and a speed at which this torque must be applied by the electric motor of

$$n_{\text{max rec}} = 60v_0/(2\pi r_{wst}) \approx 1910 \text{ rpm} . \tag{7.4}$$

City: The second case corresponds to a braking operation in city traffic with an assumed speed of $v = 15$ m/s. The rotating mass factor is assumed to be $\lambda = 1.2$, the overall transmission ratio $i_t = 7$. We obtain the following results for this situation: $F_{\text{max rec}} = -4269$ N, $M_{\text{max rec}} \approx 183$ Nm, $n_{\text{max rec}} \approx 3324$ rpm, $P_{\text{max rec}} \approx 64$ kW.

Country road: The third case concerns application of the brakes on a country road. (As in the case of urban transport, recovery power will be charged only at one point, i.e. at the initial velocity). The starting point is a speed of $v = 25$ m/s, $\lambda \approx 1$, $i_t = 4$. Here we obtain: $F_{\text{max rec}} = -3375$ N, $M_{\text{max rec}} \approx 235$ Nm, $n_{\text{max rec}} \approx 3183$ rpm, $P_{\text{max rec}} \approx 84$ kW.

Highway: The fourth case concerns application of the brakes on a highway. (As in the case of urban transport, recovery power will be charged only at one point, i.e. at the initial velocity). The starting point is a speed of $v = 35$ m/s, $\lambda \approx 1$, $i_t = 4$. Here, we obtain: $F_{\text{max rec}} = -3159$ N, $M_{\text{max rec}} \approx 237$ Nm, $n_{\text{max rec}} \approx 4456$ rpm, $P_{\text{max rec}} \approx 111$ kW.

Applying these values in a diagram for an electric motor, see Figure 7.16 (the forces are depicted as grey bullets with a positive sign, the black bullets and the black and dark grey lines are the demand forces for NEDC), we can see that the city driving was well covered by the engine, whereas braking on the country road is at the limit, meaning that the electric machine is not able to recover the braking power at a velocity of 35 m/s.

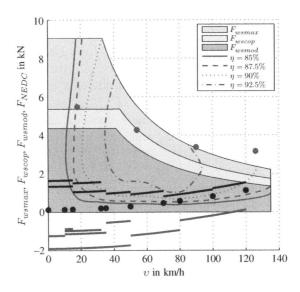

Figure 7.16   Points of energy recovery (with positive sign)

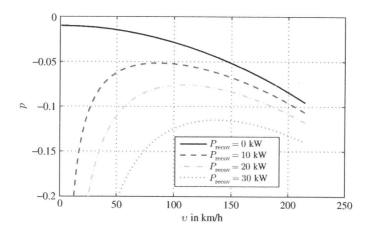

Figure 7.17   Curve of constantly recoverable power

Figure 7.17 shows the values of a negative inclination for which energy recovery is possible from potential energy. The solid line is the limit, below which energy recovery is possible; the other three lines are (p, v) for which the recoverable power is P_{recov}, where $P_{\text{recov}} = 10$ kW, $P_{\text{recov}} = 20$ kW and $P_{\text{recov}} = 30$ kW. It is evident that significant amounts of power can only be recovered for large values of p.

In the last part, we will discuss the fact that even the theoretical amount of regenerative energy cannot currently be stored as chemical energy in the battery by each vehicle configuration. Another limiting factor in the energy recovery is that not all axles are driven in many vehicles. Since energy recovery can only take place on the driven axles; this is a limiting factor for vehicles without all-wheel drives. Furthermore, from the physical point of view of the vehicle, it is necessary for the braking forces to be larger at the front axles than at the rear axles. This immediately leads to the conclusion that the regenerative potential is greater for a front-wheel-drive vehicle than for a rear-wheel-drive vehicle.

Assuming that the braking force at the front axle is 60%, the efficiency of the electric motor is 90%, and the efficiency of the power electronics and the battery is 80%; these values will result in a regenerative potential of only 41% of the theoretical value at the front axle. Under similar assumptions of the efficiency of the electric motor and the battery, a regeneration potential of 27% is obtained at the rear. With slightly better values originating from an efficiency of 95% for the electric motor and 90% for the battery, we see a potential of 49% on the front axle, and a potential of 32% on the rear axle. With four-wheel-drive vehicles, we see a potential in each case which is composed of the sum of the two.

The constraints on the theoretical regenerative potential explained above, which relate to the driving resistances along with the regeneration of moments by an electric motor, the distribution of braking forces and the constraints resulting from efficiencies reveal that only a small portion of energy in a vehicle can actually be regenerated. This should always be considered when comparing different drive concepts.

7.4   Questions and Exercises

Remembering

1. Describe the idea of hybrid powertrains.
2. Describe different toplogies of hybrid powertrains.
3. Explain regenerative braking and charging for deceleration and for a negatively inclined roads.

Applying

1. Consider a vehicle on an inclined road $p = -0.1$. Calculate the maximum regenerative power for a braking force distribution of 60% front and 40% rear for a front-wheel-drive and a rear-wheel-drive vehicle at a velocity of $v = 20$ m/s. The parameter are: $c_d = 0.3$, $A = 2$ m^2, $\rho_a = 1.2$ kg/m^3, $f_r = 0.01$, $m_{\text{tot}} = 1500$ kg, $g = 10$ m/s^2.

8

Adaptive Cruise Control

In this chapter, active cruise control is explained. In Section 8.1, the principal components and a control algorithm are considered. Section 8.2 is devoted to the measurement of distances and relative velocities. Thereafter, the approach ability of a vehicle is discussed in Section 8.3.

8.1 Components and Control Algorithm

The abbreviation ACC stands for adaptive cruise control. ACC is an extension of cruise control (CC). CC (also known as speed control) provides the opportunity for the vehicle to be driven at a constant speed without the driver having to intervene by using the brake or accelerator pedal. Different systems are in use for CC. Often there is only one engagement in the engine, which is via the throttle position (or control of the injection quantity in a diesel engine) so that the engine torque is set to maintain a certain speed. Rapid interventions that are necessary for the traction control (ASR), such as the firing angle adjustment or the suppression of individual injection pulses, are not necessary in CC.

The starting point for a mathematical description is the basic equation of the longitudinal dynamics which contains all the driving resistances.

We derived the equation of motion for the vehicle in Chapter 3, see Equation (8.1):

$$\frac{M_{a1}}{r_{wst1}} + \frac{M_{a2}}{r_{wst2}} = c_d A \frac{\rho_a}{2} \dot{x}_v^2$$

$$+ \left(m_b + J_c \left(\frac{i_d}{(1 - S_2)\, r_{wst2}} \right)^2 + J_e \left(\frac{i_d i_g}{(1 - S_2)\, r_{wst2}} \right)^2 \right) \ddot{x}_v$$

$$+ \left(m_{a1} + m_{a2} + J_{a1} \frac{(1 - S_1)^2}{r_{wst1}^2} + J_{a2} \frac{1}{(1 - S_2)^2 r_{wst2}^2} \right) \ddot{x}_v$$

Vehicle Dynamics, First Edition. Martin Meywerk.
© 2015 John Wiley & Sons, Ltd. Published 2015 by John Wiley & Sons, Ltd.
Companion Website: www.wiley.com/go/meywerk/vehicle

$$+ G \sin \alpha$$

$$+ f_{a1} F_{z1} + f_{a2} F_{z2} \, . \tag{8.1}$$

We do not need all the details, hence we reduce Equation (8.1) to the simpler form

$$\frac{1}{r_{wst1}} M_{a1} + \frac{1}{r_{wst2}} M_{a2} = c_d A \frac{\rho_a}{2} \dot{x}_v^2 + F_i + F_g + f_{a1} F_{z1} + f_{a2} F_{z2} \, . \tag{8.2}$$

The torque is controlled so that a certain speed is maintained. A CC is relatively easy to implement because of the presence of an engine control unit in all vehicles, since it is only necessary to intervene in engine management in its simplest form. Extensions of the CC are possible if brake interventions can be performed with the existing ESP and with an additional sensor for measuring distances. In it simplest form, CC is therefore possible in all vehicles.

ACC represents a significant expansion of the CC; initially the speed set by the driver on a free roadway is maintained. If there are other vehicles in the trajectory which are driving more slowly, the velocity ahead is no longer kept constant, but a time gap or distance controller keeps the vehicle within a predetermined distance (which depends on the speed) from the vehicle ahead. A controller structure for an ACC is shown in Figure 8.1. We can observe the vehicle model and the engine model in the control structure; both are inverted. Furthermore, it can be seen that the ACC is connected to other systems, particularly to the ESP, the engine and the transmission control.

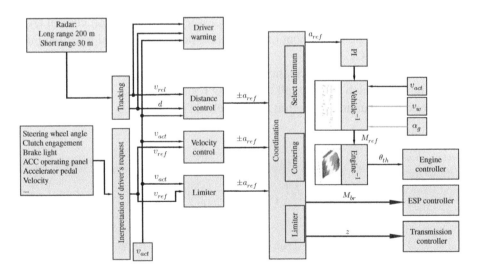

Figure 8.1   ACC control (adapted from Winner et al. 2012)

The vehicle model of longitudinal dynamics from Equation (8.2) can be used in two ways:

1. The first and straightforward way is to calculate the maximum gradeability or the maximum acceleration for a given wheel torque.
2. The second way is to prescribe one or more resistances and then to calculate the necessary torque.

Since the first way is often used as a conventional, straightforward calculation, the second, i.e. the opposite way, is referred to as an inverted model of longitudinal dynamics. This is the reason for the expression 'Vehicle^{-1}' in Figure 8.1.

The functionality of the ACC can be summarized in the steps:

1. The driver's request is interpreted: the reference velocity, v_{ref}, which is requested by the driver and which the driver has fixed via the ACC operating panel or lever, is transmitted to the velocity controller (and the limiter for acceleration and the driver warning device, usually a visual signal). This transmission is disrupted, for example, when the driver requests a higher acceleration than that specified for the ACC (kick down), when the driver brakes, when the clutch is disengaged, etc.
2. The next level of control consists of three units (besides the warning device, which has no electronic control but a human control function serving, for example, to motivate the driver to slow down):

    The velocity (or speed) controller sets the desired acceleration in order to reach the reference velocity, v_{ref}. For this purpose, the reference velocity, v_{ref}, is compared with the actual velocity, v_{act}. A function of the difference $v_{ref} - v_{act}$ yields the desired acceleration a_{ref}: $a_{ref} < 0$, if $v_{ref} - v_{act} < 0$, or $a_{ref} > 0$, if $v_{ref} - v_{act} > 0$. The velocity controller should choose a_{ref} such that it is comfortable for the driver (e.g. not too high, so as not to unsettle the driver, not too low, so as not to make the driver impatient). The velocity controller could fulfil another task: the reference acceleration, a_{ref}, yields a desired torque for the engine; it is therefore possible to operate the engine in an economical state.

    The speed limiter could, for example, consider legal requirements or ISO standards. Figure 8.2 shows the acceleration limits and the jerk limit from ISO 22179. In the first option, basic ACC control strategy in accordance with ISO 15622, the ACC switches off below $v_{low,max} = 5$ m/s. The minimum set speed is $v_{set,min} = 7$ m/s. The maximum or minimum accelerations are speed dependent. The requirements of ISO 22179 (FSRA: Full speed range adaptive cruise control; the system is designed for standstill to maximum speed) prescribe different limits, but there is also the possibility of regulating to a standstill.

    The last control unit at this level is the distance controller (or time gap controller). The input values of distance controller are the signals from the radar sensor (relative velocity of the vehicle travelling ahead and the distance to this vehicle) and the actual velocity, v_{act}. These data are used to derive the desired acceleration; while the reference acceleration, a_{ref}, of the velocity controller is mainly positive, the reference acceleration of the distance controller is negative if the distance is too small, or it alternates in follow-up control.

3. The reference values for acceleration are coordinated in the next unit, which considers special situations such as cornering. If the desired negative acceleration, a_{ref}, is larger (in terms of its absolute value) than the acceleration which the engine is able to deliver from drag torque, additional braking torque from ESP has to be applied to the wheels. If the ACC operates over the whole velocity range, an automatic or automated transmission is necessary in order to change the gear during velocity changes. A switchover from speed to distance control, and vice versa, takes place automatically, so that the driver does not have to intervene.

4. The last unit consists of a PI controller, which is described below.

In order to determine the reference wheel torque, M_{ref}, which is necessary to achieve the reference acceleration, a_{ref}, we set a_{ref}, the actual velocity, v_{act}, and the gear ratio, z, in Equation (8.1). Using these three variables, it is possible to determine the acceleration resistance, F_i, and air resistance, F_a, (with zero velocity of the wind). Since the inclination of the road cannot easily be measured (the inclination can, for example, be estimated from a longitudinal acceleration sensor, which has to be corrected by the vehicle's acceleration, which is the result of the estimated

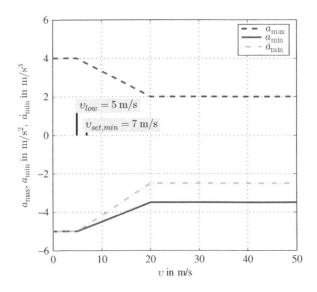

Figure 8.2   Acceleration limits for ACC: ISO 15622 and ISO 22179

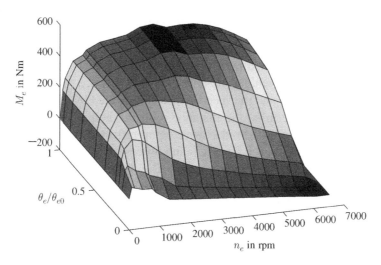

Figure 8.3 Engine map: engine torque, M_e, depending on the speed, n_e, and the relative throttle position, θ_e/θ_{e0}

engine torque) and the current set of tyres and tyre inflation pressure are not known, rolling and gradient resistance cannot be determined but merely estimated. The sum of acceleration and air resistance can be used to estimate the desired wheel torque, $M_{w\ \text{ref}}$, which yields the desired torque at the engine, M_{ref} (z is the engaged gear, i_d the gear ratio of the differential, i_z the gear ratio of the transmission):

$$M_{\text{ref}} = \frac{1}{i_d i_z} M_{w\ \text{ref}} \, . \tag{8.3}$$

Using the desired torque, M_{ref}, and the actual speed, $n_{\text{act}} = i_d i_z n_w$, of the engine ($n_w$ is the speed of the wheels) it is possible to determine the throttle angle, θ_e, which is necessary to deliver the desired engine torque. An engine map such as that in Figure 8.3 is used for this purpose. Looking at the engine map, $M_e = M_e(n_e, \theta_e)$, as a function of the two variables speed, n_e, and throttle angle, θ_e, we see that the determination of the throttle angle θ_e is a kind of inversion of M_e. That is the reason why some published sources and Figure 8.1 refer to $\theta_e = M_e^{-1}(M_{\text{ref}}, n_{\text{act}})$ or engine^{-1}.

The distance between the two vehicles which is necessary for safe operation depends on the speed. The lower boundary for the distance is generally determined by a minimum time gap $\tau_{\min} = 1\,$s. This is the time it takes for the ACC vehicle to travel the distance to the vehicle ahead. The minimum distance, $d_{\min}$, is

$$d_{\min} = \tau_{\min} v \, . \tag{8.4}$$

In ISO 15622 the time gap is recommended between 1.5 s and 2.2 s.

8.2 Measurement of Distances and Relative Velocities

The difficulty with ACC is to determine, in addition to the distance, the relative speed of the vehicle ahead, while distinguishing between relevant vehicles driving in the same lane and non-relevant vehicles travelling in other lanes.

Determining the distance and the speed is often performed by distance radar. There are different ways to determine the distance with the help of radar. One measuring principle involves an emitted radar wave being reflected from a metallic object, with the reflected (electromagnetic) wave being processed by the receiver signal, thus indicating the distance and the relative speed. The relative distance is based on the determination of the time delay required for a radiated signal to return to the receiver. The relative velocity can be determined from the Doppler shift of the frequency.

Since direct distance determination using the time delay of electromagnetic waves and direct determination of the velocity via the Doppler effect are relatively expensive (you need high sampling rates so as to detect the time delay and the Doppler shift)[1], ACC frequently uses distance radars with indirect determination by means of a so-called FMCW (frequency modulated continuous wave). In this method, the frequency of a sinusoidal signal is changed linearly (a frequency ramp) with both negative and positive slopes. The frequency shift of the received signal is based firstly on the Doppler effect due to the relative speed and, secondly, on the fact that the received signal was transmitted at a time when a different frequency was being transmitted to that being transmitted at the time of reception (the latter effect is the result of the time delay).

Let us assume that f_s is the frequency of the transmitted signal varying linearly according to the following formula ($f_0 = 76, 5\,\text{GHz}$ or $f_0 = 24\,\text{GHz}$)

$$f_s(t) = f_0 + m_1 t .\tag{8.5}$$

We obtain the frequency of the received signal

$$f_e(t) = f_0 + m_1\left(t - \frac{2d}{c}\right) - 2\left(f_0 + m_1\left(t - \frac{2d}{c}\right)\right)\frac{v_{\text{rel}}}{c}.\tag{8.6}$$

The second term of the formula is the frequency shift due to the time delay, the third term represents the frequency shift due to the Doppler effect.

The received signal with frequency f_e and the transmitted signal with frequency f_s are mixed by adding the signals; as the difference of the frequencies is small, the result is a signal that includes a beat with a relatively low frequency (of half the frequency difference). The second frequency component (the mean value of frequencies) of this mixed signal is large. After low pass filtering and a fast Fourier transformation, we obtain the frequency difference, Δf, by measurement. This frequency difference can

---

[1] If a resolution of 5 m is needed for the distance between the vehicles, the time delay difference is $10\,\text{m}/3 \times 10^8\,\text{m/s} = 1/3 \times 10^{-7}\,\text{s}$, the magnitude of the relative frequency shift is approximately $\Delta v_{\text{rel}}/c = 1/3 \times 10^{-8}\,\text{s}$ for a velocity resolution of 1 m/s.

be calculated from the two signals f_s and f_e at a time t; we obtain

$$\Delta f(t) = \frac{1}{2}|f_e - f_s|$$

$$= \left| -\frac{dm_1}{c} - \left(f_0 + m_1 \left(t - \frac{2d}{c} \right) \right) \frac{v_{\text{rel}}}{c} \right|. \tag{8.7}$$

If Δf_1 is the measured frequency of the beat and if we substitute this in Equation (8.7), we obtain

$$\Delta f_1 = -\frac{dm_1}{c} - \underbrace{\left(f_0 + m_1 \left(t - \frac{2d}{c} \right) \right)}_{=\Delta f \ll f_0} \frac{v_{\text{rel}}}{c}. \tag{8.8}$$

Equation (8.8) is a non-linear function for two unknown variables, the distance, d, and the relative velocity, v_{rel}. This means that the distance, d, and relative velocity, v_{rel}, cannot be derived from one measurement of Δf_1. At least one more measurement is necessary. To obtain good results (with respect to errors from measurements), it is necessary to have a second measurement with a different slope m_2 for frequency change. If the slope is changed to m_2 and then the frequency of the beat, Δf_2, is again measured, a second equation for d and v_{rel} is obtained

$$\Delta f_2 = -\frac{dm_2}{c} - \left(f_0 + m_2 \left(t - \frac{2d}{c} \right) \right) \frac{v_{\text{rel}}}{c}. \tag{8.9}$$

Considering the intersection of Equations (8.8) and (8.9) in the graph, where the time of measurement has to be chosen for the time t, we obtain the desired distance, d, and the desired relative velocity, v_{rel}. Assuming that the frequency changes are relatively small, Δf_i, $(i = 1, i = 2)$, with respect to the fundamental frequency, f_0, we can neglect the non-linear term in Equations (8.8) and (8.9), whereby the determination of d and v_{rel} reduces to the simple solution of a linear system of two equations.

Figure 8.4 shows how the ramped frequencies are changed for a radar sensor. Here, we observe more than two slopes. These slopes, which vary in both sign and magnitude can be used to determine the relative speed and distance more accurately.

We now consider the various frequency ramps with different slopes, m_i $(i = 1, \ldots , 4)$, as shown in Figure 8.4. A slowly rising ramp will now be added to this figure (same absolute value of the slopes). If we assume that the measurements of the beat frequency, Δf, have an error of $\pm 1\%$, then the possible distances, d, and relative velocities, v_{rel}, fulfil the equation

$$(1 \pm 1\%)\Delta f = -\frac{dm_i}{c} - \left(f_0 + m_i \left(t - \frac{2d}{c} \right) \right) \frac{v_{\text{rel}}}{c}. \tag{8.10}$$

The limits are shown in Figure 8.5. The parameters for this figure are: distance $d = 120\,\text{m}$ and relative velocities $v_{\text{rel}} = 10\,\text{m/s}$. Ideally, without any error, the curves

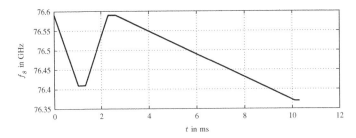

Figure 8.4 Frequency changes with different slopes to improve the relative velocity determination (adapted from Winner et al. 2012)

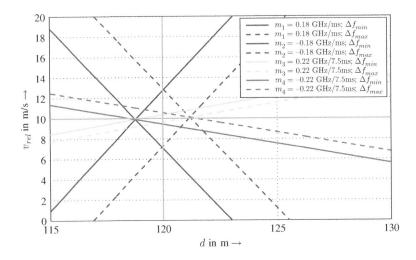

Figure 8.5 Frequency determination for different frequency changes, errors for the beat frequencies, each $\pm 1\%$

would intersect at the point $(d, v_{\text{rel}}) = (120\,\text{m}, \ 10\,\text{m/s})$. It can be seen that the inaccuracy of this method, which uses only two fast ramps, is higher than that using two slow ramps. Hence, using the third, slowly increasing function increases the accuracy.

Multiple antennas and receivers are further used to predict the direction from which the reflected beam arrives. Figures 8.6–8.9 depict the detection ranges for each of three different transmitters. A number of devices sometimes also work with four transmitting and receiving units.

Figures 8.6–8.9 illustrate sample configurations that need to be recognized by an ACC. The examples include situations in which a vehicle leaves its lane and another vehicle moves into the detection area, as shown in Figure 8.6. In this situation, the slow-moving vehicle 1 is shown briefly between the two points in time within the detection range of the ACC but then leaves it again on the right-hand side. The faster vehicle 2 enters from the left. The ACC has to distinguish these two vehicles, which

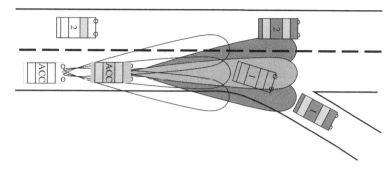

Figure 8.6   Slow vehicle 1 is currently in the detection range of the ACC and leaves it again; and vehicle 2 is in the field of coverage

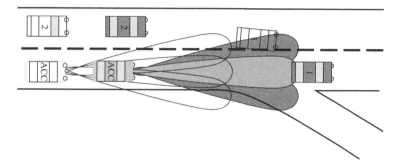

Figure 8.7   Vehicle 1 enters the detection range of the ACC vehicle

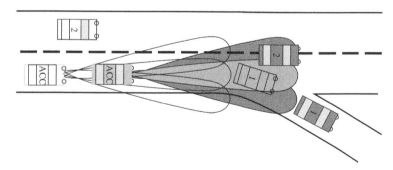

Figure 8.8   Vehicle 2 moves into the detection range of the ACC vehicle, vehicle 1 exits the range

is not complicated in the situation shown due to the different positions and different speeds.

Figure 8.7 illustrates a case in which the ACC switches from speed control to distance control because vehicle 1 cuts into the safe distance. In this situation, there

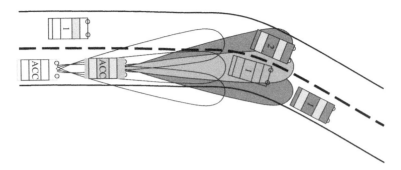

Figure 8.9 Vehicle circumstances as shown in Figure 8.8, the ACC may not be changed due to the faster vehicle travelling in the other lane on distance control

would be a reduction in speed of the ACC vehicle. If vehicle 1 was very slow, then the ACC vehicle would also brake sharply. The circumstances shown in Figure 8.8 illustrate a case similar to that in Figure 8.6. In contrast to 8.6, the ACC has to intervene in Figure 8.8 because the faster vehicle 2 is in the middle detection area and thus in the same lane as the ACC vehicle. The fact that the slow vehicle 1 was previously in this region means that ACC signals have to distinguish between the two vehicles from the side detecting sections and from the measured speeds of the vehicles. For this reason, ACC systems work with tracking algorithms to calculate the trajectories of these vehicles.

Comparing the situations in Figures 8.8 and 8.9, we can see that the locations of the vehicles involved are the same and only the directions differ. In the case of Figure 8.9, the ACC may not interfere with the circumstances. Since ACC systems may also operate in a low speed range to some extent, the situation shows an increased requirement for the ACC.

Since the ACC is connected to the ESP, automatic or automated transmission and the engine control unit, all signals of these control devices are available. It is therefore also possible for the ACC to recognize cornering due to the lateral acceleration sensor of ESP. For such a situation, i.e. the detection of curves and of vehicles travelling on a curved path ahead, it is possible to use these sensor signals. Again, there are situations such as that shown in Figure 8.9, where allocation is clearly not possible. Consequently, tracking algorithms are implemented in the ACC equipment to track the vehicles on their prospective tracks.

The functionality of the ACC is restricted or is completely lost if there is a failure of one of the control devices on which the ACC is based (engine control unit, transmission control and ESP). This is accepted as the ACC can be implemented much more economically by relying on the existing actuators and sensors of these three systems than if it were installed with its own separate set of sensors and actuators.

8.3   Approach Ability

If a slow-moving vehicle comes in front of an ACC vehicle, the ACC vehicle has to reduce its speed. The so-called approximation capability is the maximum velocity difference, Δv_0, that can just be controlled by the ACC.

Here it is assumed that the maximum acceleration, $a_{\max}$, is achieved linearly from the value 0 (see Figure 8.10). The slope is $\gamma_{\max}$ (this is the jerk, which should be limited for comfort reasons, cf. Figure 8.2), the time which is necessary to build up the maximum acceleration is τ_{up}. Consequently, the acceleration is given by

$$a(t) = \begin{cases} \gamma_{\max} t & t \le \tau_{\mathrm{up}} \\ & \text{for} \\ a_{\max} & t > \tau_{\mathrm{up}} \end{cases} .$$

(8.11)

As a result, the time to build up the maximum acceleration, τ_{up}, fulfils the following relationship:

$$\tau_{\mathrm{up}} = \frac{a_{\max}}{\gamma_{\max}} .$$

(8.12)

By integrating the acceleration, we obtain the relative velocity, $\Delta v(t)$. The relative velocity, $\Delta v(t)$, starts with the initial differential velocity, Δv_0, then first decreases parabolically and then linearly to 0. The value 0 is reached at time τ_e. We hereafter

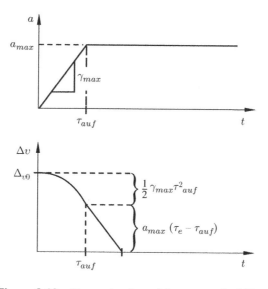

Figure 8.10   Determination of the approach ability

assume that this time τ_e occurs after the maximum acceleration has been attained, so $\tau_e > \tau_{up}$. The relative velocity curve is then given by

$$\Delta v(t) = \Delta v_0 + \frac{1}{2}\gamma_{max}\tau_{up}^2 + a_{max}(t - \tau_{up})$$

$$= \Delta v_0 - \frac{a_{max}^2}{2\gamma_{max}} + a_{max}t \ . \tag{8.13}$$

The condition that the difference in velocity must disappear, $\Delta v(\tau_e) = 0$, gives us the equation for the time τ_e. Integrating the velocity curve, we obtain the curve of the distance, $\Delta s(t)$. This starts with the distance d_0, which is present between the vehicles when they merge, and decreases with time.

This distance must not fall below a certain minimum distance, d_{min}. Substituting τ_e in the formula for the differential distance, Δs, we obtain $\Delta s(\tau_e) = d_{min}$.

Transformation yields a relationship between the minimum distance, d_{min}, and the relative velocity.

8.4   Questions and Exercises

Remembering

1. Name the two control modes on which the ACC is generally based.
2. Give the order of magnitude for the acceleration limits of ACC.
3. On what mechatronic system components does the ACC rely?
4. What happens if one of these system components fails?

Understanding

1. Explain the functioning of ACC.
2. Explain the relationship between the longitudinal dynamics, the engine characteristic map and ACC control algorithms.
3. Explain how the distance and velocity of vehicles travelling ahead are determined with FMCW.
4. How should an ACC system be equipped by sensors in order to detect the complete longitudinal dynamics?
5. Describe driving situations that illustrate the need for a vehicle tracking system.
6. In the context of longitudinal vehicle dynamics, what in general influences the presence of disturbance variables on ACC systems.

Applying

1. Calculate the frequency of the beat resulting from the Doppler effect and of the time for wave propagation (distance between the cars is 150 m) for the signal, which is transmitted at the time $t = 0$ s.

The frequency of the transmitted signal is $f_s = f_0 + mt$ ($f_0 = 74\,\text{GHz}$), where t is the time and $m = 100\,\text{MHz/s}$.

The relative velocity of the cars is $v_{\text{rel}} = 3\,\text{m/s}$. Please use $c = 3 \times 10^8\,\text{m/s}$ for velocity of electromagnetic waves. The frequency shift can be approximated by: $\Delta f = 2v_{\text{rel}} f_s/c$.

Hint: The beat frequency of the transmitted frequency f_s and the received frequency f_r is $\Delta f = |(f_s - f_r)/2|$.

- Calculate the propagation time T_p for both ways (ACC car to run ahead car and back to ACC car). For this, neglect the change in distance between the cars during wave propagation.
- Calculate the beat frequency from wave propagation time between transmitted signal and received signal for $t = T_p$.
- Calculate the beat frequency from the Doppler effect between transmitted signal and received signal.

9

Ride Dynamics

The quality of ride is an important aspect to comfort for customers, and it is one criterion in deciding whether or not to buy a car. The perception of ride quality cannot be measured as an objective value, but depends on the customer's perception and experience and so there are a great deal of factors influencing the customer's judgement of comfort: age, gender, whether the customer is used to drive high-quality or low-quality cars, perhaps even the customer's health. Besides these intrinsic factors, which cannot be influenced by the car manufacturer, there are extrinsic factors: noise, vibration, heat and flow of air. All these latter factors can be influenced. In this chapter, we concentrate mainly on vibration, the sources of vibrations and how they can be reduced or influenced.

As ride quality is an important factor for a customer to purchase a specific vehicle, it is an important aspect for an original equipment manufacturer (OEM), too. Although ride quality is subjective, there are many methods of measuring or judging ride in order to obtain an objective assessment.

The first stage of modern vehicle development is nowadays purely virtual, which means that the entire vehicle exists only on various computers as a CAD model, as a model for judging the stiffness of the body, as a model for the driving dynamics of the vehicle, as a crash model for passenger safety, as a vibrational model for the powertrain and, depending on the OEM strategy in the virtual development process, there may be other models as well. This means that the ride quality as one aspect of comfort can also be judged in the early stages of development with the aid of models.

The main difference between these virtual methods is whether a human being is present to judge the ride quality or whether there is a perceptual model of the human being.

The method with a human being requires a driving or comfort simulator; this simulator describes the dynamic behaviour of the car by means of multi-body systems (MBS). In the method without the human being, a procedure is required to compute the human being's perception.

Vehicle Dynamics, First Edition. Martin Meywerk.
© 2015 John Wiley & Sons, Ltd. Published 2015 by John Wiley & Sons, Ltd.
Companion Website: www.wiley.com/go/meywerk/vehicle

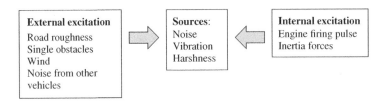

Figure 9.1   Internal and external sources

We concentrate in this chapter on sources of noise and vibration and on the fundamental means of influencing the noise and vibration in the environment of a passenger, i.e. the vibration of the seat, the steering wheel and the area where the feet are placed (e.g. pedals) and the noise received by the ear.

The excitation of vibrations and noise can be roughly divided into internal and external sources. Apart from the sources of excitation, the transfer paths from the source to the passenger are also important in order to reduce noise and vibration arriving at the passenger.

The external sources (cf. Figure 9.1) are the result of uneven roads (road roughness and single obstacles such as bumps), from the headwind and from noise generated by other vehicles. Uneven roads and noise from other vehicles cannot be influenced, they are determined by the environment. Wind from the weather is predetermined, too, while the noise generated by protruding parts of vehicles can be influenced by an OEM. The avoidance of such protruding or turbulence-generating parts is an issue of aerodynamic optimization of a vehicle, and this area of vehicle development is not dealt with in this consideration.

The internal sources are mainly the engine and the rotating parts of the whole powertrain (transmission, Cardan shaft, differential, drive shafts and wheels). These parts are taken into consideration in the development of a new car. The internal sources are caused by the engine firing pulses (in the case of an internal combustion engine), by reciprocating pistons and by mass imbalances in all rotating parts.

In addition to the sources, the transfer paths also play a vital part in judging and optimizing vibrations. Apart from the sources, Figure 9.2 illustrates some possible transfer paths (not all). The transfer paths are divided into those belonging to the external and those belonging to the internal sources.

One crucial factor is joints between sources (e.g. internal combustion engine, mass imbalance of a shaft) and chassis, for example, the centre bearing of the Cardan shaft, the mounts of the powerplant (engine and transmission), the bushings between the suspension and the chassis or between the subframe and the chassis or the hangers of the exhaust system.

A huge effort is made to isolate oscillations of the powerplant from the chassis, especially in designing the mounts. Frequently, the powerplant is fixed in two mounts at the

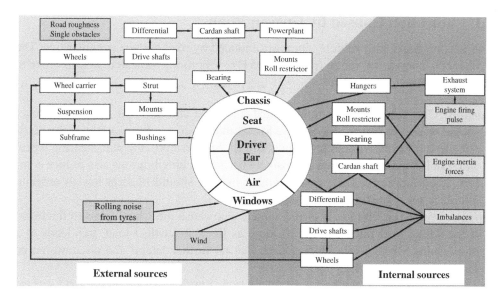

Figure 9.2   Transfer paths of oscillations

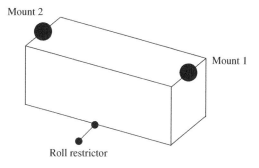

Figure 9.3   Mount at the powerplant

chassis (sometimes called an engine and a transmission mount) and an additional roll restrictor (to fix the last degree of freedom; cf. Figure 9.3). The mounts at the power-plants can be designed as hydrodynamic rubber bushings with a special characteristic in reducing distinct oscillation for special frequency intervals.

There are different transfer paths for the oscillations and for the noise. Headwind noise, for example, enters the car through the windows and chassis, and, from these parts, the vibrations cause oscillations of the air cavity inside the vehicle, which results in noise in the ears of passengers or the driver.

In the next two sections, we look at details concerning the reduction of torsional vibrations of the powertrain and concerning vibrations caused by rough roads.

9.1 Vibration Caused by Uneven Roads

This section deals with vehicle vibrations induced by road surface irregularities. The frequencies considered range from 0 to 25 Hz. Vibrations are calculated in order to answer a number of questions.

Comfort: One necessary requirement that a vehicle should fulfil is that the driver should be able to operate the vehicle for a lengthy period of time without any detriment to health and without becoming unwell. Furthermore, the seat, steering wheel and pedals should not vibrate excessively because such vibrations also reduce comfort. The vibrations of the entire vehicle interior should be kept low in order to minimize the acoustic strain on the occupants.

Driving safety: The vibrations of the vehicle also cause the wheel loads to fluctuate. If the wheel load fluctuation is as large as the static wheel load, this leads to a diminishing wheel load, which means that no further lateral and circumferential forces can be transferred.

Road surface strain: The wheel load fluctuations impose a strain on the road surface in addition to the static wheel loads.

Engineering strength: The strength is reduced due to the vibrating loads. The consequence of this is a lower operating lifetime of vehicle parts.

Package space: Vehicle vibrations must not result in parts colliding with each other, as this firstly produces noise disturbance and secondly leads to a large strain on the parts or to their destruction. For this purpose, the designed space has to be calculated under a vibrating load.

9.1.1 Damped Harmonic Oscillator

The damped harmonic oscillator shown in Figure 9.4 serves as an introduction to and as revision of the basic principles of vibration theory.

The massless wheel travels over an uneven road surface. The mass, m, represents one quarter of the total vehicle. Between the mass, m, and the wheel is a shock absorber (damper constant b) and a spring (spring constant k). The single mass oscillator moves at a constant velocity v. The parameters sought are the movement $z(t)$, of the mass m and the dynamic wheel load fluctuation F_{zdyn}.

From the free-body diagram, we obtain

$$m\ddot{z} + F_d + F_s = 0. \tag{9.1}$$

We assume that the spring and shock absorber forces are linearly dependent upon the distances and velocities.

Consequently, the result is

$$F_d = b(\dot{z} - \dot{h}) , \tag{9.2}$$

$$F_s = k(z - h) . \tag{9.3}$$

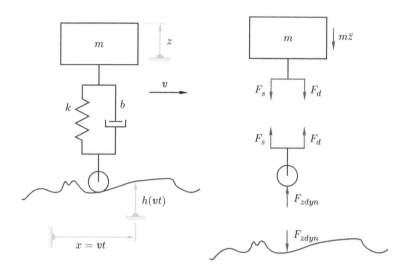

Figure 9.4 Damped harmonic oscillator

From this, we obtain the equation of motion of the mass m

$$m\ddot{z} + b\dot{z} + kz = b\dot{h} + kh \tag{9.4}$$

and the dynamic wheel load fluctuation

$$F_{\text{zdyn}} = -F_s - F_d \ . \tag{9.5}$$

We first consider the natural vibrations of the system. These can be obtained by solving the homogeneous differential equation

$$m\ddot{z} + b\dot{z} + kz = 0 \ . \tag{9.6}$$

With the aid of the abbreviations $\sigma = \frac{b}{2m}$ and $\nu^2 = \frac{k}{m}$, we obtain with an $e^{\lambda t}$-approach of $z_{\text{hom}} = \hat{z}e^{\lambda t}$ the characteristic polynomial

$$\lambda^2 + 2\sigma\lambda + \nu^2 = 0 \ . \tag{9.7}$$

From Equation (9.7), we obtain two eigenvalues:

$$\lambda_{1,2} = -\sigma \pm \sqrt{\sigma^2 - \nu^2} \ , \tag{9.8}$$

from which the solution of the homogeneous differential equation is obtained:

$$z_{\text{hom}} = \hat{z}_1 e^{-\sigma t} e^{j\sqrt{\nu^2 - \sigma^2}t} + \hat{z}_2 e^{-\sigma t} e^{-j\sqrt{\nu^2 - \sigma^2}t} \ . \tag{9.9}$$

Here $\hat{z}_1$ undergoes complex conjugation to $\hat{z}_2$. If the system is not too strongly damped (this is the case for $\nu^2 > \sigma^2$), then σ is the decay constant and $\sqrt{\nu^2 - \sigma^2}$ the natural frequency of the system.

We determine a particular solution of the equation of motion for the special case where the road surface irregularity, h, is a cosine function. To simplify the calculation, we set $h(x)$ as complex:

$$h(x) = \hat{h}e^{j\kappa x} . \tag{9.10}$$

Ultimately, it is the real part that is of interest for the solution. Inserting this into the equation of motion, we obtain with $x = vt$:

$$m\ddot{z}_{\text{part}} + b\dot{z}_{\text{part}} + kz_{\text{part}} = b\hat{h}j\kappa v e^{j\kappa vt} + k\hat{h}e^{j\kappa vt} . \tag{9.11}$$

The free parameter κ here is not the eigenvalue as above but, instead, the angular wavenumber:

$$\kappa = \frac{2\pi}{L} , \tag{9.12}$$

where L is the wavelength of the harmonic road surface irregularity. Inserting expression (9.13) for the particular solution z_{part} into the equations of motion

$$z_{\text{part}} = \hat{z}_{\text{part}} e^{j\kappa vt} , \tag{9.13}$$

yields

$$\hat{z}_{\text{part}} = \hat{h}\frac{jb\kappa v + k}{-m(\kappa v)^2 + jb\kappa v + k} . \tag{9.14}$$

The expression

$$\kappa v = \omega \tag{9.15}$$

is the angular frequency with which the system is excited. The expression

$$\frac{\hat{z}_{\text{part}}}{\hat{h}} = \frac{bj\omega + k}{-m\omega^2 + bj\omega + k} \tag{9.16}$$

is called the transfer function.

If the frequency ratio of the excited frequency to the natural angular frequency of the undamped system, $\eta = \omega/\sqrt{k/m}$, and the damping constant $D = b/(2m\sqrt{k/m})$ are introduced, the magnification or transfer function (9.16) can be rewritten as

$$\frac{\hat{z}_{\text{part}}}{\hat{h}} = \frac{1 + j2D\eta}{(1 - \eta^2) + j2D\eta} . \tag{9.17}$$

The transfer function is therefore merely dependent upon the damping constant, D, and the frequency ratio, η. Interest is often only directed towards the magnitude of the transfer function, the phase plays a subordinate role.

For many issues, it is also the acceleration and not the deflection or the path that is of greater interest. Figure 9.5 is a plot of the acceleration of the mass, m. The function is obtained from the complex magnification functions by forming the absolute value:

$$\left|\frac{\ddot{z}}{\hat{h}}\right| = \nu^2\eta^2\sqrt{\frac{1 + 4D^2\eta^2}{(1 - \eta^2)^2 + 4D^2\eta^2}} . \tag{9.18}$$

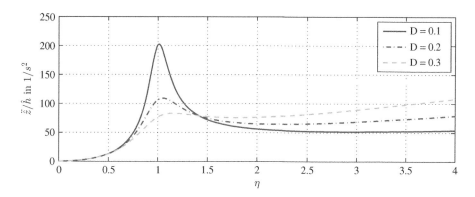

Figure 9.5 Magnification function for the acceleration

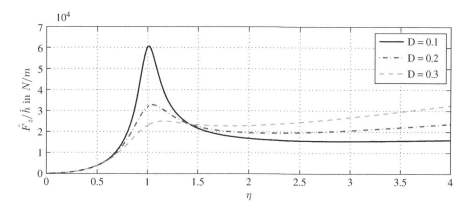

Figure 9.6 Magnification functions for the wheel load fluctuation

Here $\nu = \sqrt{\frac{k}{m}}$ is the natural angular frequency of the undamped system. Figures 9.5 and 9.6 show the magnification functions for a natural frequency of 1 Hz (hence: $\nu = 2\pi\frac{1}{s}$) and for the damping factors $D = 0.1$, $D = 0.2$ and $D = 0.3$. The lower graph shows the wheel load fluctuation for a mass of $m = 300$ kg. If an amplitude of $\hat{h} = 0.1$ m is assumed for the excitation, it can be seen that the wheel load between $D = 0.1$ and $D = 0.2$ becomes zero by a value of $\eta \approx 1$. The magnification function for the force in Figure 9.6 can be derived for this simple mechanical system by multiplying the function from Figure 9.5 by the mass $m = 300$ kg.

If the road surface is periodically uneven, this irregularity can be developed into a Fourier series. The vibration response of the system is then simply a superimposition of the individual responses.

We can write h as

$$h(t) = \sum_{i=-n}^{n} \hat{h}_i e^{ji\omega t} \ . \tag{9.19}$$

Here $h(t)$ is a real variable if $\hat{h}_i = \overline{\hat{h}}_{-i}$ is demanded; hence, if the ith complex coefficient is equal to the complexly conjugated $(-i)$th coefficient. The response of the system can then be written as

$$z_{\text{part}} = \sum_{i=-n}^{n} \hat{h}_i \frac{bji\omega + k}{-m(i\omega)^2 + bji\omega + k} e^{ji\omega t} . \tag{9.20}$$

These Fourier series can be expanded to Fourier integrals if stochastic irregularities are present. The irregularity of the road surface can then be written as

$$h(t) = \int_{-\infty}^{\infty} \hat{h}(\omega) e^{j\omega t} d\omega . \tag{9.21}$$

We obtain the response of the system as above

$$z_{\text{part}} = \int_{-\infty}^{\infty} \left(\frac{\hat{z}_{\text{part}}}{\hat{h}} \right) (\omega) \hat{h}(\omega) e^{j\omega t} d\omega . \tag{9.22}$$

The function $(\hat{z}_{\text{part}}/\hat{h})(\omega)$ is the transfer function (or magnification function) of the system. Statistical parameters are frequently used to characterize random vibrations. Initially, the mean is an obvious choice

$$\overline{z}_{\text{part}} = \frac{1}{T} \int_{0}^{T} z_{\text{part}}(t) \, dt , \tag{9.23}$$

where a sufficiently large value of T must be selected. Another important value is the standard deviation:

$$\sigma_z = \sqrt{\frac{1}{T} \int_{0}^{T} \left[z_{\text{part}}(t) - \overline{z}_{\text{part}} \right]^2 dt} . \tag{9.24}$$

This standard deviation can of course be also determined for the wheel load fluctuation and the acceleration.

The importance of the standard deviation is illustrated by the following example. Let the mean of the wheel load be $\overline{F} = 3000$ N, the standard deviation $\sigma_F = 300$ N. The probability that the wheel load is above $\overline{F} + \sigma_F = 3300$ N or below $\overline{F} - \sigma_F = 2700$ N is 31.7%. The probability that the wheel load is above $\overline{F} + 2\sigma_F = 3600$ N or below $\overline{F} - 2\sigma_F = 2400$ N is 4.6%. The probability that the wheel load will be below $\overline{F} - 3\sigma_F = 2100$ N or above $\overline{F} + 3\sigma_F = 3900$ N is 0.3%.

9.1.2 Assessment Criteria

This section explains some assessment criteria for oscillations.

Rebound clearance
The rebound clearance is the distance between the body and the wheel. This must not exceed a certain value in order to avoid collisions between the wheel and the wheel housing.

In order to determine the maximum rebound clearance, we first start on the basis of an unladen vehicle. The static spring deflection is then

$$z_{\text{stat unlad}} = \frac{m_{\text{unlad}}g}{k} \; ; \tag{9.25}$$

here m_{unlad} is the mass of the unladen vehicle, g the acceleration due to gravity and k the total stiffness of the springs. The spring deflection fluctuates due to the road irregularities. Let the standard deviation be $\sigma_{z\text{unlad}}$ (cf. Figure 9.7). The following applies to the laden vehicle:

$$z_{\text{stat lad}} = \frac{m_{\text{lad}}g}{k} \; ; \tag{9.26}$$

here m_{lad} is the mass of the laden vehicle, where $m_{\text{lad}} = m_{\text{unlad}} + \Delta m$. The static spring deflection between the unladen and laden vehicle is therefore

$$\begin{aligned}
\Delta z_{\text{stat}} &= \frac{m_{\text{lad}}g}{k} - \frac{m_{\text{unlad}}g}{k} \\
&= \frac{\Delta m}{m_{\text{unlad}}} \frac{g}{\nu_{\text{unlad}}^2} .
\end{aligned} \tag{9.27}$$

Here $\nu_{\text{unlad}} = \sqrt{k/m_{\text{unlad}}}$ is the natural angular frequency of the unladen vehicle.

In specifying the maximum travel, it is common not to use the maximum ranges because they are relatively unlikely to occur. It is therefore common practice to use three times the standard deviation instead of the maximum values. In addition to the static spring deflections, Figure 9.7 also shows the fluctuations due to irregularities. The broken lines show the respective deviations from $3\sigma_{z\text{ unlad}}$ and $3\sigma_{z\text{ lad}}$. The difference then yields the maximum spring deflection:

$$z_{\text{max}} = \frac{\Delta m}{m_{\text{unlad}}} \frac{g}{\nu_{\text{unlad}}^2} + 3(\sigma_{z\text{ lad}} + \sigma_{z\text{ unlad}}) . \tag{9.28}$$

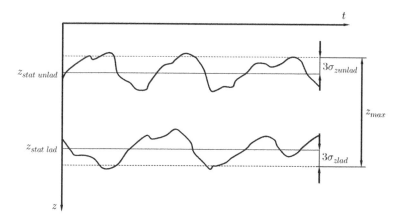

Figure 9.7   Spring deflection

Example 9.1 If we start with a passenger car having an unladen weight of $m_{\text{unlad}} = 1200$ kg and a load of five persons (80 kg per person) and 60 kg luggage, then $\Delta m = 460$ kg. If the natural frequency is $f_{\text{unlad}} = 1$ Hz, the static change in spring deflection results from Equation (9.27):

$$\Delta z_{\text{stat}} = \frac{460 \text{ kg}}{1200 \text{ kg}} \frac{9,81\frac{\text{m}}{\text{s}^2}}{\left(2\pi\frac{1}{\text{s}}\right)^2} = 95 \text{ mm} . \tag{9.29}$$

Wheel Load Impact Factor

In order to assess the wheels and the wheel bearings, the so-called wheel load impact factor, n, is often introduced. This is the ratio of the maximum wheel load to the static wheel load:

$$n = \frac{F_{z \text{ max}}}{F_{z \text{ stat}}} = 1 + \frac{F_{z \text{ dynmax}}}{F_{z \text{ stat}}} . \tag{9.30}$$

Frequently $F_{z \text{ dynmax}}$ is replaced by three times the value of the standard deviation.

$$n = 1 + \frac{3\sigma_F}{F_{z \text{ stat}}} . \tag{9.31}$$

9.1.3   *Stochastic Irregularities*

If the random irregularity of the road is described by a Fourier integral, the time dependent, stochastic responses of the vehicle as a system can also be described by Fourier integrals. However, interest here is often not directed towards the precise sequence of, for example, a wheel load or acceleration, but as a general rule merely towards the statistical parameters such as the root mean square. For this reason, we will therefore not specify the Fourier transforms of the corresponding variables when describing stochastic road irregularities and stochastic responses, but the so-called spectral densities instead.

If there is interest in the spectral density of a vehicle variable as a function of the spectral density of the road, these values can be interconverted easily. Let $q(t)$ be any vehicle-specific variable (e.g. an acceleration). Then the response function over time, $q(t)$, can initially be expressed as

$$q(t) = \int_{-\infty}^{\infty} \left(\frac{\hat{q}}{\hat{h}}\right) (\omega) \, \hat{h} \, (\omega) \, e^{j\omega t} \, d\omega , \tag{9.32}$$

with the quotient $\frac{\hat{q}}{\hat{h}}$ representing the response behaviour of the vehicle as a system to the excitation by an irregular road surface.

The integrands (apart from the exponential function)

$$\hat{q}(\omega) = \left(\frac{\hat{q}}{\hat{h}}\right) (\omega) \, \hat{h} \, (\omega) \tag{9.33}$$

can be regarded as Fourier transforms of the corresponding response (e.g. acceleration). If means of the stochastic functions are considered, they generally do not give any deeper insight into the behaviour of the system or into the road surface irregularities. Consequently, the mean of the wheel load is then exactly the static wheel load and the mean road surface irregularity is 0.

More meaningful information comes from the root mean squares (also called effective values):

$$\tilde{q}(T) = \sqrt{\frac{1}{T} \int_0^T q^2(T) \, dt} \; . \tag{9.34}$$

If the stochastic irregularities are described by normal distributions (Gaussian distributions), certain deductions can be drawn with the aid of the root mean squares. Taking account of the limiting transformation $\lim\limits_{T \to \infty}$ for the root mean square and replacing the time-dependent function by the corresponding Fourier integral, we obtain

$$\tilde{q} = \sqrt{\int_0^\infty \left(\frac{\hat{q}}{\hat{h}}\right)^2 (\omega) \underbrace{\lim_{T \to \infty} \frac{4\pi}{T} (\hat{h}(\omega))^2 d\omega}_{\Phi_h(\omega)}} \; . \tag{9.35}$$

It can be seen that the expression $\Phi_h(\omega)$ multiplied by the response behaviour of the vehicle yields the root mean square for $\tilde{q}$. The operation that leads to this result is a single integration over the square of the response function multiplied by the function $\Phi_h(\omega)$. This function is called the spectral density, and in this case it is the spectral density of the stochastic road surface irregularity. With the aid of the response function and the spectral density, we can thus use formation of the integral to determine the root mean square values for any variables of the vehicle.

Measurements of real road irregularities have shown that they can essentially be described by three parameters. Here it makes sense to use the spectral density, although this is often not specified for the time-dependent range but for the spatial range instead:

$$\Phi_h(\Omega) = v \, \Phi_h(\omega) \tag{9.36}$$

where v represents the driving speed of the vehicle. After the measurements, road irregularities can be represented by

$$\Phi_h(\Omega) = \Phi_h(\Omega_0) \left(\frac{\Omega}{\Omega_0}\right)^{-w} . \tag{9.37}$$

Figure 9.8 gives the spectral irregularity densities for a number of roads. It can be seen in the double-logarithmic plot that the spectral density essentially takes a linear path. This is also reflected in the straightforward representation of the spectral density by a simple power function.

The factor $\Phi_h(\Omega_0)$ is also called the irregularity constant, coefficient of unevenness or coefficient of roughness and w the waviness. Table 9.1 provides a compilation of some mean irregularity constants and waviness values for different types of road (the reference wavenumber for this table is $\Omega_0 = \frac{2\pi}{L_0} = 1 \text{ rad/m}$).

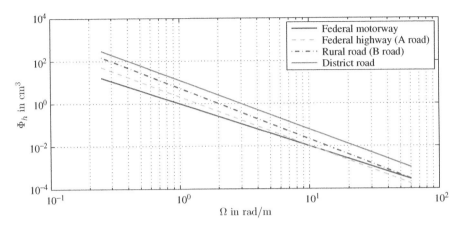

Figure 9.8 Examples of spectral irregularity densities

Table 9.1 Typical values for uneven road parameters

Road	$\Phi_h(\Omega_0)/\mathrm{cm}^3$	w
Federal motorway	1	2.0
Federal highway (A road)	2.1	2.3
Rural road (B road)	5.3	2.4
District road	12.2	2.3

$\Omega_0 = 1$ rad/m

With the aid of the spectral irregularity density for the road surface together with the response functions of a vehicle, it is possible to determine any spectral densities of vehicle variables. This is the starting point for comfort and safety investigations, which are described in the next subsection. One measure of comfort here is provided by the weighted acceleration values and a measure of safety by the wheel load.

9.1.4 Conflict between Safety and Comfort

The vehicle vibrations are transmitted to the driver via the seat, steering wheel and the floor. Depending on their frequency and amplitude and according to the site at which they occur, the driver perceives them as disruptive to a greater or lesser extent[1].

The vibrations of the seat serve as a main criterion for assessing the comfort. In order to assess these vibrations as objectively as possible, test subjects were exposed to vertical vibrations. The subjects had to classify the vibrations as not perceptible,

---

[1] The considerations closely follow the monograph of Mitschke and Wallentowitz 2004.

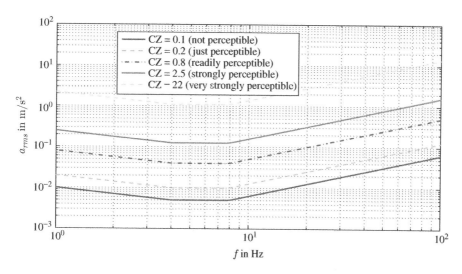

Figure 9.9 Evaluated vibration strengths

just perceptible, strongly perceptible and very strongly perceptible. These experiments showed that the classification depends on the amplitude of the seat vibration and on its frequency. Figure 9.9 shows an example of a number of curves (cf. VDI Guideline 2057) of the same classification.

It can be seen from the curves that humans react sensitively to seat vibrations in the range from 4 Hz to 8 Hz because this is the range in which the r.m.s. of acceleration is lowest. These so-called CZ values are obtained from the r.m.s. values of the accelerations in m/s² by multiplying the numerical values of the minimum of the curves by 20.

The term CZ comes from comfort and its effect in the z direction. As there are also C values for the hands and feet, we use the abbreviations C_{seat}, C_{hand} and C_{foot} in the following.

Investigations have revealed that humans react sensitively to vibrations of the steering wheel in the range from 8 Hz to 16 Hz.

Frequently it is not the comfort value that is used for assessment but, instead, the evaluation function E

$$E_{\text{seat}} = \frac{K_{\text{seat}}}{\ddot{z}_{\text{seat,eff}}} \, . \tag{9.38}$$

Figure 9.10 shows the evaluation functions for the seat.

In order to assess the comfort of vehicles, it is not sufficient to assume one transmission point to the body, but, instead, several transmission points have to be considered simultaneously. In this respect, it has proven useful to first treat the transmission points of hand, foot and seat separately from one another, and then combine them with the aid of a suitable weighting scheme to produce an overall comfort evaluation.

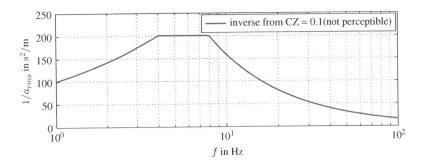

Figure 9.10   Evaluation functions for the seat

Figure 9.14 shows three different additional evaluation functions for harmonic exci-tations in the z direction for seat, foot and hand, which are used in the following comfort assessment. These evaluation functions, E_i, for harmonic excitations can be converted into evaluation functions for stochastic excitations by scaling.

$$E_{\text{stoch }i} = G_{\text{stoch }i}E_i . \tag{9.39}$$

The evaluation functions, E_i and $E_{\text{stoch }i}$, depend on the natural angular frequency, ω. Here the index i is intended to represent the information on the transmission point and the direction of transmission. The weighting factors for the different transmission points in the z direction are

$$G_{\text{stoch }z\text{ seat}} = G_{\text{stoch }z\text{ foot}} = G_{\text{stoch }z\text{ hand}} = 1,26 . \tag{9.40}$$

A comfort value of $C_{\text{stoch }z\text{ seat}}$, e.g. for stochastic seat vibrations in the z direction, results from:

$$C^2_{\text{stoch }z\text{ seat}} = \int_0^\infty E^2_{\text{stoch }z\text{ seat}}(\omega)\phi_{\ddot{z}\text{ seat}}(\omega)\,d\omega . \tag{9.41}$$

Here $\Phi_{\ddot{z}\text{ seat}}$ seat is the spectral density of the seat acceleration in the z direction. The corresponding comfort evaluations are obtained in a similar manner for the vibra-tions from the steering wheel (hand) and from the floor (foot). If the human body is simultaneously excited at several transmission points, these stochastic comfort values are weighted and added to obtain the total comfort value C_{tot}:

$$C^2_{\text{tot}} = 1,1^2\,C^2_{\text{stoch }z\text{ seat}} + 0,75^2\,C^2_{\text{stoch }z\text{ hand}} + 1,3^2\,C^2_{\text{stoch }z\text{ foot}} . \tag{9.42}$$

The C values for foot and hand again result from the integral over the evaluation function multiplied by the spectral density of the acceleration at the corresponding transmission points. This method yields an overall evaluation of the vibration comfort, although it should be noted that both vehicle and road characteristics are contained in the respective spectral densities.

The following shows the procedure for a simple three-mass model as a means of determining the overall comfort factor. Driving safety continues to be assessed using this model by calculating the standard deviation of the dynamic wheel loads. During the course of this section, it becomes clear that a conflict exists between driving safety and comfort. The starting point is the three-mass oscillator shown in Figure 9.11.

The mass m_1 corresponds to a wheel mass plus mass fractions of movable wheel suspension parts (hub carriers, control arms, etc.), mass m_2 represents one quarter of the body mass in addition to the masses for legs and arms of a driver, with mass m_3 accounting for the remainder of the driver's mass (without legs and arms). Only one-half of the driver is considered in this process. If we assume a mass of 18 kg for arms and legs with a total mass for the driver of 74 kg, we obtain a value of $m_3 = 28$ kg for the residual mass of the driver. The following compilation includes the data for the reference model under consideration:

$$m_1 = 31 \text{ kg}, \qquad m_2 = 229 \text{ kg}, \qquad m_3 = 28 \text{ kg},$$
$$k_1 = 128 \text{ kN/m}, \quad k_2 = 20.2 \text{ kN/m}, \quad k_3 = 9.9 \text{ kN/m},$$
$$b_2 = 1.14 \text{ kNs/m}, \quad b_3 = 0.26 \text{ kNs/m}.$$

Ignoring small stiffness values enables the natural frequencies of the system to be estimated by

$$\frac{\nu_1}{2\pi} \approx \frac{1}{2\pi}\sqrt{\frac{k_1}{m_1}} \approx 10 \text{ Hz},$$

$$\frac{\nu_2}{2\pi} \approx \frac{1}{2\pi}\sqrt{\frac{k_2}{m_2}} \approx 1,5 \text{ Hz}, \qquad\qquad (9.43)$$

$$\frac{\nu_3}{2\pi} \approx \frac{1}{2\pi}\sqrt{\frac{k_3}{m_3}} \approx 3 \text{ Hz} .$$

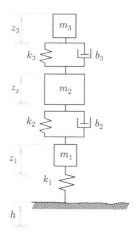

Figure 9.11 Quarter-vehicle model with driver mass

The equations of motion for the system are

$$m_3 \ddot{z}_3 + b_3(\dot{z}_3 - \dot{z}_2) + k_3(z_3 - z_2) = 0 \tag{9.44}$$

$$m_2 \ddot{z}_2 + b_3(\dot{z}_2 - \dot{z}_3) + b_2(\dot{z}_2 - \dot{z}_1)$$
$$+ k_3(z_2 - z_3) + k_2(z_2 - z_1) = 0 \tag{9.45}$$

$$m_1 \ddot{z}_1 + b_2(\dot{z}_1 - \dot{z}_2) + k_2(z_1 - z_2) + k_1 z_1 = k_1 h . \tag{9.46}$$

Inserting a harmonic excitation for the height profile, h, into this equation yields response functions for z_1, z_2 and z_3. Similarly, these response functions can be used to obtain the corresponding response functions for the accelerations. If $\Phi_h(\omega)$ is the spectral density of the road surface plus the driving speed, the spectral densities for the accelerations of hand, foot and seat can be obtained, namely $\Phi_{\ddot{z}\text{ hand}}$, $\Phi_{\ddot{z}\text{ foot}}$ and $\Phi_{\ddot{z}\text{ seat}}$, respectively. The spectral densities of the respective comfort values as well as the spectral density for the wheel load can be calculated with the following equations:

$$\Phi_{C\text{ seat}}(\omega) = 1.26^2 E_{\text{seat}}^2(\omega) \underbrace{\left(\frac{\hat{\ddot{z}}_3}{\hat{h}}\right)^2 (\omega) \, \Phi_h(\omega)}_{\Phi_{\ddot{z}\text{ seat}}(\omega)} , \tag{9.47}$$

$$\Phi_{C\text{ hand}}(\omega) = 1.26^2 E_{\text{hand}}^2(\omega) \underbrace{\left(\frac{\hat{\ddot{z}}_2}{\hat{h}}\right)^2 (\omega) \Phi_h(\omega)}_{\Phi_{\ddot{z}\text{ hand}}(\omega)} , \tag{9.48}$$

$$\Phi_{C\text{ foot}}(\omega) = 1.26^2 E_{\text{foot}}^2(\omega) \underbrace{\left(\frac{\hat{\ddot{z}}_2}{\hat{h}}\right)^2 (\omega) \, \Phi_h(\omega)}_{\Phi_{\ddot{z}\text{ foot}}(\omega)} , \tag{9.49}$$

$$\Phi_F(\omega) = \left(\frac{\hat{F}_z}{\hat{h}}\right)^2 (\omega) \, \Phi_h(\omega) , \tag{9.50}$$

where the following apply:

$$\frac{\hat{F}_z}{\hat{h}}(\omega) = \omega^2 \left(m_1 \frac{\hat{z}_1}{\hat{h}} + m_2 \frac{\hat{z}_2}{\hat{h}} + m_3 \frac{\hat{z}_3}{\hat{h}}\right) , \tag{9.51}$$

$$\left(\frac{\sigma_F}{F_{z\text{ stat}}}\right)^2 = \frac{1}{F_{z\text{ stat}}^2} \int_0^\infty \Phi_F(\omega) \, d\omega . \tag{9.52}$$

The reference vehicle described above is compared with another vehicle (denoted by Vehicle 2 in the following, whereas the reference vehicle is denoted by Vehicle 1) in which only the shock absorber is changed. The shock absorber of Vehicle 2 has the damping constant $b_2 = 1.54$ kNs/m. The road excitations for both vehicles are identical. The spectral densities for the road (i.e. as a function of the wavenumber) and the spectral density of the road together with the driving speed (as a function of the excitation frequency) are shown in Figures 9.12 and 9.13.

It can be seen that the spectral densities, for both road irregularity and for excitation (road and velocity) closely resemble each other. Here the same road excitation parameters have been selected for Vehicle 1 and Vehicle 2:

$$\Phi(\Omega_0) = 4 \times 10^{-6} \text{ m}^3 \,,$$

$$\Omega_0 = 1 \frac{\text{rad}}{m} \,,$$

$$w = 2 \,.$$

The velocity for both vehicles, v, is 20 m/s. From the graph for the spectral density of the road, we arrive at the graph for spectral density of the road plus driving speed by multiplying the abscissa axis by v and dividing the ordinate axis by v.

The following example shows how the gradient can be estimated from the graph in Figure 9.12. We assume that the two angular spatial frequencies of $\Omega_1 = 2\pi \times 0.1\frac{1}{m}$ and $\Omega_2 = 2\pi \times 0.3\frac{1}{m}$. From the graphs, we can read off $\Phi(\Omega_1) \approx 10^{-5}$ m^3 and $\Phi(\Omega_2) \approx 10^{-6}$ m^3. The result is

$$\frac{\Phi(\Omega_2)}{\Phi(\Omega_1)} = 0.1 = \left(\frac{2\pi \times 0.1}{2\pi \times 0.3} \right)^w . \tag{9.53}$$

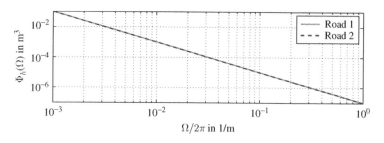

Figure 9.12   Spectral density of the road

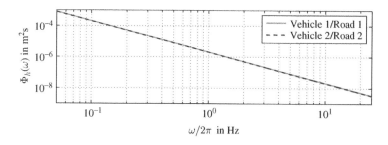

Figure 9.13   Spectral density of the road with driving speed

Taking the logarithms and rearranging, we obtain

$$w = \frac{\log(0.1)}{\log(0.1/0.3)} \approx 2.095 \ . \tag{9.54}$$

If we consider the vertical evaluation functions (cf. Cucuz 1993) in Figure 9.14, we can see that the evaluation function for the vertical seat acceleration is approximately a factor of three above the evaluation functions for the accelerations of foot and hand. It is obvious that the seat accelerations have a greater influence on the overall comfort. For this reason, in the following we merely consider the corresponding functions for the seat in the graphs. However, all comfort evaluations will be included in the calculation of the overall comfort grade at the end of this chapter.

In Figure 9.15, we can see the magnification functions of the seat $m_{a\ seat}$ vibrations for the two vehicles. Considering Vehicle 1, we can clearly see that the two lowest natural frequencies at approximately 1.5 and 3 Hz. Both natural frequencies are distinguished by clear resonance rises. If the magnification function of Vehicle 1 is compared with that of Vehicle 2, it becomes clear that the natural oscillation of the body of the vehicle (frequency at approx. 1.5 Hz) no longer becomes as clearly visible, which can be explained by the greater body damping b_2. However, the second natural frequency at approx. 3 Hz (natural frequency of seat) becomes significantly greater in Vehicle 2.

If, however, we compare the magnification functions with the corresponding spectral densities for the seat in Figure 9.16, we can see that the magnification functions are so distorted by the corresponding spectral densities of the road and the driving speed that the body acceleration is more strongly emphasized than the seat acceleration. The reason for this lies in the fact that in the case of road irregularities smaller excitation frequencies with a greater amplitude occur as larger frequencies. Nevertheless, in a comparison of the two vehicles in terms of the spectral density of the seat

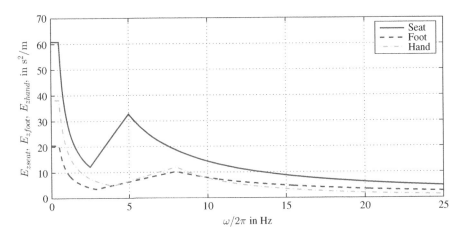

Figure 9.14   Evaluation function after Cucuz 1993

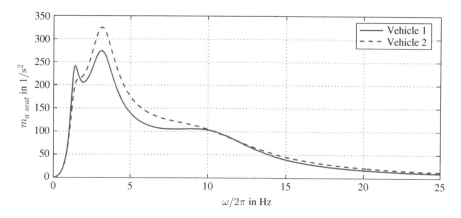

Figure 9.15 Magnification function for seat acceleration

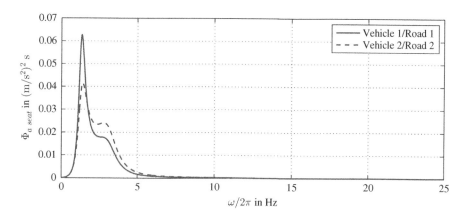

Figure 9.16 Spectral density of the seat vibrations

acceleration, the larger seat acceleration is still distinct in the natural frequency of the seat.

In evaluating the comfort of the seat vibrations, the evaluation function is of importance as well as the spectral density for the vertical seat vibrations. As the evaluation function exhibits a maximum at 5 Hz, the natural frequency of the seat is highlighted more than the natural oscillation of the body. This becomes apparent in Figure 9.17. Comparing the two spectral densities of the seat comfort evaluation in this graph reveals that the natural frequency of the seat for Vehicle 2 is significantly more apparent than for Vehicle 1 and the resonance rise in the natural frequency of the seat is more clearly visible than in the spectral density of the seat acceleration.

The comfort evaluation results from the integral over the spectral density of the seat acceleration multiplied by the corresponding evaluation function (Figure 9.17). In the comparison of the two vehicles in this figure, the value of the function of Vehicle 1 is

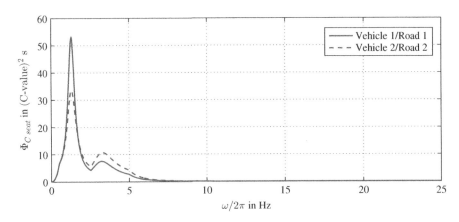

Figure 9.17 Comfort evaluation function

greater for the body vibration, while in the case of Vehicle 2 it is greater for the seat vibration. We can only determine what effects this has on the overall comfort value after integration over the function. The influence of the body damping is examined more closely in the last part of this section. On the basis of the comparison shown here, it is not possible to make a clear statement on whether the comfort value rises or falls with increasing damping.

Figure 9.18 shows the spectral density of the wheel load. In the comparison of the two vehicles, it can be seen that the resonance in the natural oscillation of the body and the resonance of the natural frequency of the wheel (approx. 11 Hz) are lower in Vehicle 2 with the higher body damping than in Vehicle 1. As the spectral density of the wheel load is included in the calculation of the standard deviation, σ_{Fz}, it can be expected that the wheel load fluctuations in Vehicle 2 will be lower than in Vehicle 1. However, certainty is not ultimately achieved until integration of the corresponding

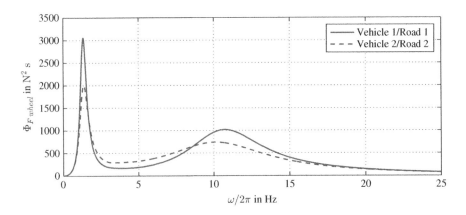

Figure 9.18 Spectral density of the wheel load vibrations

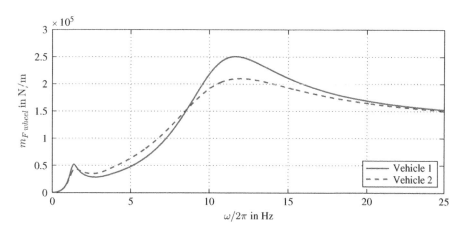

Figure 9.19   Magnification function of wheel load

integral of the magnification function multiplied by the road excitation (the spectral density of the wheel load results from the product of the square of the magnification function of the wheel load with the spectral density of the road plus the driving speed).

The magnification function of the wheel load is shown in Figure 9.19. Here, it can be seen that the body damping has a small influence on the natural oscillation of the body and a significant influence on the natural frequency of the wheel. Comparing the magnification function of the wheel load with the spectral density of the wheel load, it can be seen, as was previously the case with the seat vibrations that smaller frequencies are evaluated significantly more highly than larger frequencies. This can be seen by the fact that the ratios of the amplitudes of the resonance points of the magnification function are inverted in the spectral density. Ultimately, however, the consideration allows the conclusion to be drawn that the wheel load fluctuation is most probably lower in Vehicle 2.

The influence on the wheel load fluctuation or on the standard deviation of the wheel load is described in greater detail in the following.

In Figures 9.20 and 9.21, the overall comfort evaluations (i.e. seat, hand and foot accelerations) are shown as a function of the related standard deviation of the wheel load (the standard deviation refers to the static wheel load). Figure 9.20 shows the relationship of C_{tot} as a function of $\sigma_z/F_{z \text{ stat}}$ for five different values of the body spring stiffness. The body damping is varied within each of the curves. Essentially, the curves proceed with increasing body damping from large values for the standard deviation of the wheel load to small values. However, in the range for very large body damping, there is a minimum in relation to the standard deviation for the wheel load and this increases again, with the total comfort value C_{tot} increasing again as well (the total comfort decreases).

Figure 9.21 shows the corresponding curves for five different body dampings. Here the body stiffness changes within the curves. The comfort values rise with increasing

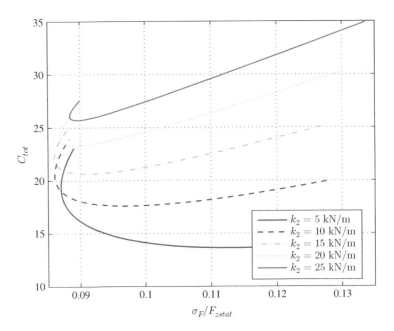

Figure 9.20 Conflict: safety and comfort; b_2 increases starting from upper right to the end point in the left part

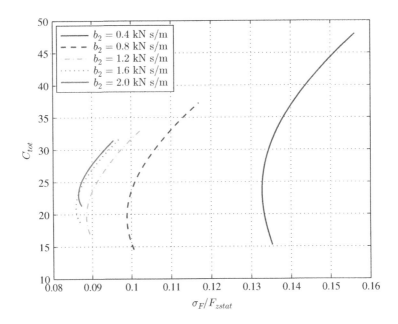

Figure 9.21 Conflict: comfort–safety

body spring stiffness. As high values for the total comfort value, C_{tot}, mean poor comfort, it may be deduced in the cases shown here that the comfort decreases with rising stiffness of the body springs.

The ranges for the comfort in relation to the body damping cannot be readily interpreted from Figure 9.20. However, it can be seen that the standard deviation for the wheel load mainly decreases as damping increases; for very high values of the damping there is again an increase.

Figure 9.22 shows curves with constant body spring and varying body damping (these curves do not differ substantially from those in Figure 9.20). However, these curves do not indicate the stiffness of the body spring but the natural frequency of the body in estimated form. It can be seen that in each case there are regions in which the standard deviation for the wheel load decreases, but the total comfort value increases. The boundaries of these regions are each marked by two dots. Here the dots at the horizontal of the curves indicate the maximum comfort for the corresponding natural frequency of the body whereas the dots at the vertical show the maximum value for safety (the smallest standard deviation for the wheel load). Between these two dots, an improvement in either comfort or safety can only be achieved by a deterioration in the respective other value. Where such a relation exists, we also refer to solutions with Pareto optimality. It becomes apparent that a compromise for the design of a vehicle has to be sought amongst the solutions with Pareto optimality.

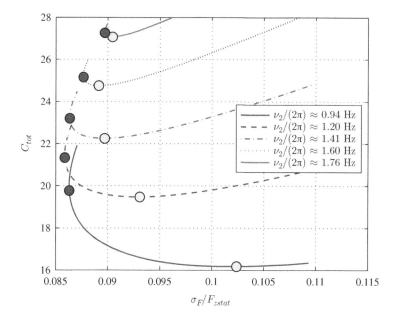

Figure 9.22 Pareto frontiers: comfort–safety

9.2   Oscillations of Powertrains

Most vehicles are driven by internal combustion reciprocal engines. Due to the combustion process and due to geometrical non-linearities (from the conversion of translation to rotation) deviation in torque and angular velocity occurs naturally. These oscillations can influence the comfort of the vehicle. Consequently, some devices are integrated in drive trains in order to reduce these oscillations.

To understand these devices, it is necessary to understand the torsional vibrations of powertrains. The first subsection of this section explains the theory of torsional oscillators similar to dual mass flywheels, while the second subsection takes a closer look at centrifugal pendulum vibration absorbers (CPVAs). The third subsection gives hints of examples.

9.2.1   Torsional Oscillators

We start with a simple torsional oscillator shown in Figure 9.23, left.

The oscillator consists of one rotating mass (moment of inertia J_1) and one rod simplified by a torsional spring (spring constant c_{T1}). For a more precise consideration, the rod has to be described by at least one partial differential equation.

The equation of motion is (φ_1 is the angle of torsion):

$$J_1\ddot{\varphi}_1 + c_{T1}\varphi_1 = 0 \ . \tag{9.55}$$

Now we can consider the free vibration of the system, which is characterized by the natural frequency (or eigenfrequency):

$$\omega_1 = \sqrt{\frac{c_{T1}}{J_1}} \ . \tag{9.56}$$

In the case of forced vibrations of the system, it is necessary to introduce a damping, b_{T1}, in order to limit resonance amplitudes (cf. Figure 9.23, right). Then the equation of motion reads

$$J_1\ddot{\varphi}_1 + b_{T1}\dot{\varphi}_1 + c_{T1}\varphi_1 = M_1 \ . \tag{9.57}$$

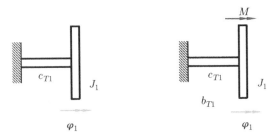

Figure 9.23   Torsional oscillators with one degree of freedom

To calculate the solution of this equation of motion in the case of harmonic excitation, an approach with complex amplitudes prove to be straightforward

$$M_1 = \hat{M}_1 \, e^{j\omega t} \,, \tag{9.58}$$

here $\hat{M}_1$ is the complex amplitude of the excitation torque, j is the imaginary unit (i.e. $j^2 = -1$), ω is the circular frequency of the excitation, and t is the time. The solution q_1 is the sum of the general solution of the homogeneous equation and a particular integral

$$q_1(t) = \left(A \, e^{\, j \, \sqrt{\frac{c_{T1}}{J_1} - \frac{b_{T1}^2}{4 J_1^2}} \, t} + \bar{A} \, e^{-j \, \sqrt{\frac{c_{T1}}{J_1} - \frac{b_{T1}^2}{4 J_1^2}} \, t} \right) e^{-\frac{b_{T1}}{2 J_1} t}$$

$$+ \frac{\hat{M}_1}{-\omega^2 J_1 + j\omega \, b_{T1} + c_{T1}} \, e^{j\omega t} \,. \tag{9.59}$$

For a harmonic excitation

$$M_0 \sin(\omega t) = \frac{M_0}{2j} (e^{j\omega t} - e^{-j\omega t}) \,, \tag{9.60}$$

$$M_0 \cos(\omega t) = \frac{M_0}{2} (e^{j\omega t} + e^{-j\omega t}) \tag{9.61}$$

two complex particular integrals have to be combined.

Damped absorbers are applied in order to reduce torsional oscillations. A simple model of a damped absorber is depicted in Figure 9.24. The equations of motion are

$$J_1 \ddot{\varphi}_1 + b_{T2} \, (\dot{\varphi}_1 - \dot{\varphi}_2) + c_{T2}(\varphi_1 - \varphi_2) + c_{T1} \, \varphi_1 = M_1 \,, \tag{9.62}$$

$$J_2 \ddot{\varphi}_2 + b_{T2} \, (\dot{\varphi}_2 - \dot{\varphi}_1) + c_{T2}(\varphi_2 - \varphi_1) = 0 \,. \tag{9.63}$$

Assuming harmonic excitation (in notation with complex amplitudes)

$$M_1 = \hat{M}_1 e^{j\omega t} \tag{9.64}$$

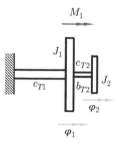

Figure 9.24   Torsional oscillator with two degrees of freedom

we can write the equations with matrices

$$\underbrace{\begin{pmatrix} -J_1\omega^2 + j\omega b_{T2} + c_{T1} + c_{T2} & -j\omega b_{T2} - c_{T2} \\ -j\omega b_{T2} - c_{T2} & -J_2\omega^2 + j\omega b_{T2} + c_{T2} \end{pmatrix}}_{\underline{A}} \begin{pmatrix} \hat{\varphi}_1 \\ \hat{\varphi}_2 \end{pmatrix} e^{j\omega t} \quad (9.65)$$

$$= \begin{pmatrix} \hat{M}_1 \\ 0 \end{pmatrix} e^{j\omega t}$$

The determinant of $\underline{A}$ is

$$\begin{aligned} \det\left(\underline{A}\right) &= (-J_1\omega^2 + j\omega b_{T2} + c_{T1} + c_{T2})(-J_2\omega^2 + j\omega\, b_{T2} + c_{T2}) \\ &\quad - (j\omega\, b_{T2} + c_{T2})^2 \\ &= (-J_1\omega^2 + c_{T1})(-J_2\omega^2 + c_{T2}) - J_2\, c_{T2}\omega^2 \\ &\quad + j\omega b_{T2}(-J_1\omega^2 - J_2\omega^2 + c_{T1}). \end{aligned} \quad (9.66)$$

With the inverse of $\underline{A}$

$$\underline{A}^{-1} = \frac{1}{\det(\underline{A})} \begin{pmatrix} -J_2\omega^2 + j\omega b_{T2} + c_{T2} & j\omega b_{T2} + c_{T2} \\ j\omega b_{T2} + c_{T2} & -J_1\omega^2 + j\omega b_{T2} + c_{T1} + c_{T2} \end{pmatrix} \quad (9.67)$$

we obtain the complex amplitude of the mass J_1

$$\hat{\varphi}_1 = \hat{M}_1 \frac{-J_2\omega^2 + j\omega b_{T2} + c_{T2}}{\det(\underline{A})}, \quad (9.68)$$

which yields the amplitude when an absolute value is applied

$$|\hat{\varphi}_1| = |\hat{M}_1| \frac{\sqrt{(-J_2\omega^2 + c_{T2})^2 + \omega^2 b_{T2}^2}}{|\det(\underline{A})|}. \quad (9.69)$$

For an undamped absorber (i.e. no damping $b_{T2} = 0$ Nms and resonance of the additional inertia J_2: $\omega_r{}^2 = c_{T2}/J_2$) the amplitude of the main inertia J_1 becomes zero: $\hat{\varphi}_1 = 0$ (cf. Figure 9.25). In this case, two infinite resonances occur for $\hat{\varphi}_1$: one below and another above ω_r. Figure 9.25 shows the amplitude of inertia J_1 for different values of the damping. The parameters are: $J_1 = 0.215$ kg m^2, $c_{T1} = 1600$ Nm, $J_2 = 0.00215$ kgm^2, $|\hat{M}_1| = 1$ Nm and

$$c_{T2} = \frac{J_2\, c_{T1}}{(1 + J_2/J_1)^2 J_1}. \quad (9.70)$$

The values for the damping are shown in the graph.

The torsional stiffness, c_{T2}, is chosen in order to obtain the same magnitudes of the two intersection points of all curves (cf. Dresig and Holzweissig 2010).

The choice $b_{T2} = 0$ Nms, i.e. a pure absorber, is not advisable for a powertrain of a vehicle because two resonances occur and because the frequency of the excitation is

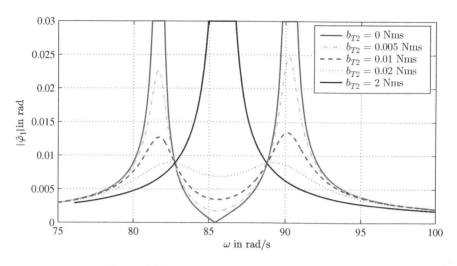

Figure 9.25 Amplitudes of a damped absorber

not constant in vehicles. A damped absorber should therefore be used, and the damping should not be too high in order to avoid high amplitudes of the inertia.

9.2.2 Centrifugal Pendulum Vibration Absorbers

Damped absorbers with a fixed area in the frequency domain with good damping characteristics are appropriate for excitation with fixed frequency, e.g. at engine idling speed.

In internal combustion engines, the frequency of excitation depends on the angular velocity of the engine itself. Consequently, damping devices with absorbing frequency which are proportional to angular velocity of the engine are useful.

One device with this characteristic is known as the centrifugal pendulum.

Several authors have dealt with this type of absorber in the literature, e.g. Salomon, Sarazin and Chilton.

Figure 9.26 depicts some possibilities.

In order to obtain the equations of motion we consider Figure 9.27, where J_1 is driven by the moment M_e.

Lagrange's equations are helpful in deriving the equations of motion for this system with two degrees of freedom: $L_L = T - V$

$$\frac{\mathrm{d}}{\mathrm{d}t} \frac{\partial L_L}{\partial \dot{\varphi}_1} - \frac{\partial L_L}{\partial \varphi_1} = M_e \tag{9.71}$$

$$\frac{\mathrm{d}}{\mathrm{d}t} \frac{\partial L_L}{\partial \dot{\varphi}_2} - \frac{\partial L_L}{\partial \varphi_2} = 0 \tag{9.72}$$

We neglect gravitational forces, i.e. $V = 0$.

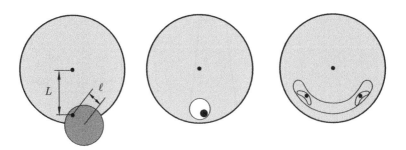

Figure 9.26 Different possibilities of pendulum absorber

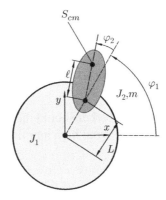

Figure 9.27 Centrifugal pendulum absorber

The coordinates x and y of the centre of mass S_{cm} of the pendulum (inertias J_2, m) are

$$x = L\cos\varphi_1 + \ell\cos(\varphi_1 + \varphi_2) \tag{9.73}$$

$$y = L\sin\varphi_1 + \ell\sin(\varphi_1 + \varphi_2) \tag{9.74}$$

Derivation with respect to time yields

$$\dot{x} = -L\dot{\varphi}_1\sin\varphi_1 - \ell(\dot{\varphi}_1 + \dot{\varphi}_2)\sin(\varphi_1 + \varphi_2) \tag{9.75}$$

$$\dot{y} = L\dot{\varphi}_1\cos\varphi_1 + \ell(\dot{\varphi}_1 + \dot{\varphi}_2)\cos(\varphi_1 + \varphi_2) \tag{9.76}$$

and these equations result in the velocity

$$\begin{aligned}
\dot{x}^2 + \dot{y}^2 = {} & L^2\dot{\varphi}_1^2\sin^2\varphi_1 + \ell^2(\dot{\varphi}_1 + \dot{\varphi}_2)^2\sin^2(\varphi_1 + \varphi_2) \\
& + 2L\ell\dot{\varphi}_1(\dot{\varphi}_1 + \dot{\varphi}_2)\sin\varphi_1\sin(\varphi_1 + \varphi_2) \\
& + L^2\dot{\varphi}_1^2\cos^2\varphi_1 + \ell^2(\dot{\varphi}_1 + \dot{\varphi}_2)^2\cos^2(\varphi_1 + \varphi_2) \\
& + 2L\ell\dot{\varphi}_1(\dot{\varphi}_1 + \dot{\varphi}_2)\cos\varphi_1\cos(\varphi_1 + \varphi_2).
\end{aligned} \tag{9.77}$$

Using $\sin^2\alpha + \cos^2\alpha = 1$ and $\cos\alpha\cos\beta + \sin\alpha\sin\beta = \cos(\alpha - \beta)$, we can simplify the equation

$$\dot{x}^2 + \dot{y}^2 = L^2\dot{\varphi}_1^2 + \ell^2(\dot{\varphi}_1 + \dot{\varphi}_2)^2 + 2L\ell\dot{\varphi}_1(\dot{\varphi}_1 + \dot{\varphi}_2)\cos\varphi_2 . \tag{9.78}$$

The kinetic energy of the whole system is then

$$T = \frac{1}{2}J_1\dot{\varphi}_1^2 + \frac{1}{2}m(\dot{x}^2 + \dot{y}^2) + \frac{1}{2}J_2(\dot{\varphi}_1 + \dot{\varphi}_2)^2 \tag{9.79}$$

and application of Lagrange's equations yields

$$\frac{\mathrm{d}}{\mathrm{d}t}\frac{\partial T}{\partial\dot{\varphi}_1} = \frac{\mathrm{d}}{\mathrm{d}t}(J_1\dot{\varphi}_1 + mL^2\dot{\varphi}_1 + m\ell^2(\dot{\varphi}_1 + \dot{\varphi}_2)$$

$$+ m\ L\ell(2\dot{\varphi}_1 + \dot{\varphi}_2)\cos\varphi_2 + J_2(\dot{\varphi}_1 + \dot{\varphi}_2))$$

$$= \ddot{\varphi}_1(J_1 + mL^2 + m\ell^2 + J_2) + (m\ell^2 + J_2)\ddot{\varphi}_2$$

$$+ mL\ell((2\ddot{\varphi}_1 + \ddot{\varphi}_2)\cos\varphi_2 - (2\dot{\varphi}_1 + \dot{\varphi}_2)\dot{\varphi}_2\sin\varphi_2) , \tag{9.80}$$

$$\frac{\mathrm{d}}{\mathrm{d}t}\frac{\partial T}{\partial\dot{\varphi}_2} = \frac{\mathrm{d}}{\mathrm{d}t}(m\ell^2(\dot{\varphi}_1 + \dot{\varphi}_2) + mL\ell\ \cos\varphi_1\cos\varphi_2$$

$$+ J_2(\dot{\varphi}_1 + \dot{\varphi}_2))$$

$$= \ddot{\varphi}_1(m\ell^2 + mL\ell\ \cos\varphi_2 + J_2)$$

$$+ \ddot{\varphi}_2(m\ell^2 + J_2) - mL\ell\ \dot{\varphi}_1\ \dot{\varphi}_2\ \sin\varphi_2 \tag{9.81}$$

$$-\frac{\partial T}{\partial\varphi_2} = mL\ell\ \dot{\varphi}_1\ (\dot{\varphi}_1 + \dot{\varphi}_2)\sin\varphi_2 \tag{9.82}$$

$$\frac{\mathrm{d}}{\mathrm{d}t}\frac{\partial T}{\partial\dot{\varphi}_2} - \frac{\partial T}{\partial\varphi_2} = \ddot{\varphi}_1(m\ell^2 + mL\ell\ \cos\ \varphi_2 + J_2)$$

$$+ \ddot{\varphi}_2(m\ell^2 + J_2) + mL\ell\ \dot{\varphi}_1^2\ \sin\varphi_2 . \tag{9.83}$$

Simplifying the trigonometric functions (assuming $\varphi_2 \ll 1$, $\cos\varphi_2 \approx 1$, $\sin\varphi_2 \approx \varphi_2$) yields the equations (additionally the external torque M_e was introduced):

$$\ddot{\varphi}_1(J_1 + J_2 + m(L + \ell)^2) + \ddot{\varphi}_2(m\ell^2 + J_2 + mL\ell)$$

$$- mL\ell(2\dot{\varphi}_1 + \dot{\varphi}_2)\dot{\varphi}_2\ \varphi_2 = M_e \tag{9.84}$$

$$\ddot{\varphi}_2(J_2 + m\ell^2) + \ddot{\varphi}_1(m\ell^2 + J_2 + mL\ell) + mL\ell\ \dot{\varphi}_1^2\ \varphi_2 = 0 . \tag{9.85}$$

Assuming small amplitudes of φ_2 and small oscillations $\hat{\varphi}_1 e^{i\omega t}$ of a stationary movement $\varphi_1 = \omega_o t$ of φ_1, we can linearize these non-linear equations:

$$\varphi_1 = \omega_o t + \hat{\varphi}_1\ e^{j\omega t} , \tag{9.86}$$

$$\varphi_2 = \hat{\varphi}_2\ e^{j\omega t} , \tag{9.87}$$

$$M_e = \hat{M}_e\ e^{j\omega t} , \tag{9.88}$$

$$\hat{M}_e = -\omega^2 \hat{\varphi}_1 (J_1 + J_2 + m(L + \ell)^2) - \omega^2 \hat{\varphi}_2 (m\ell^2 + J_2 + mL\ell) , \qquad (9.89)$$

$$0 = -\omega^2 \hat{\varphi}_2 (J_2 + m\ell^2) - \omega^2 \hat{\varphi}_1 (m\ell^2 + J_2 + mL\ell) + mL\ell \, \omega_0^2 \, \hat{\varphi}_2 . \quad (9.90)$$

Introducing

$$J_{11} = J_1 + J_2 + m(L + \ell)^2 \qquad (9.91)$$

$$J_{22} = J_2 + m \, \ell^2 \qquad (9.92)$$

$$J_{12} = J_{21} = m \, \ell^2 + J_2 + mL\ell \qquad (9.93)$$

allows the equations to be written as

$$\begin{pmatrix} -\omega^2 J_{11} & -\omega^2 J_{12} \\ -\omega^2 J_{21} & -\omega^2 J_{22} + mL\ell\omega_0^2 \end{pmatrix} \begin{pmatrix} \hat{\varphi}_1 \\ \hat{\varphi}_2 \end{pmatrix} = \begin{pmatrix} \hat{M}_e \\ 0 \end{pmatrix} . \qquad (9.94)$$

In the case of resonance excitation of the pendulum

$$\omega = \omega_0 \sqrt{\frac{mL\ell}{J_{22}}} \qquad (9.95)$$

the solution is

$$\hat{\varphi}_1 = 0 \qquad (9.96)$$

$$\hat{\varphi}_2 = -\hat{M}_e \frac{J_{22}}{\omega_0^2 J_{12} mL\ell} . \qquad (9.97)$$

Thus it is possible to adjust the parameters of the pendulum in such a way that the oscillation of the main inertia J_1 vanishes.

One remarkable feature is that the resonance frequency ω in Equation (9.95) is proportional to the excitation frequency ω_0. As one excitation frequency in reciprocating internal combustion machines is proportional to the angular velocity of the crankshaft, the pendulum absorber is ideal because its absorbing frequency is proportional to the angular velocity.

We look now at the special choice of the parameters; for simplicity we neglect $J_2 = 0$ and consider a mathematical pendulum. Then for the resonance (or absorbing) frequency we have

$$\omega = \omega_0 \sqrt{\frac{L}{\ell}} . \qquad (9.98)$$

If the goal was to eliminate the nth order of the crankshaft vibration, i.e.

$$\omega = n\omega_0 \qquad (9.99)$$

then it holds that

$$n^2 = \frac{L}{\ell} . \qquad (9.100)$$

The length, L, depends on the diameter of the inertia, J_1, which would be the flywheel; this means that $L < R_{\text{flywheel}}$. If, for example, the fourth order is to be absorbed and the flywheel has a radius of $R_{\text{flywheel}} = 0.12\,\text{m}$, then $\ell = 0.12\,\text{m}/16 \approx 7.5\,\text{mm}$, which is very small.

The linearized equations of motion are valid for small amplitudes of the pendulum. This means that the frequency depends on the amplitudes. To avoid the dependency, the pendulum has to be modified in such a way that the curve on which the inertia J_2, m moves is not a circle. In Denman 1992 or Nester 2004, the so-called epicycloids or tautochrones are investigated.

9.3   Examples

In this section, we will have a closer look at examples of devices to reduce torsional vibrations.

The first one is a classical spring-constrained damper located at the free end of a crankshaft (cf. Figure 9.28). For example, annular plates or rings are fixed with a rubber spring to the crankshaft. As rubber is very stiff, the mass has to be high, which is a disadvantage when we are trying to integrate the damper in the crankcase. One even bigger disadvantage is the temperature dependence of the rubber stiffness. Another possibility is to integrate the mass into the crankweb (cf. Figure 9.28), thus it becomes an internal crankshaft damper.

To reduce the space, which is necessary and compensate for an imbalance in mass, the damper does not have an annular shape but is formed like a horseshoe. The springs

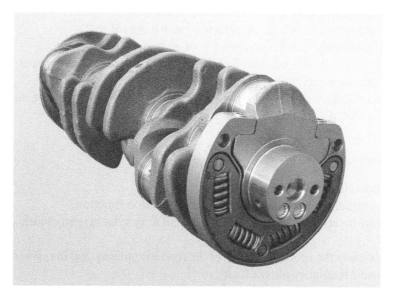

Figure 9.28   Internal crankshaft damper (reproduced with permissions of Schaeffler)

are steel coil springs and the damping arises from friction between the rotating mass and the plastic bearing on the inner side of the housing. This means that the damping is caused by dry, nearly velocity-independent Coulomb friction.

Because Coulomb friction depends on the normal force, the angular velocity of the crankshaft and the oscillation velocity enter indirectly into the damping force by centrifugal forces.

The convergence is that the damping moment increases with the angular velocity of the crankshaft. To avoid clearance in the system, the springs are preloaded. One secondary effect of preloading is a preload-normal force and therefore a constant friction force without rotation of the crankshaft. The interaction between centrifugal forces and preload and their influence on friction forces in the oscillator have an advantageous effect on the damping properties.

More example are shown in Section 17.1, for example, a dual-mass flywheel with centrifugal pendulum vibration absorber in Figure 17.8 or a clutch disc with torsional damper and centrifugal pendulum vibration absorber in Figure 17.6.

9.4   Questions and Exercises

Remembering

1. In which issues do vibrations play a role?
2. What is a magnification or transfer function?
3. In which frequency range do humans exhibit a sensitive reaction to seat vibrations, and in which frequency range do they exhibit a sensitive reaction to the vibrations of the steering wheel?
4. Which parameters have a great influence on comfort?
5. Which parameters have a great influence on driving safety?
6. Give a typical value for the waviness w of an uneven road.
7. In which ranges do the maxima lie for the evaluation functions for seat, hand and foot accelerations according to Cucuz?
8. Which accelerations – seat, hand or foot – generally have the greatest influence on the total comfort value?

Understanding

1. What are evaluated vibration intensities?
2. How do natural frequencies affect the magnification function?
3. How do we determine the response of an oscillatory system to stochastic excitation stimuli?
4. Explain clearly the spectral density of the road irregularity and the spectral density of the road irregularity plus velocity.

5. What influence do respective increases in the masses m_1, m_2 and m_3 have on the magnification functions?
6. How do different waviness w affect the spectral densities of seat acceleration and wheel load?
7. Explain the conflict between safety and comfort.
8. Why a frequency-dependent absorber is important for an IC engine?
9. Explain the differences between a simple vibration absorber and a CPVA?

10

Vehicle Substitute Models

Section 10.1 presents a very simple vehicle substitute model known as the quarter-vehicle model as described already in Chapter 9. It is described with the help of a two-mass oscillator. With the aid of this model, it is possible to describe the first natural frequency of a vehicle with which the body mainly oscillates in the vertical direction, and the second one, with which the wheel mainly oscillates. Section 10.2 is dedicated to a two-axle vehicle with a single-track excitation (in single-track excitation, the left-hand and right-hand wheels of an axle are subjected to identical excitation). With the aid of this model, which exhibits five degrees of freedom, it is possible to investigate pitch oscillations as well as vertical oscillations. Finally, Section 10.3 deals with the effects of non-linear characteristic curves for springs and shock absorbers.

10.1   Two-mass Substitute System

The single-mass substitute system dealt with in Chapter 9 is very simple and is of little assistance in designing spring and damper properties.

Quarter-vehicle model: The quarter-vehicle model (two-mass substitute system, Figure 10.1) is the simplest substitute system that already exhibits essential features of a vehicle in terms of vertical dynamics. The substitute system consists of the two masses, m_b (in this case m_b is one-fourth of the body mass) and m_w (this is the wheel mass). The body springs and shock absorbers are located between the masses (spring stiffness k_b, damping constant b_b). A spring–damper system (stiffness k_w, damping constant b_w) also acts between the wheel mass, m_w, and the uneven road surface. Dividing the wheel into the components of wheel mass, m_w, wheel stiffness, k_w, and wheel damping, b_w, is a simplified model that permits a good reproduction of the wheel properties.

Vehicle Dynamics, First Edition. Martin Meywerk.
© 2015 John Wiley & Sons, Ltd. Published 2015 by John Wiley & Sons, Ltd.
Companion Website: www.wiley.com/go/meywerk/vehicle

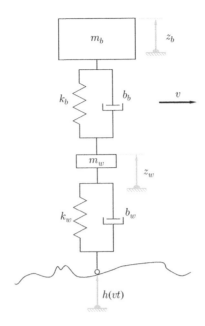

Figure 10.1 Two-mass substitute systems

As a two-mass system is involved, there are four eigenvalues, with pairs of them being complex conjugates. This means that there are two natural frequencies. In the following, we first determine the equations of motion for the system. We then consider the eigenvalues and the magnification functions.

For systems such as that shown in Figure 10.1, which are composed of masses and spring–shock absorber or damper elements, one possible approach is to derive the equations of motion with the aid of the Lagrange formalism. We apply the Lagrange formalism to the undamped system. The Lagrange function, L, is the difference between kinetic energy, T, and potential energy, V:

$$L = T - V$$
$$= \frac{1}{2}m_b\dot{z}_b^2 + \frac{1}{2}m_w\dot{z}_w^2 - \frac{1}{2}k_b(z_b - z_w)^2 - \frac{1}{2}k_w(z_w - h)^2 . \tag{10.1}$$

If we set $q_1 = z_b$ and $q_2 = z_w$, the Lagrange equations become

$$\frac{\mathrm{d}}{\mathrm{dt}}\left(\frac{\partial L}{\partial \dot{q}_j}\right) - \frac{\partial L}{\partial q_j} = 0 , \; j = 1, 2 . \tag{10.2}$$

By differentiation, we thus obtain the two equations of motion as

$$m_b\ddot{z}_b + k_b z_b - k_b z_w = 0 , \tag{10.3}$$
$$m_w\ddot{z}_w + (k_b + k_w)z_w - k_b z_b = k_w h . \tag{10.4}$$

The damping systems can now be inserted directly in a similar way to the stiffness terms into (10.3) and (10.4):

$$m_b \ddot{z}_b + b_b \dot{z}_b + k_b z_b - b_b \dot{z}_w - k_b z_w = 0 \ , \tag{10.5}$$

$$m_w \ddot{z}_w + (b_b + b_w) \dot{z}_w + (k_b + k_w) z_w \tag{10.6}$$
$$- b_b \dot{z}_b - k_b z_b = b_w \dot{h} + k_w h \ .$$

We first consider the natural frequencies of the undamped homogeneous system. With the aid of an $e^{\lambda t}$ approach, we obtain the eigenvalue equation

$$\begin{pmatrix} z_b \\ z_w \end{pmatrix} = \begin{pmatrix} \hat{z}_b \\ \hat{z}_w \end{pmatrix} e^{\lambda t}. \tag{10.7}$$

If we insert (10.7) into Equations (10.5) and (10.6) and divide by $e^{\lambda t}$ (for all complex values of λ and for all t the following applies: $e^{\lambda t} \neq 0$), it follows that

$$\begin{pmatrix} \lambda^2 m_b + k_b & -k_b \\ -k_b & \lambda^2 m_w + k_b + k_w \end{pmatrix} \begin{pmatrix} \hat{z}_b \\ \hat{z}_w \end{pmatrix} = \begin{pmatrix} 0 \\ 0 \end{pmatrix} . \tag{10.8}$$

The terms $b_w \dot{h}$ and $k_w h$ represent an external excitation and therefore have no influence on the natural frequencies. We obtain the characteristic equation from the condition that the determinant of the 2×2 matrix from (10.8) disappears:

$$\lambda^4 + \lambda^2 \frac{m_w k_b + m_b (k_b + k_w)}{m_b m_w} + \frac{k_b k_w}{m_b m_w} = 0 \ . \tag{10.9}$$

The solutions of (10.9) for λ^2 are

$$\lambda^2_{1,2} = -\frac{m_w k_b + m_b (k_b + k_w)}{2 m_b m_w} \tag{10.10}$$
$$\pm \sqrt{\left(\frac{m_w k_b + m_b (k_b + k_w)}{2 m_b m_w} \right)^2 - \frac{k_b k_w}{m_b m_w}} \ .$$

The values for $\lambda^2_{1,2}$ are purely real and negative; this means that the four solutions $\lambda_1, \ldots, \lambda_4$ from (10.9) are purely imaginary. The natural frequencies result from the values of $\lambda_1, \ldots, \lambda_4$, with pairs of the values being identical; hence, two natural frequencies, f_1 and f_2, result.

Example 10.1 We assume the following values for the parameters of the quarter-vehicle model: $m_b = 300\,\text{kg}$, $m_w = 30\,\text{kg}$, $k_w = 120\,000\,\text{N/m}$ and consider the natural frequencies as a function of the body spring stiffness, k_b. The graphs in Figure 10.2 show the natural frequencies f_1 and f_2 for values of k_b in a range from $12\,000\,\text{N/m}$ to $44\,000\,\text{N/m}$ of the quarter-vehicle model.

The first natural frequency lies in the range from approximately 1 Hz to 1.7 Hz, and the second from 10 Hz to 12 Hz, in modern vehicles up to 16 Hz. The body natural

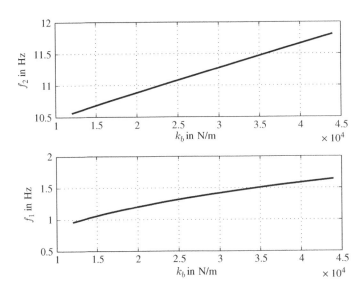

Figure 10.2   First and second natural frequency as a function of the body spring stiffness

frequency (hence the first natural frequency) lies significantly below the range of 4 Hz to 8 Hz, in which humans react sensitively to vibrations. The second natural frequency lies above this range.

10.2   Two-axle Vehicle, Single-track Excitation

The vehicle model presented in this section comprises four masses. It provides a good reproduction of the actual vibrations of a vehicle if we make some constraining assumptions.

1. The irregularities of the left-hand and right-hand wheel tracks are identical. We call this type of excitation the single-track excitation. Let us further assume that the vehicle is symmetrical in terms of its inertia characteristics with respect to the $\vec{e}_{vx} - \vec{e}_{vz}$ plane. As a result, no rolling or sliding movements occur. As the vehicle is driving in a straight line, there are also no yaw movements.
2. The rear wheels travel in the same track as the front wheels. This means that the excitations on the rear wheels are identical to those on the front wheels, although a phase shift does occur.

Figure 10.3 shows the vehicle model. It consists of four masses. The two masses of the wheels, m_{w1} and m_{w2}, support themselves against the road surface via the substitute stiffnesses, k_{w1} and k_{w2}. The spring–shock absorber pairs, k_{b1}, b_{b1} and k_{b2}, b_{b2}, act between the body mass (mass, m_b, mass moment, J_b) and the wheel masses. The seat–human system is located on the body. In the former, the mass of the driver,

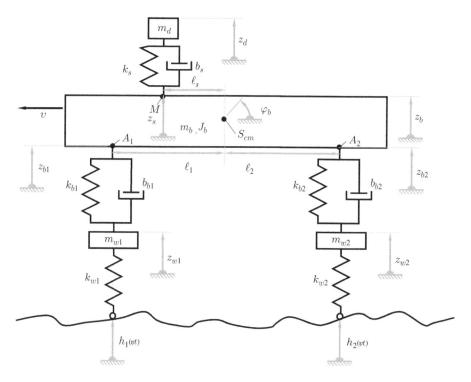

Figure 10.3   Four-mass oscillator as a model of a two-axle vehicle

m_d, is supported against the body via the spring–damper pair of the seat k_s, b_s. The deflections of the wheel masses are z_{w1} and z_{w2}, those of the body and the seat–human system z_b and z_d, respectively. The body has an additional rotational degree of freedom (pitch movement). The respective angle of rotation about the centre of mass, S_{cm}, of the body is φ_b.

We consider merely small oscillations about the rest position. This means that trigonometric functions dependent upon the pitch angle, φ_b, can be linearized. The base of the seat–human vibration system moves as a function of the pitch angle, φ_b, and the rise, z_b. The following applies for the coordinates, z_s, of the base, M:

$$z_s = z_b - \ell_s \varphi_b \ . \tag{10.11}$$

In this equation, the sine function $\sin \varphi$ has been linearized. In addition to a knowledge of the auxiliary variables, z_d, the z-coordinates z_{b1} and z_{b2} of the points A1 and A2 are also helpful in establishing the equations of motion:

$$z_{b1} = z_b - \ell_1 \varphi_b \ , \tag{10.12}$$

$$z_{b2} = z_b + \ell_2 \varphi_b \ . \tag{10.13}$$

The Lagrange function for the undamped system can be readily established with the help of the variables z_b, φ_b, z_{w1}, z_{w2}, z_d and the auxiliary variables z_{b1}, z_{b2}.

$$
\begin{aligned}
L &= T - V \\
&= \frac{1}{2}m_d\dot{z}_d^2 + \frac{1}{2}m_b\dot{z}_b^2 + \frac{1}{2}J_b\dot{\varphi}_b^2 + \frac{1}{2}m_{w1}\dot{z}_{w1}^2 + \frac{1}{2}m_{w2}\dot{z}_{w2}^2 \\
&\quad - \left(\frac{1}{2}k_s(z_d - (z_b - \ell_s\varphi_b))^2 \right. \\
&\quad + \frac{1}{2}k_{b1}(z_{w1} - (z_b - \ell_1\varphi_b))^2 + \frac{1}{2}k_{b2}(z_{w2} - (z_b + \ell_2\varphi_b))^2 \\
&\quad \left. + \frac{1}{2}k_{w1}(z_{w1} - h_1)^2 + \frac{1}{2}k_{w2}(z_{w2} - h_2)^2\right) \ .
\end{aligned}
\tag{10.14}
$$

If the following abbreviations are introduced for the variables: $q_1 = z_d$, $q_2 = z_b$, $q_3 = \varphi_b$, $q_4 = z_{w1}$, $q_5 = z_{w2}$, the equations of motion for the undamped system become

$$
\frac{\mathrm{d}}{\mathrm{d}t}\left(\frac{\partial L}{\partial \dot{q}_j}\right) - \frac{\partial L}{\partial q_j} = 0 \ , \ j = 1, \ldots, 5 \ .
\tag{10.15}
$$

We obtain

$$
m_d\ddot{z}_d + k_s(z_d - (z_b - \ell_s\varphi_b)) = 0 \ ,
\tag{10.16}
$$

$$
m_b\ddot{z}_b + k_s(z_b - \ell_s\varphi_b - z_d)
\tag{10.17}
$$
$$
+ k_{b1}(z_b - \ell_1\varphi_b - z_{w1})
$$
$$
+ k_{b2}(z_b + \ell_2\varphi_b - z_{w2}) = 0 \ ,
$$

$$
J_b\ddot{\varphi}_b + k_s\ell_s(\ell_s\varphi_b - z_b + z_d)
\tag{10.18}
$$
$$
+ k_{b1}\ell_1(\ell_1\varphi_b - z_b + z_{w1})
$$
$$
+ k_{b2}\ell_2(\ell_2\varphi_b + z_b - z_{w2}) = 0,
$$

$$
m_{w1}\ddot{z}_{w1} + k_{b1}(z_{w1} - (z_b - \ell_1\varphi_b)) + k_{w1}z_{w1} = k_{w1}h_1 \ ,
$$

$$
m_{w2}\ddot{z}_{w2} + k_{b2}(z_{w2} - (z_b + \ell_2\varphi_b)) + k_{w2}z_{w2} = k_{w2}h_2 \ .
$$

The damping systems can be introduced into the equations in a similar manner to the two-mass substitute system. In the following, we consider the special case of a symmetrical vehicle. The following applies to this $\ell_s = 0$, $\ell_1 = \ell_2$, $k_{b1} = k_{b2}$, $k_{w1} = k_{w2}$, $m_{w1} = m_{w2}$, $b_{b1} = b_{b2}$.

In the matrix form, the allocated eigenvalue problem can be written with the help of an $e^{\lambda t}$ approach

$$
(z_d, z_b, \varphi_b, z_{w1}, z_{w2})^T = (\hat{z}_d, \hat{z}_b, \hat{\varphi}_b, \hat{z}_{w1}, \hat{z}_{w2})^T e^{\lambda t}
\tag{10.19}
$$

as

$$\underline{M}(\hat{z}_d, \hat{z}_b, \hat{\varphi}_b, \hat{z}_{w1}, \hat{z}_{w2})^T = (0, 0, 0, k_{w1}h_1, k_{w2}h_2)^T, \tag{10.20}$$

with the matrix $\underline{M}$ having the following form:

$$\underline{M} = \begin{bmatrix} m_d\lambda^2 + k_s & -k_s & 0 \\ -k_s & m_b\lambda^2 + k_s + 2k_{b1} & 0 \\ 0 & 0 & J_b\lambda^2 + 2k_{b1}\ell_1^2 \\ 0 & -k_{b1} & k_{b1}\ell_1 \\ 0 & -k_{b1} & -k_{b1}\ell_1 \end{bmatrix}$$

$$\begin{bmatrix} 0 & 0 \\ -k_{b1} & -k_{b1} \\ k_{b1}\ell_1 & -k_{b1}\ell_1 \\ m_{w1}\lambda^2 + k_{b1} + k_{w1} & 0 \\ 0 & m_{w1}\lambda^2 + k_{b1} + k_{w1} \end{bmatrix} \tag{10.21}$$

If we multiply the third row of $\underline{M}$ with

$$(m_{w1}\lambda^2 + k_{b1} + k_{w1})/(k_{b1}\ell_1) \tag{10.22}$$

and then subtract the fourth row from the third and add the fifth, everything in the third row disappears apart from the entry in the third column. This entry is

$$\frac{J_b m_{w1}\lambda^4 + \lambda^2(J_b(k_{b1} + k_{w1}) + 2k_{b1}\ell_1^2 m_{w1})}{k_{b1}\ell_1} + 2k_{w1}\ell_1 . \tag{10.23}$$

From the condition that this entry (10.23) disappears, we obtain four of the total of 10 eigenvalues. The following applies to the squares of the eigenvalues that we obtain from (10.23):

$$\lambda_{1,2}^2 = -\frac{J_b(k_{b1} + k_{w1}) + 2k_{b1}\ell_1^2 m_{w1}}{2J_b m_{w1}} \tag{10.24}$$

$$\pm \sqrt{\left(\frac{J_b(k_{b1} + k_{w1}) + 2k_{b1}\ell_1^2 m_{w1}}{2J_b m_{w1}}\right)^2 - \frac{2k_{w1}\ell_1^2 k_{b1}}{J_b m_{w1}}}.$$

Setting (10.24) in $m_{w1}\lambda^2 + k_{b1} + k_{w1}$, we obtain

$$m_{w1}\lambda^2 + k_{b1} + k_{w1} = \frac{k_{b1} + k_{w1}}{2} - \frac{k_{b1}\ell_1^2 m_{w1}}{J_b} \tag{10.25}$$

$$\pm \frac{1}{2J_b}\sqrt{(J_b(k_{b1} + k_{w1}) + 2k_{b1}\ell_1^2 m_{w1})^2 - 8J_b m_{w1}k_{w1}k_{b1}\ell_1^2}.$$

If the two fractions and the square root expression on the right-hand side of (10.25) are each squared separately, we can see that the value of the two fractions is smaller than the square root expression by

$$k_{b1} \sqrt{\frac{\ell_1^2 m_{w1}}{J_b}} \tag{10.26}$$

The expression

$$m_{w1}\lambda^2 + k_{b1} + k_{w1} \tag{10.27}$$

is therefore greater than zero for the positive sign preceding the square root and less than zero for the negative sign. Consequently, the two accompanying eigenvectors

$$\underline{e}_1 = \left(0, 0, 1, -\frac{k_{b1}\ell_1}{m_{w1}\lambda_1^2 + k_{b1} + k_{w1}}, \frac{k_{b1}\ell_1}{m_{w1}\lambda_1^2 + k_{b1} + k_{w1}}\right)^T, \tag{10.28}$$

$$\underline{e}_2 = \left(0, 0, 1, -\frac{k_{b1}\ell_1}{m_{w1}\lambda_2^2 + k_{b1} + k_{w1}}, \frac{k_{b1}\ell_1}{m_{w1}\lambda_2^2 + k_{b1} + k_{w1}}\right)^T, \tag{10.29}$$

have components preceded by the following signs:

$$\underline{e}_1 : (0, 0, +, -, +) , \tag{10.30}$$

$$\underline{e}_2 : (0, 0, +, +, -) . \tag{10.31}$$

The first eigenmode is therefore equivalent to an oscillation in which the ends of the body and the wheel masses oscillate in the same phase. In the second eigenmode, the ends of the body and the wheel masses vibrate in opposite phases. The eigenmode of the remaining six natural vibrations correspond to purely translatory movements in the z direction.

In automotive engineering, it is common to define a substitute for the model presented previously in which merely masses with translational motion occur. The substitute system is shown in Figure 10.4. The body mass in the substitute system is replaced by three masses. These masses, m_{b1}, m_{b2} and m_c (coupling mass) are connected by a rigid, massless beam. The three masses are treated as point masses (without mass moment of inertia). To ensure that the inertia characteristics of the three masses agree with those of the body, three conditions have to be fulfilled:

$$m_{b1} + m_{b2} + m_c = m_b , \tag{10.32}$$

$$\ell_1 m_{b1} - \ell_2 m_{b2} = 0 , \tag{10.33}$$

$$\ell_1^2 m_{b1} + \ell_2^2 m_{b2} = J_b . \tag{10.34}$$

The above equations guarantee equality of the total mass (10.32), the centre of mass, (10.33) and the mass moment (10.34). In the form of a system of equations, these

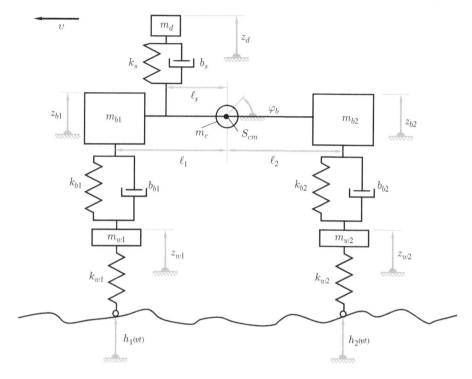

Figure 10.4   Two-axle vehicle with coupling mass

conditions have the following structure:

$$
\begin{bmatrix} 1 & 1 & 1 \\ \ell_1 & 0 & -\ell_2 \\ \ell_1^2 & 0 & \ell_2^2 \end{bmatrix} \begin{bmatrix} m_{b1} \\ m_c \\ m_{b2} \end{bmatrix} = \begin{bmatrix} m_b \\ 0 \\ J_b \end{bmatrix}.
\tag{10.35}
$$

The determinant of the matrix is $\ell_1\ell_2(\ell_1 + \ell_2)$. This means that the system of equations can always be solved provided that the centre of mass does not coincide with A1 or A2. However, there may be solutions for which $m_c < 0$ applies; from the engineering mechanics point of view, a negative mass makes no sense. As the equations of motion can be arranged in a similar manner to the first model and the solutions of the equations of motion result in the same motions as in the first model, we shall dispense with a closer consideration of this second model.

Remark 10.1 The excitations h_1 and h_2 due to the irregular road surface appear in the last two differential equations. The different designations indicate that these functions and hence the excitations are independent of each other. However, as we assume that the front and rear tyres travel in one track, the excitation at the rear, h_2, is the same as that at the front but shifted in phase ($\ell = \ell_1 + \ell_2$):

$$h_2(vt) = h_1 \left(v \left(t - \frac{\ell}{v} \right) \right) . \tag{10.36}$$

If we assume that h_1 can be written as a Fourier series:

$$h_1(vt) = \sum_{i=-N}^{N} \hat{h}_i e^{ji\omega t} , \tag{10.37}$$

then

$$h_2(vt) = \sum_{i=-N}^{N} e^{-ji\omega \frac{\ell}{v}} \hat{h}_i e^{ji\omega t} . \tag{10.38}$$

The excitation angular frequencies, ω, are dependent upon the wavenumbers of the irregularity of the road surface

$$h(x) = \sum_{i=-N}^{N} \hat{h}_i e^{ji\kappa_w x} . \tag{10.39}$$

The following applies:

$$\kappa_w v = \omega . \tag{10.40}$$

From this, the excitation on the rear wheels results in the following:

$$h_2(vt) = \sum_{i=-N}^{N} e^{-ji\kappa_w \ell} \hat{h}_i e^{ji\omega t} . \tag{10.41}$$

The fixed phase shift, $i\kappa_w \ell$, which is independent of the driving speed, v, has the consequence that

- a harmonic component of the irregularity of the road surface excites the pitch and vertical vibrations in a certain ratio independent on the driving speed;
- there may be harmonic components of the road irregularity that exclusively excite either pitch or vertical vibrations.

The phase shift depends on the wheelbase, ℓ, of the vehicle and the wavenumbers, κ_w. This has the consequence that in the case of two different vehicles travelling over a certain test circuit it will only be possible to stimulate vertical vibrations in one of the vehicles and only pitch vibrations in the other. During the planning of test circuits and in the comparison, it is therefore necessary to make sure that a certain wavelength distribution is present.

10.3 Non-linear Characteristic Curves

To conclude this chapter, we examine non-linearities in characteristic curves in springs and shock absorbers. As a simple example, we consider the single-mass oscillator with non-linear characteristic curve for springs:

$$\ddot{x} + 2D\dot{x} + f(x) = p_0 \cos \eta t \ . \tag{10.42}$$

We restrict ourselves here to a dimensionless notation. We obtain the periodic solutions of this differential equation with the help of a Fourier series. In the following, however, we merely consider the first member of this series. The essential properties can be identified with support of this approximate solution:

$$x = Q \cos(\eta t - \alpha) \ . \tag{10.43}$$

Let the non-linearity be

$$f(x) = x + 0.05x^3 \ . \tag{10.44}$$

This is the non-linear stiffness which is called a Duffing oscillator.

Figure 10.5 shows the solutions for the undamped system (hence $D = 0$).

We can see the relation between the amplitude, Q, and the frequency, η^2, of the autonomous system (hence no external excitation: $p_0 = 0$). In contrast to the linear single-mass oscillator, there is no fixed natural frequency, but, instead, the frequency, η, with which the system oscillates depends upon the amplitude, Q. We can see that the frequency with which the system vibrates is very close to the natural angular frequency, $\eta = 1$, of the linear single-mass oscillator for small amplitudes Q. The line for $p_0 = 0$ is called the backbone curve.

The amplitude for the enforced vibrations can also be seen in Figure 10.5. With a given amplitude, p_0, of excitation, there are three possible amplitudes for the vibration

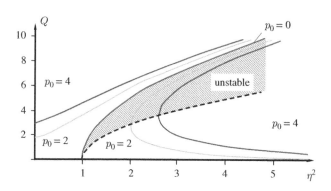

Figure 10.5 Relation of amplitude to frequency (undamped Duffing oscillator)

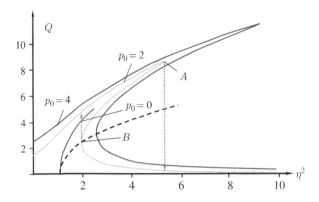

Figure 10.6   Relation of amplitude to frequency (damped Duffing oscillator)

that becomes established. The following will examine their significance with the support of a damped system. Figure 10.6 shows the amplitudes and the frequencies for a damped system:

$$\ddot{x} + 0.1\dot{x} + x + 0.05x^3 = p_0 \cos(\eta t) . \tag{10.45}$$

Here, too, we can see the backbone curve. However, here the amplitude of the oscillations is not shown as the oscillations of the free system decrease because of the damping. Yet, in contrast to the undamped system, the backbone curve ends at a certain frequency. Unlike the undamped system, the curves for $p_0 \neq 0$ in the damped system are closed. In the damped case, the backbone curve merely has the function of a separation line in the amplitude–frequency graph.

We consider the curve for $p_0 = 2$ and raise the frequency beginning from zero. If the frequency reaches the reversal point, A, the amplitude falls to the section of the curve with the negative gradient (Figure 10.6).

If the frequency is reduced starting from a high frequency, at the reversal point, B, the amplitude jumps to the higher branch for $p_0 = 2$. The solutions for the middle

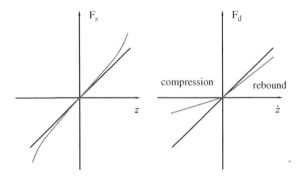

Figure 10.7   Non-linear characteristic curves

branch are unstable and therefore practically play no role. The leap from one branch to the other always takes place at the points with a vertical tangent.

In summary, this means that there may be more than one solution for forced vibrations in non-linear systems.

Next to the non-linear characteristic curve of the spring (linear and cubic part) Figure 10.7 shows the path of a non-linear hydraulic shock absorber. The bilinear path is characteristic. Here, the gradient is lower in the compression range than in the rebound range.

10.4   Questions and Exercises

Remembering

1. What does a very simple vehicle substitute model look like?
2. In what order of magnitude does the first body natural frequency lie, and in which does the second lie?
3. Draw a model to investigate the vertical and pitch vibrations of a vehicle.
4. How many natural frequencies does this model have?
5. What are the main natural modes that exist in this model?

Understanding

1. What is a coupling mass?
2. Which conditions have to be fulfilled for the introduction of a coupling mass?

11

Single-track Model, Tyre Slip Angle, Steering

In this chapter, we present the main concepts, technical terms and inter-relationships of the lateral dynamics. The lateral dynamics plays a central role in cornering. Section 11.1 presents important technical terms such as the single-track model for understanding the cornering and deriving the underlying equations of motion. Section 11.2 is devoted to the central element of cornering, the cornering of the tyres and the tangential stress distribution in the contact patch. The steering and the steering angle and concepts necessary to understand oversteering and understeering are the content of Section 11.3. Thereafter, the linearized equations of motion of the single-track model are derived in Section 11.4; these equations are the most important outcomes of this chapter. Section 11.5 discusses the relationship between the longitudinal and the lateral forces of the tyre, the effect of differential gears on cornering is discuss in Section 11.6.

11.1 Equations of Motion of the Single-track Model

In the following section, we deal with driving a vehicle in the plane along a trajectory. The forces arising during cornering and the effect of the tyres are also studied. Figure 11.1 shows the model with two axles and four wheels. We assume that the centre of mass, S_{cm}, lies in the plane on which the vehicle travels[1]. As a result, no wheel load transfer occurs (either during cornering or during acceleration or deceleration). The model can therefore be reduced to a single-track model (Figure 11.1).

---

[1] Of course, this an essential simplification, but it allows simple equations of motion to be derived and hence provides a clear overview of some important phenomena.

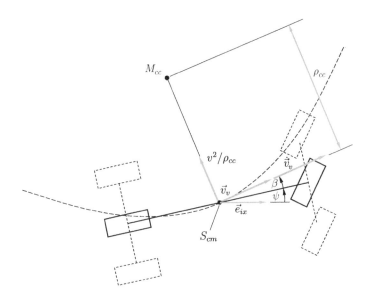

Figure 11.1   Single-track model

Single-track model: The single-track model is a key model in the lateral dynamics of a
vehicle which allows important parameter dependencies to be considered and conclusions to be drawn on the lateral dynamics. The single-track model often forms
the basis of simple ESP systems. One important assumption of the single-track
model is that the centre of mass of the vehicle is on the road, which means that the
distance of the centre of mass to the road plane is zero: $h_{\rm cm} = 0$. This simplification
limits the applicability of the single-track model.

   The centre of mass $S_{\rm cm}$ of the model moves along the trajectory, the velocity, $\vec{v} =$
$\vec{v}_v$, of the centre of mass is always tangential to the trajectory. The angle between the
$\vec{e}_{ix}$-axis and the vehicle longitudinal $\vec{e}_{vx}$-axis is the yaw angle ψ (see also Figure 1.8;
we mainly omit the index v in the following, $v = v_v$ etc.).

Vehicle sideslip angle: The angle between the direction of motion of the vehicle's centre of mass and the vehicle's longitudinal axis is called the vehicle sideslip angle β.
The sum of the yaw angle and the vehicle sideslip angle is the course angle.
Circle of curvature: The circle of curvature is a purely geometric object which approximates the trajectory locally at one point. In other words, the circle of curvature
exists even when there is no vehicle moving along the trajectory; it is a characteristic of the trajectory. The centre of the circle of curvature is $M_{\rm cc}$.

   The circle of curvature can be calculated by means of a limiting process as shown in
Figure 11.2. Both points marked by the crosses move against the point in the middle

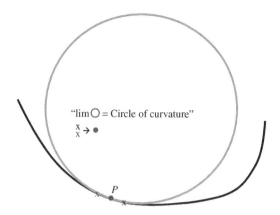

"lim $\bigcirc$ = Circle of curvature"

$\underset{x \to \bullet}{\overset{x}{x}}$

Figure 11.2   Determination of the circle of curvature using a limiting process

of the crosses. Exactly one circle is defined by the crosses and the point. The circle which arises in this limiting process is the circle of curvature at the point P.

If the trajectory is given by mathematical functions in the form of parameterized curves

$$\vec{r} = (x_v(\zeta), y_v(\zeta)) \begin{pmatrix} \vec{e}_{ix} \\ \vec{e}_{iy} \end{pmatrix} \ , \tag{11.1}$$

(here ζ is a curve parameter without a dimension) the curve radius ρ_{cc} can be calculated by

$$\rho_{cc} = \left| \frac{\left((x_v')^2 + (y_v')^2\right)^{\frac{3}{2}}}{x_v' y_v'' - x_v'' y_v'} \right| \ . \tag{11.2}$$

For a motion on a straight line the radius is infinite, $\rho_{cc} = \infty$, or the curvature κ_{cc} is zero:

$$\kappa_{cc} = \frac{1}{\rho_{cc}} = 0 \frac{1}{\mathrm{m}} \ . \tag{11.3}$$

This means that the transition from a straight line motion to a motion on a circle with the radius ρ_{cc} results in discontinuous change in the lateral acceleration

$$a_y = \frac{v^2}{\rho_{cc}} \ . \tag{11.4}$$

To avoid this discontinuity, in the planning of roads straight lines are usually not connected to parts of a circle, but special curves, so-called clothoids (or Euler spirals) are used. These special track transition curves can be described by so-called Fresnel integrals (cf. Abramowitz 1984):

$$\begin{pmatrix} x_v(\zeta) \\ y_v(\zeta) \end{pmatrix} = A_c \sqrt{\pi} \int_0^\zeta \begin{pmatrix} \cos\left(\frac{\pi \xi^2}{2}\right) \\ \sin\left(\frac{\pi \xi^2}{2}\right) \end{pmatrix} \, \mathrm{d}\xi \ . \tag{11.5}$$

A closed form solution is not possible, but series expansions and approximate numerical functions exists. Substituting these integrals in Equation (11.2) yields

$$\rho_{cc} = \frac{A_c}{\zeta\sqrt{\pi}} \ . \tag{11.6}$$

Thus, the lateral acceleration (and therefore the centrifugal forces) depends linearly on ζ:

$$a_c = \frac{v^2}{\rho_{cc}} \tag{11.7}$$

$$= \frac{v^2\sqrt{\pi}}{A_c}\zeta \ . \tag{11.8}$$

This is the reason why the clothoid is suitable for transition from straight line motion to motion on a circle. The length of a clothoid is

$$L = A\sqrt{\pi}\zeta \ , \tag{11.9}$$

which means that the curvature $\kappa_{cc} = 1/\rho_{cc}$ increases linearly with the length.

Clothoids can be used for transition from two straight line motion or for transition from a straight line to circular motion and vice versa.

The centripetal acceleration (or radial acceleration), $a_c = \frac{v^2}{\rho_{cc}}$, is directed towards the centre of curvature, M_{cc}, of the trajectory ($v = |\vec{v}|$ is the absolute value of the velocity vector $\vec{v} = \vec{v}_v$ and ρ_{cc} is the radius of the circle of curvature). The tangential acceleration, $\dot{v}$, is directed tangentially to the trajectory (and tangentially to the circle of curvature). The free-body diagram is shown in Figure 11.3.

In addition to the tangential inertial forces, $F_t = m\dot{v}$, and the centrifugal force, $F_c = \frac{mv^2}{\rho_{cc}}$, it also shows the air forces, F_{ax} and F_{ay}, and the forces in the contact patches of the front and rear wheels. The front wheel is turned by the steering angle, δ_1.

From the free-body diagram of Figure 11.3, we obtain three equations of motion (these form the basis for further investigations):

- Equilibrium of forces in the longitudinal direction of the vehicle:

$$m\frac{v^2}{\rho_{cc}}\sin\beta - m\dot{v}\cos\beta + F_{x2} - F_{ax} + F_{x1}\cos\delta_1 - F_{y1}\sin\delta_1 = 0 \ , \tag{11.10}$$

- Equilibrium of forces perpendicular to the longitudinal direction of the vehicle:

$$m\frac{v^2}{\rho_{cc}}\cos\beta + m\dot{v}\sin\beta - F_{y2} + F_{ay} - F_{x1}\sin\delta_1 - F_{y1}\cos\delta_1 = 0 \ , \tag{11.11}$$

- Moment equilibrium about S_{cm}:

$$J_z\ddot{\psi} - (F_{y1}\cos\delta_1 + F_{x1}\sin\delta_1)\ell_1 + F_{y2}\ell_2 + F_{ay}\ell_{cm} = 0 \ . \tag{11.12}$$

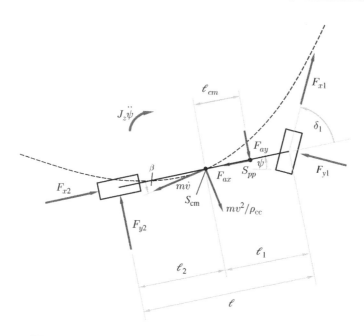

Figure 11.3 Free-body diagram of the single-track model

The axle loads are

$$F_{z1} = G\frac{\ell_2}{\ell} - F_{az1} \,, \tag{11.13}$$

$$F_{z2} = G\frac{\ell_1}{\ell} - F_{az2} \,. \tag{11.14}$$

Here, F_{az1} and F_{az2} are the aerodynamic lift forces at the front and rear axles. The effects of the rolling resistances and of inertia effects of rotating parts (especially an engine with a lateral rotation axis) on the axle load are neglected. The static parts from a gradient of the road or the dynamic parts from acceleration or braking do not result in a moment because we assume that the centre of mass lies on the road.

The non-linear equations of motion (11.10)–(11.12) are later linearized in order to obtain linear equations of motion of the single-track model. We first discuss the movement of the vehicle in terms of the curvature centre point, M_{cc}, (or the centre of the circle of curvature) of the trajectory, the centre of mass, S_{cm}, and instantaneous centre of rotation, M_{cr}, of the vehicle movement. The centre of curvature, M_{cc}, of the trajectory is a purely geometric object, which allows the centre of mass acceleration to be interpreted and split into radial and tangential components. The instantaneous centre of rotation, M_{cr}, also contributes to the calculation of the rotation of the vehicle (the yaw) about the centre of mass.

In the following, we derive the relations for the distances of the instantaneous centre of rotation, M_{cr}, and the centre of curvature, M_{cc}, from the centre of mass, S_{cm}. To derive the relationship for the instantaneous centre of rotation, we start from Figure 11.4.

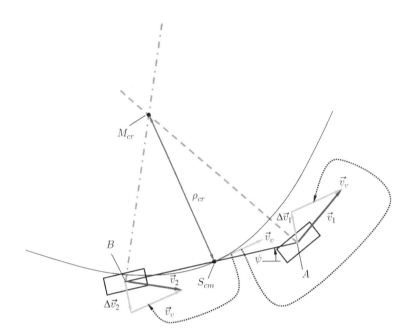

Figure 11.4   Instantaneous centre of rotation of the vehicle in motionl

Instantaneous centre of rotation: The instantaneous centre of rotation is an imaginary
point. The vehicle rotates around this point at a particular moment. If we imagine
an infinitely large rigid plate which is fixed to the vehicle and parallel to the road,
the instantaneous centre of rotation is that point of the plate which does not move,
i.e. the velocity of this point vanishes. The instantaneous centre of rotation, M_{cr},
is the intersection of two normals of two arbitrary velocity vectors in two different
points of the vehicle.

Figure 11.4 shows an example of the two points A and B. The velocity, $\vec{v}_1$, of point
A is the sum of the velocity vector of point S_{cm} and the velocity vector, $\Delta\vec{v}_1$, due to
the yaw motion. The vector $\Delta\vec{v}_1$ is derived from the following Equation (11.15):

$$\Delta\vec{v}_1 = \dot{\vec{\psi}} \times (\vec{r}_A - \vec{r}_{cm}) \ . \tag{11.15}$$

Here $\vec{r}_A$ is the vector to the point A, $\vec{r}_{cm}$ is the vector to the centre of mass S_{cm}, and
$\dot{\vec{\psi}}$ is the yaw angle velocity vector. Therefore

$$\vec{v}_1 = \vec{v}_v + \Delta\vec{v}_1 \ , \tag{11.16}$$

$$\vec{v}_2 = \vec{v}_v + \Delta\vec{v}_2 \ \text{where} \tag{11.17}$$

$$\Delta\vec{v}_2 = \dot{\vec{\psi}} \times (\vec{r}_B - \vec{r}_{cm}) \ . \tag{11.18}$$

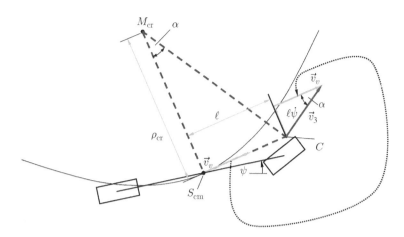

Figure 11.5 Distance of the instantaneous centre of rotation from the centre of gravity

The dashed and chain dotted lines in Figure 11.4 are perpendicular to the velocities $\vec{v}_1$ and $\vec{v}_2$, respectively, and the points A and B are on the lines. They are described by the following two equations in the normal form as

$$(\vec{r} - \vec{r}_A) \cdot \vec{v}_1 = 0 , \tag{11.19}$$

$$(\vec{r} - \vec{r}_B) \cdot \vec{v}_2 = 0 , \tag{11.20}$$

where $\cdot$ denotes the scalar product and $\vec{r}$ is the vector to each point of the line. From Equations (11.19) and (11.20), we obtain the position of the instantaneous centre of rotation, which is the intersection of the two lines.

Often we are interested only in the distance, ρ_{cr}, of the instantaneous centre of rotation, M_{cr}, from the centre of mass, S_{cm}. We obtain this by simple geometrical observations from Figure 11.5. Without any loss of generality, we choose an arbitrary point, C, on the line through the centre of gravity, S_{cm}, with the direction of the velocity, $\vec{v}_v$. For this point C we determine the velocity $\vec{v}_3$ (cf. Figure 11.5). The velocity $\vec{v}_3$ is the sum of $\vec{v}_v$ and of the portion perpendicular to the $\vec{v}_v$-direction from yaw $\ell\dot{\psi}$. The angle α in the velocity triangle is given by ($v_v = |\vec{v}_v|$):

$$\tan \alpha = \frac{\ell\dot{\psi}}{v_v} . \tag{11.21}$$

The dashed-line triangle is similar (in the mathematical sense of similar triangles) to the velocity triangle; hence, the angles at the vertex M_{cr} is the same as the angle between $\vec{v}_v$ and $\vec{v}_3$ in the velocity triangle. We obtain from the dashed line triangle:

$$\tan \alpha = \frac{\ell}{\rho_{\text{cr}}} . \tag{11.22}$$

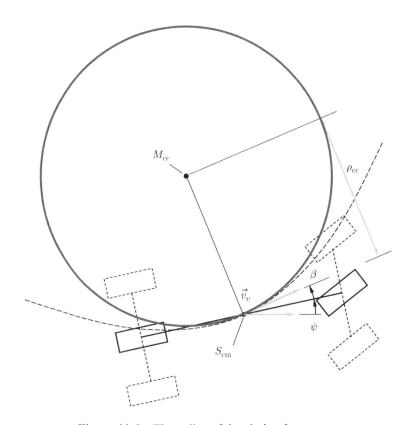

Figure 11.6   The radius of the circle of curvature

From Equations (11.21) and (11.22) together, we have

$$\rho_{\mathrm{cr}} = \frac{v_v}{\dot{\psi}} \; . \tag{11.23}$$

The radius, ρ_{cc}, of the circle of curvature is associated with the centre of mass velocity and the course angular velocity, $\dot{\beta} + \dot{\psi}$. The relationship can be easily determined with the help of Figure 11.6.

The circle of curvature touches the trajectory at one point. The velocity, $\vec{v}_v$, is a tangent to the trajectory and to the circle of curvature, hence the circle of curvature is an instantaneous approximation of the trajectory. The velocity of S_{cm} on this circle is equal to the velocity of the movement on the trajectory. The direction of $\vec{v}_v$ changes with the course angular velocity $\dot{\beta} + \dot{\psi}$. We therefore obtain

$$|\vec{v}_v| = \rho_{\mathrm{cc}}(\dot{\beta} + \dot{\psi}) \; , \tag{11.24}$$

which yields (with $v_v = |\vec{v}_v|$)

$$\rho_{\mathrm{cc}} = \frac{v_v}{\dot{\beta} + \dot{\psi}} \; . \tag{11.25}$$

It may seem surprising at this point that just the angles β and ψ, associated with the movement of the vehicle, appear in the formula for ρ_{cc}, the instantaneous radius of the trajectory, since these angles have nothing to do with the trajectory. This is because the course angle is merely expressed by the yaw and the vehicle sideslip angle.

11.2   Slip Angle

The following section is devoted to the slip angle of the tyre. Figure 11.7 shows three views of a tyre: in Figure 11.7(a) the front view, in Figure 11.7(b) the bottom view and in Figure 11.7(c) the side view.

Slip angle: Lateral slip occurs in a tyre when the x_w-direction (i.e. the longitudinal direction in the tyre coordinate system) does not coincide with the direction of motion ($\vec{v}_w$-direction in Figure 11.7(b)). Between the x_w-direction and $\vec{v}_w$-direction, we call this angle the slip angle α.

If the x_w and $\vec{v}_w$-direction do not coincide, this results in lateral deformations of the tyre (indicated by dashed lines in Figure 11.7(a) and (b)), it also leads to a force F_y in the tyre contact patch acting in the y_w-direction.

Caster: The point of application of the force F_y does not lie in the symmetry plane of the tyre, but is shifted against n_{tc} in the x_w-direction. We call n_{tc} the tyre caster trail (see Figure 11.7(b))[2].

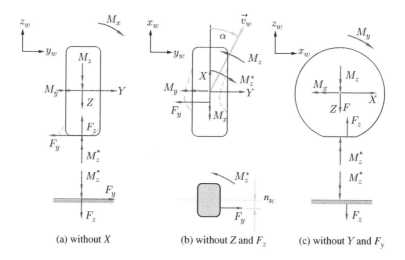

| (a) without X | (b) without Z and F_z | (c) without Y and F_y |

Figure 11.7   Forces on a tyre with lateral slip

---

[2] In the literature the technical term *pneumatic trail* is often used, in Reimpell et al. 2001 the term is *tyre caster*; as the lever arm occurs at solid tyres, too, the term *pneumatic trail* may be confusing, and therefore the term *tyre caster trail* is used in this book.

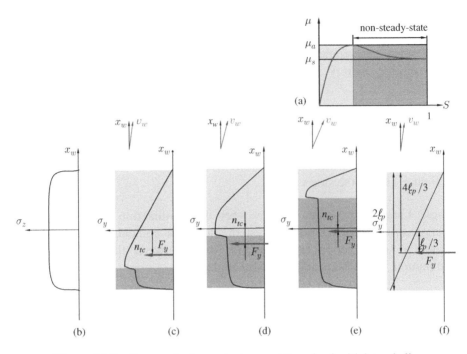

Figure 11.8 Stresses in the contact area at the wheel with lateral slip

Figure 11.7 lists the section forces. Considering Figure 11.7(b), it is obvious that, due to the tyre caster trail n_{tc}, a moment $M_z = n_{\text{tc}}F_y$ is necessary for the conditions of equilibrium to be fulfilled (here the moment M_z^* was set to zero). The moment M_z^*, which was introduced here only for the sake of completeness, is due to the angular velocity of the wheel about the z_w-axis, which may occur, for example, during parking manoeuvre. Such motion results in a moment which is generally small for a rolling tyre. Since the steady slip is considered here, we set $M_z^* = 0$.

The moment M_z counteracts an increase of the slip angle (self-aligning moment)[3]. An explanation of the tyre caster trail is illustrated with the help of Figure 11.8, in which the lateral stresses, σ_y, are shown schematically. Figure 11.8(a) shows the longitudinal force coefficient, μ, as a function of the slip. Essential features for the following explanation are the division in the adhesion area (at the front of the contact patch) and the sliding area (at the end of the contact patch). It is important that the lateral stresses should fall off rapidly in the transition from adhesion to sliding, while μ should fall rapidly from μ_a to μ_s. Figure 11.8(b) shows the normal stress, σ_z, in the longitudinal symmetry plane of the contact area as a function of x_w. The asymmetrical stress distribution due to the rolling has been omitted for the sake of clarity. Figure 11.8(c)–(e) shows the tangential stresses, σ_y, in the y_w-direction for three different slip angles.

---

[3] Longitudinal forces in the contact patch from braking or from traction could result in moments, if they do not act in the centre of the contact patch.

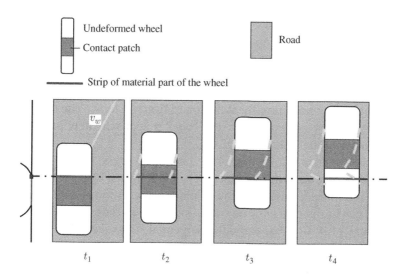

Figure 11.9 Development of lateral forces

The slip angle α is the smallest in Figure 11.8(c) and is the greatest in Figure 11.8(e). What happens when lateral slip occurs in a wheel can be first explained with the help of Figure 11.9, before we return to the stresses. The four sub-images of Figure 11.9 show a wheel in motion on the road for four consecutive points in time, t_1, t_2, t_3, t_4. The wheel moves from bottom left to top right. Wheel and contact area are shown undeformed in the diagram. We now concentrate on the small strip of material which is part of the tyre. At $t = t_1$, this strip runs in the contact patch and touches the road for the first time (the small image on the left shows a section in the x_w-z_w plane; the strip in this section is the small rectangle). At time $t = t_2$, the tyre has moved a little towards the $\vec{v}_w$-direction. However, the strip adheres substantially to the road and therefore does not change its position. The tyre is therefore deformed; the deformed tyre is shown by the dashed line.

In the further course of the manoeuvre, the movement of the tyre continues in the $\vec{v}_w$-direction ($t = t_3$). The deformations of the tyre become greater. This in turn draws larger deformation forces or tangential stresses, which means that the tangential forces acting between the road and the strip become larger. If the forces are too large, then the strip begins to slide ($t = t_4$). This can occur in the back section of the contact patch. Consequently, the contact patch may be divided in the adhesion area, which is located at the front part of the contact patch, and in a sliding area at the rear part of the contact patch. It is evident that the contact patch no longer has a simple rectangular shape when cornering.

We now return to the consideration of the stresses in the contact area from Figure 11.8. Figure 11.8(c) shows the tangential stresses σ_y for a small slip angle. The stress increases approximately linearly (light grey area) until it comes to slide because it exceeds the coefficient of adhesion. In the adhesion area, the location of the strip on the road as in Figure 11.9, does not change. In fact, the strip changes its position only slightly, because a small amount of sliding occurs even in the so-called adhesion area. This can be seen in Figure 11.8(a). Major sliding occurs in the transition from the adhesion area to the sliding area. Although the normal stress, σ_z, changes only slightly, the tangential stresses, σ_y, drop quite abruptly, hence the friction coefficient decreases rapidly at this point from μ_a to μ_s. The tangential stress σ_y thus increases in the adhesion area and behaves approximately proportionally to the normal stresses in the sliding area.

Due to this asymmetric distribution of the tangential stress, the resultant force, F_y, is not located in the centre of the contact patch, but is shifted by n_{tc} in the direction of the transition from the adhesion area to the sliding area. An increase in the slip angle (Figure 11.8(d)) over the course of time also increases the lateral displacement of the strip from Figure 11.9. Consequently, the tangential stress increases rapidly and the transition to sliding occurs earlier. This results in a smaller adhesion area and a larger sliding area. As the tangential stress distribution approaches a symmetrical shape, the tyre caster trail, n_{tc}, becomes smaller. In Figure 11.8(e), this effect is more apparent at an even greater slip angle; here the tyre caster trail has decreased to almost zero. By extending the sliding area for very large slip angles on almost the entire contact patch, the tangential stress distribution is proportional to the normal stress distribution. If we consider the asymmetric normal stress distribution due to the tyre rolling, which involves the normal stresses increasing in the front portion of the contact patch, then theoretically a negative tyre caster trail could arise. This negative tyre caster trail would mean that the wheel becomes unstable. However, for practical driving patterns, these extreme slip angles have no meaning.

Cornering stiffness: For small slip angles (approx. $\alpha < 4°$) the lateral force F_y can be approximated by a linearized law:

$$F_y = c_\alpha \alpha \ . \tag{11.26}$$

The coefficient c_α is called the lateral force coefficient or the cornering stiffness.

For small slip angles we can approximately assume that the sliding portion vanishes and the tangential stress increases linearly from zero to the maximum value in the adhesion area of the contact patch. We then obtain the tyre caster trail simply by determining the centre point of the tangential stress triangle (cf. Figure 11.8(f)):

$$n_{tc} = \frac{1}{3}\ell_{cp} \ . \tag{11.27}$$

This is an approximate formula, which applies only under the stated conditions.

In Figure 11.10, the lateral force, F_y, the moment, M_z, and the tyre caster trail, n_{tc}, are shown as a function of slip angle α. Both the lateral force, F_y, and the tyre caster trail, n_{tc} (order of magnitude for passenger cars: $n_{tc} \approx 0.02 - 0.06$ m, $c_\alpha \approx 40 - 110$ kN/rad) depend on the vertical force, F_z. For small slip angles, α, this can

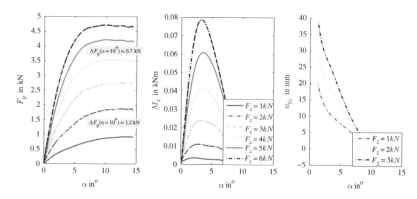

Figure 11.10   Lateral force, self-aligning moment and tyre caster trail (from measurements with slight corrections of offsets)

be approximated by polynomials ($F_{z\text{nom}}$ is a nominal vertical load):

$$c_\alpha = (\hat{c}_\alpha - \tilde{c}_\alpha F_z)F_z \ , \tag{11.28}$$

$$n_{\text{tc}} = n_{\text{tc0}}\frac{F_z}{F_{z\text{nom}}} \ . \tag{11.29}$$

Equation (11.28) is crucial in subsequent considerations.

11.3   Steering

This section derives the relationship between steering angle, δ_1, of the front wheels and the angle, δ_s, by which the driver rotates the steering wheel. We use Figure 11.11 to explain the relationship. The rotational movement of the steering wheel is converted by the rack and pinion into a translational motion of the tie rod. The tie rods are linked to the steering arms. The rotation of the steering arm results in a rotation of the wheel carrier, which rotates about the steering axis (the steering axis is the axis through the two ball joints of the wheel hub carrier; for a McPherson suspension, the steering axis is the axis through the lower ball joint and the joint between the strut and the body).

The gear ratio of the steering gear is $i_s = \ell_{sa}\delta_s/u_r^*$ (order of magnitude: $i_s \approx 16$–22; here ℓ_{sa} is the length of the steering arm and u_r^*/ℓ_{sa} is approximately the steering angle of the wheels for small angles for neglected steering compliance). In existing steering amplifiers (power steering), the steering moment M_s is increased by a factor of V_s. The steering stiffness, k_s, is introduced between the rack and the tie rods. This steering stiffness, k_s, represents all compliances (flexibilities), e.g. of the steering column, the steering gear, the tie rod and the steering arms. The wheel rotates around the steering axis. In general, this axis is not vertical but tilted. Its position is described by two tilt angles: the king pin inclination angle, σ, and the so-called caster angle τ. Due to tilting about the caster angle τ, the instantaneous centre of the steering movement of

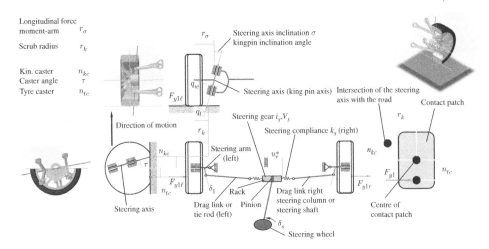

Figure 11.11 Schematic diagram of a rack and pinion steering system (adapted from Mitschke and Wallentowitz 2004)

the wheel lies in the direction of travel, the distance is the kinematic caster trail[4], n_{kc}, (order of magnitude: $n_{kc} \approx 1–30$ mm).

The steering moment at the wheels is then calculated as

$$M_s^* = (F_{y1\ell} + F_{y1r})(n_{kc} + n_{tc}) \cos(\tau) . \tag{11.30}$$

For small caster angles, τ, we can only consider the sum of the kinematic caster trail n_{kc} and the tyre caster trail n_{tc} in this formula and replace the cosine term by 1. The steering wheel torque is given by

$$M_s = \frac{M_s^*}{i_s V_s} . \tag{11.31}$$

The steering moment on the tyres depends on the difference in displacement[5] $u_{tr} = \ell_{sa}\delta_1$ of the tie rod and the displacement of the rack u_r^* (here ℓ_{sa} is the length of the steering arm):

$$M_s^* = 2k_s(u_r^* - \ell_{sa}\delta_1) . \tag{11.32}$$

The displacement of the rack is obtained from the angle of rotation of the steering wheel, δ_s, by means of the transmission ratio i_s of the steering gear:

$$u_r^* = \ell_{sa}\frac{\delta_s}{i_s} . \tag{11.33}$$

---

[4] In the literature or in ISO 8855 2011 other technical terms are used, e.g. caster offset at ground, castor trail or kinematic trail.
[5] For convenience, we neglect the angle between the steering arm and the longitudinal direction, x_v, of the vehicle as well as the angle between the tie rod and the lateral direction, y_v, of the vehicle. Furthermore, the equation $u_{tr} = \ell_{sa}\delta_1$ holds only for small angles δ_1.

By eliminating M_s^*, we obtain

$$(F_{y1\ell} + F_{y1r})(n_{kc} + n_{tc}) = 2k_s(u_r^* - \ell_{sa}\delta_1,) \tag{11.34}$$

and hence (the translational stiffness, k_s, is substituted by a rotational stiffness, $\tilde{k}_s = 2\ell_{sa}k_s$) by substituting u_r^* using Equation (11.33):

$$\delta_1 = \frac{\delta_s}{i_s} - \frac{(F_{y1\ell} + F_{y1r})(n_{kc} + n_{tc})}{\tilde{k}_s}. \tag{11.35}$$

Similar results can be obtained for other steering systems such as a lever arm steering system[6].

It should be emphasized that we have neglected non-linearities of trigonometric functions in the above derivation.

The influence of longitudinal forces in the tyre contact patch when driving or braking or when driving over obstacles during cornering can have an effect on the steering wheel torque. In general, steering arms are not arranged in parallel. During cornering, the longitudinal forces cause different moments depending on whether they act on the inside of the curve or the outside of the curve. We denote the influence factors by i_i for the inner wheel and i_o for the outer wheel. In most cases, $i_i > i_o$ (i.e. the inside wheel is turned more than the outer wheel).

During the braking process, the braking torque is applied on the one hand by a longitudinal brake force F_b in the contact patch and on the other hand by the corresponding tangential force on the brake disc. The entire brake force is therefore supported via the suspension on the body of the vehicle (see Figure 11.12). The total moment acting on the steering system due to the braking forces is therefore related to the braking forces and to the corresponding scrub radius, r_k. The scrub radius is the distance from the intersection point of the axis of symmetry of the wheel with the roadway to the intersection point of the steering axis with the roadway. The total moment is therefore given by[7]

$$M_s \approx \underbrace{(F_{bo}i_o - F_{bi}i_i)r_k\cos\sigma}_{q_T}. \tag{11.36}$$

Because the steering arms are not parallel, $i_i > i_o$, and as the braking forces at the outer and the inner wheels are of the same magnitude ($F_{bo} = F_{bi}$), the moment acting on the steering system is proportional to the scrub radius r_k. The scrub radius can be very small or even zero. To achieve this goal, the distance between steering

---

[6] The stiffness of the compliance $\tilde{k}_s$ in the equations of the rack and pinion steering system has to be substituted by k_s of a lever arm steering system. The unit of the constant k_s in the rack and pinion steering system is a unit of force, the unit of the constant k_s in a lever arm steering is a unit of moment. The moment arm ℓ_{sa} is introduced in $u_r^* = \ell_{sa}\frac{\delta_s}{i_s}$ to obtain the same equations for rack and pinion and lever arm steering systems. This moment arm ℓ_{sa}, of course, is not part of the rack and pinion steering system but it is the length of the steering arm for both steering systems.

[7] To calculate the torque of the braking forces F_{bo} and F_{bi} at the outer and inner sides, respectively, the lever arm, i.e. the distance between the kingpin axis and the centre of the contact patch, should be introduced. In the literature and in ISO 8855 2011 this is included by using the symbol q_T; here we prefer to introduce r_k, which can be used to calculate $q_T = r_k\cos\sigma$.

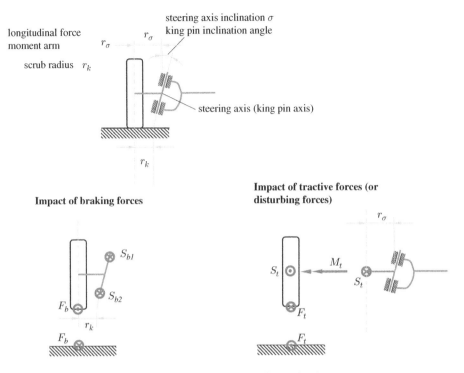

Figure 11.12   Impact of longitudinal forces in the contact area

axis and centre of the wheel and the inclination angle of the steering axis have to chosen appropriately. A negative scrub radius is also possible. During braking with ABS control, the braking forces are very high, which could result in a high moment in the steering system. It is therefore advantageous to have a small or zero scrub radius.

When considering the driven wheels, we assume drive shafts which are parallel to the lateral vehicle axis (y_v-axis). In general, drive shafts with universal joints (or constant velocity joints, CV joints) and splines for adaption to changes in lengths are used. These drive shafts are, generally not in a straight line. In such cases, the consideration is more complicated and geometric non-linearities have to be taken into account. For convenience, we restrict the following to straight drive shafts. The tractive forces act in the contact patch. Since the drive torque at the wheel cannot be supported via the wheel hub carrier and the suspension by the body of the vehicle, the corresponding tractive force, F_t, in the contact patch has to be equal to the section force, S_t, acting on the centre of the wheel. The equation $S_t = F_t$ results from the free-body diagram in the right-hand part of Figure 11.12. The moments arising from the driving forces on the inner and outer wheel are not proportional to the scrub radius, but proportional to the so-called longitudinal force moment-arm or the disturbing force lever arm radius r_σ. (On closer examination the disturbing force lever arm must be modified in accordance with the geometry of the drive shafts; in this context we speak of a driving torque arm radius). A straight drive shaft results in a moment, M_s, due tractive forces F_{xi} and F_{xo}[8]:

$$M_s \approx (F_{xi} i_i - F_{xo} i_o) \underbrace{r_\sigma \cos \sigma}_{q_W} . \tag{11.37}$$

This steering moment is proportional to the disturbing force lever arm r_σ. In the suspensions and wheel hub bearings commonly in use today, r_σ cannot be chosen to be arbitrarily small. Impact forces have a similar influence to driving forces, such as when driving over obstacles. Although these disturbing forces act in the opposite direction, their influence is similar to that of tractive forces: both of them produce a steering moment that is proportional to r_σ. The name disturbing force lever arm comes from the effect of these disturbing impact forces. By design, the scrub radius, r_k, can be zero or even negative. This became possible in passenger cars particularly through the introduction of sliding calliper brakes, which only need small package dimensions, meaning that the steering axle can be placed close to the wheel centre plane.

A negative scrub radius is advantageous for diagonally split braking systems. In these systems, the diagonally opposite wheels are each combined in one of the two brake circuits. If one circuit should fail during braking, a yaw moment occurs. This yaw moment is the result of load transfer from rear to front axle and therefore greater braking forces at the front axle. The negative scrub radius, together with a one-sided braking force at the front wheels, causes a compensatory steering (initiated by the moment) to the yaw moment from braking forces (cf. Figure 11.13(a)).

Similarly, a negative scrub radius, r_k, has a favourable effect on braking on a split-μ road. The yaw moment from different braking forces is partly compensated for by steering initiated by the steering moment (from negative scrub radius and the different braking forces at the front wheels; cf. Figure 11.13(b)). Similar compensation occurs during braking and cornering (cf. Figure 11.13(c)) where a steering to the inside of the trajectory is initiated.

11.4   Linearized Equations of Motion of the Single-track Model

In this section we linearize the equations of motion of the single-track model that were derived in Section 11.1. From $v = |\vec{v}_v|$ we obtain

$$v = \rho_{cc} \left(\dot{\beta} + \dot{\psi} \right) . \tag{11.38}$$

Multiplying this equation by v / ρ_{cc} yields the centrifugal acceleration:

$$\frac{v^2}{\rho_{cc}} = v \left(\dot{\beta} + \dot{\psi} \right) . \tag{11.39}$$

---

[8] In the literature and in ISO 8855 2011 for the lever arm, the symbol $q_W = r_\sigma \cos \sigma$ is additionally introduced in order to calculate the torque.

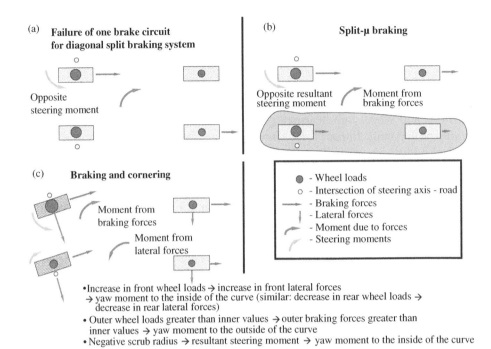

Figure 11.13   Effect of a negative scrub radius in different driving situations

From Figure 11.14, we obtain the relationships between the slip angles, α_1 and α_2, of the wheels and the vehicle sideslip angle, β. The velocity components of the velocities $\vec{v}_1$, $\vec{v}_2$, and $\vec{v}_v$ in the vehicle longitudinal direction (x_v-direction) must be the same ($v = |\vec{v}_v|$, $v_1 = |\vec{v}_1|$, $v_2 = |\vec{v}_2|$):

$$v \cos \beta = v_1 \cos (\delta_1 - \alpha_1) \,, \tag{11.40}$$

$$v \cos \beta = v_2 \cos \alpha_2 \,. \tag{11.41}$$

The velocity components in the vehicle transverse direction (y_v-direction) differ by the amount of yaw $\ell_j \dot{\psi}$, $j = 1, 2$ (see also Figure 11.3):

$$v_1 \sin(\delta_1 - \alpha_1) = \ell_1 \dot{\psi} + v \sin \beta \,, \tag{11.42}$$

$$v_2 \sin \alpha_2 = \ell_2 \dot{\psi} - v \sin \beta \,. \tag{11.43}$$

Note that the slip angles α_j were introduced against the yaw, steering and the vehicle sideslip angle. From the equations, we obtain

$$\tan(\delta_1 - \alpha_1) = \frac{\ell_1 \dot{\psi} + v \sin \beta}{v \cos \beta} \,, \tag{11.44}$$

$$\tan \alpha_2 = \frac{\ell_2 \dot{\psi} - v \sin \beta}{v \cos \beta} \,. \tag{11.45}$$

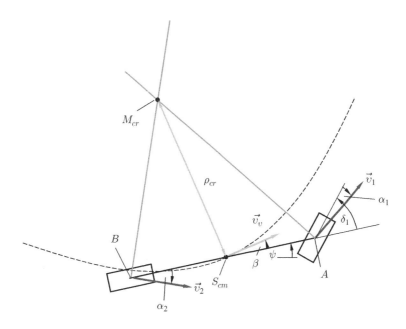

Figure 11.14   Kinematics on the single-track model

Linearizing these equations ($\tan \alpha_j \approx \alpha_j$, $j = 1, 2$, $\cos \beta \approx 1$, $\sin \beta \approx \beta$), we obtain

$$\alpha_1 = -\beta + \delta_1 - \ell_1 \frac{\dot{\psi}}{v} , \tag{11.46}$$

$$\alpha_2 = -\beta + \ell_2 \frac{\dot{\psi}}{v} . \tag{11.47}$$

Substituting the equation for the lateral forces (11.26) in a linear form in the equations of motion (11.10), (11.11) and (11.12) of Section 11.1 and linearizing them results in (here Equation (11.39) is used)):

$$m\dot{v} = F_{x1} + F_{x2} - F_{ax} , \tag{11.48}$$

$$mv(\dot{\beta} + \dot{\psi}) + m\dot{v}\beta = c_{\alpha 1}\left(-\beta + \delta_1 - \ell_1 \frac{\dot{\psi}}{v}\right) \tag{11.49}$$

$$+ c_{\alpha 2}\left(-\beta + \ell_2 \frac{\dot{\psi}}{v}\right) - F_{ay} ,$$

$$J_z\ddot{\psi} = c_{\alpha 1}\ell_1\left(-\beta + \delta_1 - \ell_1 \frac{\dot{\psi}}{v}\right) \tag{11.50}$$

$$- c_{\alpha 2}\ell_2\left(-\beta + \ell_2 \frac{\dot{\psi}}{v}\right) - F_{ay}\ell_{cm} .$$

The term $F_{x1}\delta_1$ was also neglected here. The set of equations is completed by the steering equation:

$$\delta_1 = \frac{1}{1 + \frac{c_{\alpha 1} n_c}{\tilde{k}_s}} \left(\frac{\delta_s}{i_s} + \frac{c_{\alpha 1} n_c}{\tilde{k}_s} \left(\beta + \ell_1 \frac{\dot{\psi}}{v} \right) \right). \tag{11.51}$$

Equation (11.51) is obtained from (11.35)

$$\delta_1 = \frac{\delta_s}{i_s} - \frac{(F_{y1\ell} + F_{y1r})(n_{kc} + n_{tc})}{\tilde{k}_s} \tag{11.52}$$

by substituting $F_{y1} = F_{y1\ell} + F_{y1r}$ and $F_{y1} = c_{\alpha 1}\alpha_1$ (α_1 from 11.46) and solving for δ_1. It is important that the cornering stiffness, $c_{\alpha 1}$ in this equation should be the stiffness for the whole axle, this means that $c_{\alpha 1} = 2c_{\alpha w1}$, where $c_{\alpha w1}$ is the cornering stiffness of one single wheel at the front axle.

Here the total caster trail, $n_c = n_{kc} + n_{tc}$, is the sum of the kinematic caster trail, n_{kc}, and the tyre caster trail, n_{tc}. For the limiting case $\lim_{\tilde{k}_s \to \infty}$ of a rigid steering system without compliances, we obtain $\delta_1 = \delta_s/i_s$.

Equations (11.48), (11.49), (11.50) and (11.51) are considered in more detail for special cases in subsequent chapters.

11.5   Relationship between Longitudinal Forces and Lateral Forces in the Contact Patch

When considering the lateral forces, F_y, (section force in the contact patch), we have so far assumed that no longitudinal forces, F_x, act on the tyre. In this chapter, the limiting adhesion stress is only affected by the lateral force, F_y. However, the longitudinal force, F_x, also acts perpendicularly to F_y, hence the lateral force considered by the adhesion stress is no longer independent of the longitudinal force. A crucial factor for the adhesion limit is the quotient $\sqrt{F_x^2 + F_y^2}/F_z$ between the vectorial sum of the two forces F_x and F_y and the vertical load, F_z. Since the lateral and longitudinal forces are perpendicular, the resultant force is $F_r = \sqrt{F_x^2 + F_y^2}$. The adhesive limit in the F_x–F_y-plane is described by the following equation:

$$\sqrt{F_x^2 + F_y^2} \le \mu_a F_z . \tag{11.53}$$

Here F_z is the vertical load. If we consider diagram a) in Figure 11.15, this equation means that the sum of F_x and F_y always lies inside the circle $\mu_a F_z$. We call this circle Kamm's circle.

This limitation also affects the maximum of the F_y-α curve. In Figure 11.15(b), this is reproduced qualitatively. The diagram shows the lateral force-slip angle curves for three different longitudinal forces $F_{x1} < F_{x2} < F_{x3}$. It can be seen that the maxima

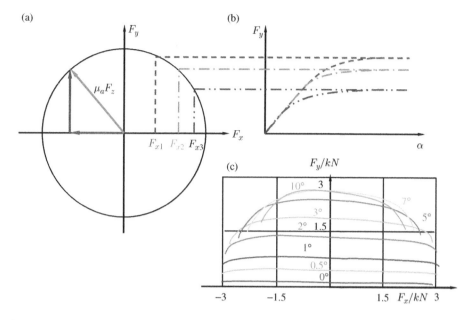

Figure 11.15   Kamm's circle and Krempel diagram ($F_x > 0$: driving; $F_x < 0$: braking)

of the lateral forces decrease with increasing longitudinal force, F_x. The reason for this is that adhesive coefficient is exceeded for smaller slip angles α by the action of the longitudinal and the transverse forces. Figure 11.15(c) shows measurements of the lateral forces as a function of the longitudinal force for different slip angles (from G. Krempel, ATZ 1967).

Again in this diagram it is obvious that the lateral forces are dependent on the longitudinal force. The decrease in the lateral force for large longitudinal forces is significant. The consequence of this relation is that the maximum cornering or lateral forces decrease if the vehicle accelerates or brakes. If a vehicle brakes or accelerates during cornering at the limit of lateral tyre forces, the tyre forces decrease and the vehicle is not able to continue driving the curve.

An asymmetry between braking ($F_x < 0$) and accelerating ($F_x > 0$) can also be seen.

11.6   Effect of Differentials when Cornering

A so-called axle differential is used in order to distribute the drive torques on the inner and outer wheels when cornering. A differential divides the input torque into two drive torques for the two driven wheels on one axle. The behaviour of this differential is described by a set of equations for the rotational speed and the torque, which cannot be altered without any external actions. It follows that the distribution of the drive torque to the drive wheels is constant. The function of a differential is firstly to avoid torsion of an axle during cornering (this torsion is caused by the different speeds of curve-inner and curve-outer wheels) and secondly to distribute the torque to the inner and the outer wheels. The most widely used differential gear design used is the bevel gear. In addition to compensating for different speeds when cornering, the differential also balances the wheels for different slip conditions between left and right wheels when driving straight ahead. Such conditions may, for example, occur on a split-μ road.

The bevel gear differential has the following characteristics:

- Same distribution of the input torque, M_i, on right and left wheel:

$$M_\ell = M_r = M_i/2 \; . \tag{11.54}$$

- Stresses between the right and left wheel when cornering or during different slip condition can be compensated for.

The disadvantage of varying adhesion conditions for the left and right wheels is that the wheel with the smaller adhesion value determines the total transferable driving force; this can result in only a very small driving force being transmitted to the road.

The so-called differential lock connects the left and right drives by friction-locking or interlocking. The simplest form of this is a claw clutch, which connects the two wheels to each other and is activated when required. Driving is then affected during cornering and there are large stresses in the axle. With the existing ASR brake system, it is possible to achieve good traction behaviour on split-μ roads without the use of differential locks. In this case, the wheel with the poorer traction performance is braked individually so that the transmitted forces at the wheel with a good traction surface are not reduced. The disadvantage of this implementation is that the brake disc at the braked wheel can heat up considerably; hence ASR with braking on one wheel is not suitable for permanent application.

Another way to allocate a higher torque to the wheel with the better traction performance can be found in systems that automatically partially restrict the speed compensation between left and right wheel when necessary. The systems worth mentioning here are also based on speed differences (e.g. Haldex) or those which work on moment sensing (Torsen). Torsen differentials are used for example in four-wheel drive vehicles in order to distribute the moments between the front and rear axles.

Another aspect of partially locked differentials is their response to load changes of the vehicle when cornering. If we consider the release of the foot from the accelerator pedal during steady-state circular cornering, then the drag moment from the combustion engine decelerates the vehicle and the wheel loads therefore decrease on the rear wheels and increase on the front wheels. Reducing wheel loads also causes the cornering forces to decrease on the rear wheels, while the cornering forces increase on the front wheels. For these reasons the vehicle turns into the curve. Now we assume an ideal, frictionless differential, so that the engine drag torque would be divided equally between inner and outer wheels. A partially locked differential, however, means that the braking torque on the outer wheel is greater than the inner wheel[9]. The varying brake force distribution between the inner and outer wheels means that there is a yaw moment, and the vehicle rotates out of path so as to counteract the inward turning as a result of the load change behaviour.

11.7 Questions and Exercises

Remembering

1. Define vehicle sideslip angle.
2. Define course angle.
3. How is the instantaneous centre of rotation defined?
4. How is the slip angle of a tyre defined?
5. How are caster angle, total caster trail, tyre caster trail (pneumatic trail) and kinematic caster trail defined?
6. By what are the longitudinal and transverse forces associated with the tyres?

Understanding

1. What key assumption determines the single-track model and what is neglected by this assumption?
2. What forces and moments act on the single-track model (including d'Alembert inertial forces and moments and section forces)?
3. Define the centre of curvature of the trajectory and explain its purpose.
4. What qualitative path do the tangential stresses follow in the contact area and how do they change as a function of the tyre slip angle?
5. What do we obtain from this process?
6. What is cornering stiffness?
7. How do lateral force, self-aligning moment and tyre caster trail depend qualitatively on slip angle?
8. What is Kamm's circle?
9. What is a Krempel diagram?

Applying

1. The following parabola is given in parametrized form:
$$x_v = A\zeta,$$
$$y_v = A\zeta^2.$$
Calculate the radius of curvature ρ_{cc}.

---

[9] We can illustrate this by imagining that the vehicle is turning about the tyre contact point on the inner wheel (in this doubtlessly unrealistic extreme case, the inner wheel no longer rotates and the outer curve would provide the complete input torque).

2. The following straight line is given:

$$x_v = A\zeta,$$
$$y_v = 2A\zeta$$

Calculate the curvature κ_{cc}.

3. The adhesion limit is $\mu_a = 1.1$, the velocity of a vehicle is $100\,\text{km/h}$ (the gravitational acceleration is $g = 9.81\ \text{m/s}^2$).

Estimate the minimal radius ρ_{cc}, which is possible for the vehicle.

4. Neglect aerodynamical forces and assume, that the centre of mass of a vehicle is centered between the axles $\ell_1 = \ell_2 = 2$ m. The cornering stiffness values for all four wheels are the same $c_\alpha = 50$ kN/rad. The vehicle runs at a steady state (that means $\dot\beta = 0$) on a circle $\rho_{cc} = 100$ m with a velocity $v = 30$ m/s. Calculate the slip angles of the tyres for the linear single-track model, the vehicle sideslip angle β and the front steering angle δ_1. As this may be a challenging task, follow the solution procedure:

• Look at the last equation of the linearized equation (without aerodynamic forces):

$$J_z\ddot\psi = c_{\alpha 1}\ell_1\left(-\beta + \delta_1 - \ell_1\frac{\dot\psi}{v}\right) - c_{\alpha 2}\,\ell_2\left(-\beta + \ell_2\frac{\dot\psi}{v}\right). \qquad (11.55)$$

Which conclusion can be drawn by considering the steady-state cornering $\ddot\psi = 0$, the equation $c_{\alpha 1} = c_{\alpha 2}$, $\ell_1 = \ell_2$ and the equations for the slip angles α_1 and α_2?

• Calculate the yaw rate $\dot\psi$.

• Look at the second equation of the linearized equation of the single-track model

$$mv(\dot\beta + \dot\psi) + m\dot v\beta = c_{\alpha 1}\alpha_1 + c_{\alpha 2}\alpha_2. \qquad (11.56)$$

Considering the steady-state condition you can now calculate the tyre slip angles α_1 and α_2.

• Using the equation for α_2:

$$\alpha_2 = -\beta + \ell_2\frac{\dot\psi}{v}$$

you can calculate the vehicle sideslip angle.

• Using the equation for α_1:

$$\alpha_1 = -\beta + \delta_1 - \ell_1\frac{\dot\psi}{v}$$

you can calculate the front steering angle δ_1.

12

Circular Driving at a Constant Speed

In this chapter, we consider steady-state driving at constant speed ($v = v_v = $ const.) on a circle with a radius of ρ_{cc}. We omit the index v in the following; thus $v = v_v$, etc. The value $\rho_{cc} = \infty$ corresponds to the case of steady-state driving in a straight line, which is also included in the following considerations.

In Section 12.1, system of algebraic equations are derived, in Section 12.2, we consider their solutions. Section 12.3 is dedicated to geometric aspects. In the Section 12.4, we discus the solutions and introduce oversteering and understeering.

12.1 Equations

This section derives the system of algebraic equations for the description of the steady-state driving.

Because of the steady state ($\dot{\psi} = $ const. and $\beta = $ const.), the following applies:

$$\dot{\beta} = 0 ,\tag{12.1}$$

$$\ddot{\psi} = 0 .\tag{12.2}$$

Furthermore, the instantaneous center of rotation M_{cr} and the center of curvature M_{cc} coincide: $M_{cr} = M_{cc}$. From the relationship between the yaw rate, $\dot{\psi}$, and the vehicle sideslip angle rate, $\dot{\beta}$, and the centripetal acceleration

$$v^2/\rho_{cc} = v(\dot{\beta} + \dot{\psi}) ,\tag{12.3}$$

we obtain

$$\frac{v}{\rho_{cc}} = \dot{\psi}\tag{12.4}$$

Vehicle Dynamics, First Edition. Martin Meywerk.
© 2015 John Wiley & Sons, Ltd. Published 2015 by John Wiley & Sons, Ltd.
Companion Website: www.wiley.com/go/meywerk/vehicle

and

$$mv(\dot{\beta} + \dot{\psi}) = \frac{mv^2}{\rho_{cc}} \ . \tag{12.5}$$

The linearized equations of motion of the single-track model (11.48)–(11.51) simplify further in steady-state circular travel. First, we use the expression for δ_1 from (11.51)

$$\delta_1 = \frac{1}{1 + \frac{c_{\alpha 1} n_c}{k_s}} \left(\frac{\delta_s}{i_s} + \frac{c_{\alpha 1} n_c}{\tilde{k}_s} \left(\beta + \ell_1 \frac{\dot{\psi}}{v} \right) \right) . \tag{12.6}$$

to substitute δ_1 in the term $c_{\alpha 1}(-\beta + \delta_1 - \ell_1 \dot{\psi}/v)$

$$c_{\alpha 1} \left(-\beta + \delta_1 - \ell_1 \frac{\dot{\psi}}{v} \right) = \beta c_{\alpha 1} \left(\frac{1}{1 + \frac{c_{\alpha 1} n_c}{k_s}} \frac{c_{\alpha 1} n_c}{\tilde{k}_s} - 1 \right)$$

$$+ \ell_1 \frac{\dot{\psi}}{v} c_{\alpha 1} \left(\frac{1}{1 + \frac{c_{\alpha 1} n_c}{k_s}} \frac{c_{\alpha 1} n_c}{\tilde{k}_s} - 1 \right)$$

$$+ \frac{c_{\alpha 1}}{1 + \frac{c_{\alpha 1} n_c}{k_s}} \frac{\delta_s}{i_s}$$

$$= c'_{\alpha 1} \left(-\beta - \ell_1 \frac{\dot{\psi}}{v} + \frac{\delta_s}{i_s} \right) \tag{12.7}$$

wherein

$$c'_{\alpha 1} = \frac{c_{\alpha 1}}{1 + \frac{c_{\alpha 1} n_c}{k_s}} \tag{12.8}$$

has been set.

The equations of motion are simplified to obtain (air forces are not taken into account):

$$(c'_{\alpha 1} + c_{\alpha 2}) \beta + (mv^2 - (c_{\alpha 2}\ell_2 - c'_{\alpha 1}\ell_1)) \frac{\dot{\psi}}{v} = c'_{\alpha 1} \frac{\delta_s}{i_s} , \tag{12.9}$$

$$-(c_{\alpha 2}\ell_2 - c'_{\alpha 1}\ell_1)\beta + (c'_{\alpha 1}\ell_1^2 + c_{\alpha 2}\ell_2^2) \frac{\dot{\psi}}{v} = c'_{\alpha 1}\ell_1 \frac{\delta_s}{i_s} . \tag{12.10}$$

Equations (12.9) and (12.10) form a linear, non-homogeneous system of algebraic equations for the unknown constants β, $\dot{\psi} = \frac{v}{\rho_{cc}}$ and δ_s. If one of these quantities is given, which is the inhomogeneity, the remaining two can be calculated by solving the system of equations.

For example, if the steering wheel angle δ_s is given for a constant velocity, then the vehicle sideslip angle, β, and the yaw angular velocity, $\dot{\psi}$, can be calculated, and the relation $\dot{\psi} = \frac{v}{\rho_{cc}}$ allows the radius ρ_{cc} of the circle to be derived.

Another example of using these equations is that for a given circle, ρ_{cc}, and a given velocity, v (which means that the angular velocity, $\dot{\psi}$, can be calculated by means of $\dot{\psi} = \frac{v}{\rho_{cc}}$ and is therefore known, too), the vehicle sideslip angle, β, and the steering angle, δ_s, can be calculated.

12.2 Solution of the Equations

In this section, the solutions of Equations (12.9) and (12.10) are derived.

Replacing $\dot{\psi}/v$ by $1/\rho_{cc}$, we then determine the variables of interest, such as the steering wheel angle, δ_s, angle of front wheels, δ_1, vehicle sideslip angle, β, or the moment on the steering wheel as a function of the centripetal acceleration divided by the acceleration due to gravity, g. We obtain the vehicle sideslip angle, β, when we multiply Equation (12.9) by ℓ_1 and then subtract Equation (12.10) (the second equation of (12.11) is a rearrangement of (11.47)):

$$\beta = \frac{\ell_2}{\rho_{cc}} - \frac{m\ell_1}{c_{\alpha 2}\ell}\frac{v^2}{\rho_{cc}} \tag{12.11}$$

$$= \frac{\ell_2}{\rho_{cc}} - \alpha_2 \ .$$

If we write this with the help of the static rear axle load, $F_{z2} = mg\ell_1/\ell$ (without influence of a gradient, $p = 0$ or $\alpha_g = 0$), we obtain ($\beta_0 = \ell_2/\rho_{cc}$):

$$\beta = \beta_0 - \frac{F_{z2}}{c_{\alpha 2}}\frac{v^2}{\rho_{cc}g} \ . \tag{12.12}$$

The value of β_0 can be interpreted graphically.

If the vehicle continues to drive on a circle of constant radius, ρ_{cc}, and reduces its speed, then the vehicle sideslip angle β approaches the value β_0:

$$\lim_{v \to 0} \beta = \beta_0 \ . \tag{12.13}$$

This means that the vehicle sideslip angle, β, approaches the value of β_0 for very small driving speeds, v. With increasing centripetal acceleration v^2/ρ_{cc}, the vehicle sideslip angle, β, decreases linearly with the centripetal acceleration, regardless of whether the vehicle is understeering or oversteering (see below for an explanation of these technical terms). The upper diagram of Figure 12.1 gives the trend for the following vehicle data: $m = 1350$ kg, $\ell_1 = 2.05$ m, $\ell_2 = 2.35$ m, $n_c = 0.051$ m, $c_{\alpha 1} = 100$ kN/rad, $c_{\alpha 2} = 90$ kN/rad, $\rho_{cc} = 100$ m, $i_s = 19$, $\tilde{k}_s = 10$ kN m/rad (Here we assume a constant value of n_c, which could be achieved for a constant tyre caster trail n_{tc}, which is only approximately valid for a small slip angle at the front wheels.).

Example 12.1 For these parameters, the angle β_0 is

$$\beta_0 = 0.0235 \text{ rad} \approx 1.35° \ . \tag{12.14}$$

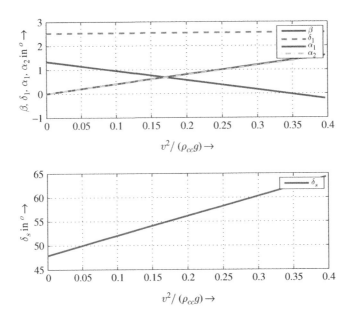

Figure 12.1   Vehicle sideslip angle, β, steering wheel angle, δ_s, tyre slip angles, α_1, α_2, and steering angle of the front wheel, δ_1, as a function of the centripetal acceleration divided by the acceleration due to gravity, g

For this angle, the vehicle's longitudinal axis is directed towards outside of the curve. The speed at which the vehicle sideslip angle becomes zero and its sign changes thus becomes:

$$v = \sqrt{\frac{\ell_2 \ell c_{\alpha 2}}{\ell_1 m}}$$

$$\approx 18.34 \text{ m/s} . \tag{12.15}$$

Substituting expression (12.11) for β in Equation (12.9) for steady-state cornering, we obtain the steering angle, δ_s:

$$\delta_s = \frac{i_s \ell}{\rho_{cc}} + m i_s \frac{c_{\alpha 2} \ell_2 - c'_{\alpha 1} \ell_1}{c'_{\alpha 1} c_{\alpha 2} \ell} \frac{v^2}{\rho_{cc}}$$

$$= \delta_{s0} + \frac{i_s \ell}{\rho_{cc}} \frac{v^2}{v_{ch}^2}$$

$$= \frac{i_s \ell}{\rho_{cc}} \left(1 + \frac{v^2}{v_{ch}^2} \right) . \tag{12.16}$$

Here

$$v_{ch}^2 = \frac{c'_{\alpha 1} c_{\alpha 2} \ell^2}{m(c_{\alpha 2} \ell_2 - c'_{\alpha 1} \ell_1)} \tag{12.17}$$

is the characteristic velocity and

$$\delta_{s0} = \lim_{v \to 0} \delta_s = \frac{i_s \ell}{\rho_{cc}} . \tag{12.18}$$

Substituting the vehicle sideslip angle (12.11) in the second linearized differential equation (11.49) (the air force is set to zero: $F_{ay} = 0$), we obtain the angle of the front wheels:

$$\delta_1 = \frac{\ell}{\rho_{cc}} + m \frac{c_{\alpha 2} \ell_2 - c_{\alpha 1} \ell_1}{c_{\alpha 1} c_{\alpha 2} \ell} \frac{v^2}{\rho_{cc}} . \tag{12.19}$$

The torque at the steering wheel is obtained by the linearized equation of motion (this is the equilibrium of forces in the y_v-direction)

$$m \frac{v^2}{\rho_{cc}} = F_{y1} + c_{\alpha 2} \left(-\beta + \ell_2 \frac{\dot{\psi}}{v} \right) \tag{12.20}$$

in which the expression for the vehicle sideslip angle

$$\beta = \frac{\ell_2}{\rho_{cc}} - \frac{m \ell_1}{c_{\alpha 2} \ell} \frac{v^2}{\rho_{cc}} \tag{12.21}$$

is substituted; it follows (with $\dot{\psi}/v = 1/\rho_{cc}$ and $\ell = \ell_1 + \ell_2$) that

$$F_{y1} = \frac{m v^2 \ell_2}{\rho_{cc} \ell} . \tag{12.22}$$

This expression is inserted into the equation for the steering wheel torque, $M_s = F_{y1} n_c / (i_s V_s)$ (cf. (11.30) and (11.31)), and the steering wheel torque is obtained as follows:

$$M_s = \frac{m n_c \ell_2}{i_s V_s \ell} \frac{v^2}{\rho_{cc}}$$

$$= \frac{F_{z1} n_c}{i_s V_s} \frac{v^2}{\rho_{cc} g} . \tag{12.23}$$

The steering wheel angle, δ_s, the tyre slip angles α_1, α_2, the vehicle sideslip angle β and the angle of the front wheels, δ_1, as a function of the centripetal acceleration divided by the gravitational acceleration can be seen in the graphs of Figure 12.1.

12.3 Geometric Aspects

In this section, some geometric aspects and interpretations of the solutions are given.

Ackermann steer angle: As described for the vehicle sideslip angle and the steering wheel angle, we introduce the front wheel angle, δ_{10}, for diminishing velocities:

$$\delta_{10} = \lim_{v \to 0} \delta_1 = \frac{\ell}{\rho_{cc}} \, . \tag{12.24}$$

We call this angle, δ_{10}, theAckermann steer angle or Ackermann angle.

The Ackermann angle, δ_{10}, and the vehicle sideslip angle, β_0, can also be illustrated geometrically (cf. Figure 12.2). For the limit of diminishing driving velocity, the lateral forces F_{y1} and F_{y2} are zero for the front and rear axles. Since these lateral forces (for small tyre slip angles) depend linearly on the tyre slip angle, this means that the tyre slip angles also disappear at the front and rear wheels. This in turn means that the trajectories of the front and rear wheels are circles, and the $\vec{e}_{wx}$ directions are tangential to the respective circle for the front and rear axle. In Figure 12.2, the circle on which the rear wheel rolls is the inner circle, the circle of the front wheel is the outer circle and the actual trajectory of the centre of mass S_{cm} is the middle circle. The angles in the triangles with the thick lines are derived from the fact that they are pairs of mutually perpendicular straight lines. For the height, h, we obtain

$$h = \sqrt{\rho_{cc}^2 - \ell_2^2}$$

$$\approx \rho_{cc} \, , \tag{12.25}$$

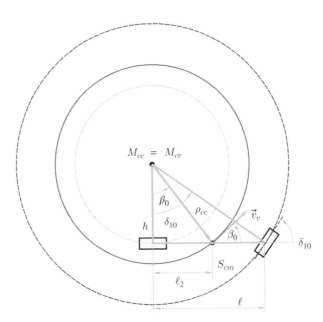

Figure 12.2 Geometric interpretation of the Ackermann angle and the vehicle sideslip angle for the disappearing tangential velocity (for a single-track model)

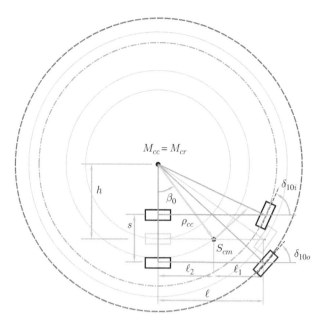

Figure 12.3 Geometric interpretation of the Ackermann angle for disappearing tangential velocity (for a double-track model)

since the angles are all small, and therefore $\rho_{cc} \gg \ell$. From $\sin \beta_0 = \ell_2/\rho_{cc}$ and $\tan \delta_{10} = \ell/h \approx \ell/\rho_{cc}$, we obtain by linearizing $\beta_0 = \ell_2/\rho_{cc}$ and $\delta_{10} = \ell/\rho_{cc}$; these are the same relationships that were derived above using the linearized equations of motion.

The geometry of a double-track model is shown in Figure 12.3. In this double-track model, the steering angle of the inner wheel δ_{10i} is greater than the angle, δ_{10o}, of the outer wheel: $\delta_{10i} > \delta_{10o}$. In this situation, the longitudinal axes are tangential to the circle on which the wheels move. These angles can be realized by a trapezoidal steering geometry as depicted in Figure 11.11. If $\delta_{10i} < \delta_{10o}$ holds, it is called anti-Ackermann steering. Anti-Ackermann steering takes into account different wheel loads (higher at the curve of the outer wheel than at the curve of the inner wheel) and therefore different transmittable lateral forces.

The angles δ_{10i} and δ_{10o} can be derived from Figure 12.3. We obtain from the rectangular triangles (with $\tan \beta_0 = \ell_2/h$):

$$\tan \delta_{10i} = \frac{2\ell \tan \beta_0}{2\ell_2 - s \tan \beta_0} \, , \tag{12.26}$$

$$\tan \delta_{10o} = \frac{2\ell \tan \beta_0}{2\ell_2 + s \tan \beta_0} \, . \tag{12.27}$$

For steady-state cornering at $\lim_{v \to 0} \cdots$, the centre of curvature and the instantaneous centre of rotation coincide: $M_{cr} = M_{cc}$.

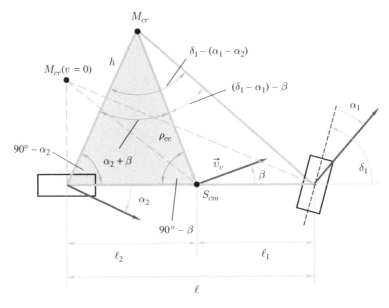

Figure 12.4   Relationship between the front wheel steering angle, δ_1, and tyre slip angles, α_1, α_2, and the vehicle sideslip angle β for steady-state cornering ($M_{\mathrm{cr}} = M_{\mathrm{cc}}$)

Figure 12.4 shows the relation between front wheel steering angle, δ_1, tyre slip angles, α_1, α_2, and the vehicle sideslip angle, β.

Using the sine theorem yields for the large triangle:

$$\frac{h}{\ell} = \frac{\sin(180° - 90° - (\delta_1 - \alpha_1))}{\sin(\delta_1 - (\alpha_1 - \alpha_2))} . \tag{12.28}$$

For small angles we have $h \approx \rho_{\mathrm{cc}}$, $\sin(90° - (\delta_1 - \alpha_1)) \approx 1$ and $\sin(\delta_1 - (\alpha_1 - \alpha_2)) \approx \delta_1 - (\alpha_1 - \alpha_2)$.

From this, we obtain

$$\delta_1 = \underbrace{\frac{\ell}{\rho_{\mathrm{cc}}}}_{\delta_{10}} + (\alpha_1 - \alpha_2) . \tag{12.29}$$

Applying the law of sines to the small grey triangle, we obtain

$$\frac{\ell_2}{h} = \frac{\sin(\alpha_2 + \beta)}{\sin(90° - \beta)} . \tag{12.30}$$

From this, we obtain

$$\beta = \frac{\ell_2}{\rho_{\mathrm{cc}}} - \alpha_2 . \tag{12.31}$$

We see:

- from (12.29) that the steering angle of the front wheel, δ_1, is dependent on the difference, $\alpha_1 - \alpha_2$, between the tyre slip angles and
- from (12.31) that the vehicle sideslip angle, β, is dependent only on the tyre slip angle, α_2, of the rear wheels.

12.4 Oversteering and Understeering

The following are the characteristic variables that are considered in more detail. We start with the steering angle, δ_s, of the steering wheel:

$$\delta_s = \frac{i_s \ell}{\rho_{cc}} + mi_s \frac{c_{\alpha 2} \ell_2 - c'_{\alpha 1} \ell_1}{c'_{\alpha 1} c_{\alpha 2} \ell} \frac{v^2}{\rho_{cc}}. \tag{12.32}$$

The dependence of the steering angle, δ_s, on the driving speed is determined by the factor

$$mi_s \frac{c_{\alpha 2} \ell_2 - c'_{\alpha 1} \ell_1}{c'_{\alpha 1} c_{\alpha 2} \ell} = \frac{i_s \ell}{v_{ch}^2} \tag{12.33}$$

The magnitude, v_{ch}^2, is the square of the so-called characteristic velocity. Since the characteristic velocity can also be purely imaginary, we should not interpret its meaning when it is complex; however, the absolute value has a meaning (cf. Chapter 13). It is important to note that v_{ch}^2 may be positive or negative (or zero).

Understeer: If $v_{ch}^2 > 0$, this means that an increase in vehicle speed, v (on a circle with radius ρ_{cc}) requires an increase in the steering wheel angle. We call this behaviour of the vehicle understeer.

An understeering vehicle behaviour is generally desirable.

Oversteer: If $v_{ch}^2 < 0$, this means that an increase in vehicle speed, v (on a circle with radius ρ_{cc}) requires an decrease in the steering wheel angle. We call this behaviour oversteer.

For oversteering vehicles, the function $\delta_s = \delta_s(v^2/\rho_{cc})$ intersects the line $\delta_s = 0$ at a certain speed; this intersection speed is called the critical velocity, v_{crit}. For $v > v_{crit}$, the vehicle has to be countersteered, so when you take a right turn the steering wheel should be turned to the left.

Generally, the vehicle understeers, if

$$\frac{\partial(\delta_s - \delta_{s0})}{\partial(v^2/\rho_{cc})} > 0 \tag{12.34}$$

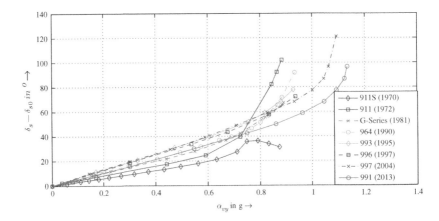

Figure 12.5 Relationship between steering angle, $\delta_s - \delta_{s0}$, and lateral acceleration, $a_{vy} = v^2/\rho_{cc}$, for several Porsche 911 models (data from Harrer et al. 2013)

and oversteers, if

$$\frac{\partial(\delta_s - \delta_{s0})}{\partial(v^2/\rho_{cc})} < 0 \ . \tag{12.35}$$

Self-steering coefficient: The following coefficient

$$\frac{1}{i_s \ell} \frac{\partial(\delta_s - \delta_{s0})}{\partial(v^2/\rho_{cc})} \tag{12.36}$$

is called the self-steering coefficient of the vehicle. Likewise, the term

$$\partial(\delta_1 - \delta_{10})/(\partial(v^2/\rho_{cc})) \tag{12.37}$$

is common, which is the self-steering coefficient without considering the steering stiffness.

For the linear model treated here, the self-steering coefficient[1] is $1/v_{ch}^2$. The linear theory applies, however, only up to centripetal accelerations of about $v^2/\rho_{cc} \approx 4 \ \mathrm{m/s^2}$. The self-steering behaviour serves to assess the vehicles. The aim is to achieve an understeering behaviour.

The linear behaviour of the single-track model is depicted in Figure 12.1. The behaviour of real vehicles differs from this ideal linear function, as shown in Figure 12.5. You can see the non-linear trend of the curves. The models of Porsche 911 from 1970 to 1990 are vehicles with a semi-trailing arm suspension, the other four from 1995 to 2013 have multi-link suspensions (cf. Chapter 16).

---

[1] The self-steering coefficient is also called the understeer/oversteer coefficient.

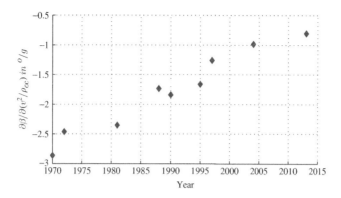

Figure 12.6 Vehicle sideslip gradient for several Porsche 911 models (data from Harrer et al. 2013)

A further criterion for the assessment of driving behaviour is the vehicle sideslip angle

$$\beta = \frac{\ell_2}{\rho_{cc}} - \frac{m\ell_1}{c_{\alpha 2}\ell} \frac{v^2}{\rho_{cc}} .$$ (12.38)

This indicates the difference in the velocity direction of the centre of mass from the direction of the longitudinal vehicle axis. Strictly speaking, the vehicle sideslip angle should be determined at the driver's position, because this is the angle observed by the driver. Since the angle at the driver's position differs slightly from the vehicle sideslip angle in the centre of mass, β is often selected as an evaluation criterion. The sideslip angle should be small and the change with the centripetal acceleration, the sideslip angle gradient

$$\frac{\partial \beta}{\partial (v^2/\rho_{cc})} = -\frac{m\ell_1}{c_{\alpha 2}\ell}$$ (12.39)

should also be small. This is achieved by a large cornering stiffness of the rear wheels.

Figure 12.6 shows gradients for models of the Porsche 911. It can be seen that the gradient has decreased over the years.

In the following, we consider the influence of the position of the centre of mass on the driving behaviour. We limit ourselves to a qualitative approach. Measurements show that an increase in the wheel load F_z does not increase the cornering stiffness, c_α, proportionally to the wheel load. The variation of the cornering stiffness in response to the wheel load can be described approximately by

$$c_\alpha = \hat{c}_\alpha F_z - \tilde{c}_\alpha F_z^2 .$$ (12.40)

The quadratic correction term shows the non-proportional dependence between wheel load and cornering stiffness. However, a static load transfer from the front to rear axle results in a reduction of the distance of the centre of mass to the rear axle (this means a reduction of ℓ_2) and an enlargement of ℓ_1. However, the cornering stiffness, $c_{\alpha 1}$, does not increase by the same amount as ℓ_1 decreases, and vice versa for $c_{\alpha 2}$ and ℓ_2. This means that the quotient

$$\frac{c_{\alpha 2}\ell_2 - c_{\alpha 1}\ell_1}{c_{\alpha 1}c_{\alpha 2}\ell} , \tag{12.41}$$

which determine the sign in the formula for the steering angle of the front wheels

$$\delta_1 = \frac{\ell}{\rho_{cc}} + \frac{c_{\alpha 2}\ell_2 - c_{\alpha 1}\ell_1}{c_{\alpha 1}c_{\alpha 2}\ell} m\frac{v^2}{\rho_{cc}} , \tag{12.42}$$

decreases with increasing rear axle load, or even changes the sign. Taking into account the simple equations for the wheel loads

$$F_{z1} = \frac{G}{\ell}\ell_2 , \tag{12.43}$$

$$F_{z2} = \frac{G}{\ell}\ell_1 \tag{12.44}$$

we obtain (using (12.40))

$$c_{\alpha 2}\ell_2 - c_{\alpha 1}\ell_1 = \tilde{c}_\alpha G^2 \frac{\ell_1\ell_2}{\ell^2}(\ell_2 - \ell_1) . \tag{12.45}$$

On condition that the front and rear axles have the same tyres, $c_{\alpha 2}\ell_2 - c_{\alpha 1}\ell_1 < 0$ applies for the back-heavy vehicle, and $c_{\alpha 2}\ell_2 - c_{\alpha 1}\ell_1 > 0$ for the front-heavy vehicle. This means that the steering angle of the front wheels decreases with increasing speed for a back-heavy vehicle, while it increases for the front-heavy vehicle. However, that does not necessarily mean that the back-heavy vehicle exhibits oversteering behaviour because the behaviour of the car depends on the quotient

$$\frac{c_{\alpha 2}\ell_2 - c'_{\alpha 1}\ell_1}{c'_{\alpha 1}c_{\alpha 2}\ell} , \tag{12.46}$$

where the corrected cornering stiffness, c'_α, enters into the quotient. Nevertheless, a back-heavy vehicle is generally less likely to understeer than a front-heavy vehicle. Loading a vehicle may result in an unfavourable change in the driving behaviour, i.e. a shift from understeering to oversteering behaviour. It used to be common to refer to vehicle understeer if $\alpha_1 - \alpha_2 > 0$ and to oversteer if $\alpha_1 - \alpha_2 < 0$. This determination is based on Equation (12.29) for the steering angle of the front wheels:

$$\delta_1 = \frac{\ell}{\rho} + (\alpha_1 - \alpha_2) . \tag{12.47}$$

As the driver turns the steering wheel directly, but the front wheels are only indirectly affected, this definition has become less important. However, it can still be also found in the recent literature. The difference between the definition used in this book and the older version, based on (12.47), is that the version used in this book takes into account the steering stiffness, $\tilde{k}_s$, and the total caster trail, n_c.

12.5 Questions and Exercises

Remembering

1. Define the Ackermann angle.
2. How can we geometrically interpret the Ackermann angle and the sideslip angle for the limiting case of diminishing driving speed?
3. How do the changes in the vehicle sideslip angle, the steering angle, the front wheel steering angle and the steering wheel torque depend on the centripetal acceleration?
4. When is a vehicle said to understeer and when is it said to oversteer?
5. Which quotient determines the sign of the relationship between the steering angle of the front wheels and the centrifugal force?

Understanding

1. Imagine you are driving on a curve with constant velocity and the radius of the curve becomes smaller. How do you change the steering wheel angle?
2. What conclusions can be drawn from this?
3. Define the self-steering coefficient.
4. What are the signs of the self-steering coefficient for oversteering and understeering in driving characteristics?
5. What might occur when loading a vehicle with respect to the self-steering behaviour?
6. What influence could passengers have on the self-steering behaviour?
7. What influence could a negative aerodynamic lift coefficient on the front axle, $c_{l,1} < 0$, or on the rear axle, $c_{l,2} < 0$, have on the self-steering behaviour and on the vehicle sideslip gradient?

Applying

The following parameters are given: $c_{\alpha 1} = 60$ kN/rad, $c_{\alpha 2} = 50$ kN/rad, $\ell_1 = 2.1$ m, $\ell_2 = 2.2$ m, $\rho_{cc} = 100$ m, $i_s = 19$, $\tilde{k}_s = 10$ kN m/rad (for a rack and pinion steering system), $m = 1350$ kg.

1. Which parameter is still needed in order to determine the self-steering coefficient, $\partial \delta_s / \partial(v^2/\rho_{cc})$ for the linear single-track model?
2. Calculate this missing parameter in order to obtain neutral steering (in other words, a self-steering coefficient of zero: $\partial \delta_s / \partial(v^2/\rho_{cc}) = 0$)!

Analysing

1. For a neutral steering vehicle $c_{\alpha 2}\ell_2 - c'_{\alpha 1}\ell_1 = 0$ with rack and pinion steering system the following equation holds for the total caster trail:

$$n_c = \tilde{k}_s \frac{\ell_1 c_{\alpha 1} - \ell_2 c_{\alpha 2}}{\ell_2 c_{\alpha 1} c_{\alpha 2}} . \tag{12.48}$$

Is it possible to change the total caster trail in order to achieve neutral steering for an understeering or for an oversteering vehicle?

2. The simplified axle loads during braking can calculated by (cf. (6.42) and (6.43))

$$F_{z1} = \frac{G}{\ell}(\ell_2 + \mathcal{Z}h) \,, \tag{12.49}$$

$$F_{z2} = \frac{G}{\ell}(\ell_1 - \mathcal{Z}h) \,. \tag{12.50}$$

Substituting the axle loads in Equation (12.40) we obtain (same tyres at the front and rear axle):

$$c_{\alpha 2}\ell_2 - c_{\alpha 1}\ell_1 = -\mathcal{Z}hG\left(\hat{c}_\alpha - \frac{4G\ell_1\ell_2}{\ell^2}\right) + \tilde{c}_\alpha \frac{G^2(\ell_2 - \ell_1)}{\ell^2}(\ell_1\ell_2 - \mathcal{Z}^2h^2) \,. \tag{12.51}$$

For an unbraked vehicle with $\ell_1 = \ell_2$ it is $c_{\alpha 1} = c_{\alpha 2}$, and therefore

$$c_{\alpha 2}\ell_2 - c'_{\alpha 1}\ell_1 > 0 \,. \tag{12.52}$$

Is this true for a braked vehicle with $\ell_1 = \ell_2$ for any values of $\mathcal{Z}$ and $\tilde{k}_s$?

13

Dynamic Behaviour

In this chapter, we consider the dynamic behaviour of a vehicle. In Section 13.1, we proceed to present the stability under steady-state driving conditions. In Section 13.2, we consider the handling of the vehicle in the event of certain changes of the steering angle. Thereafter, the role of aerodynamic side forces will be discussed in Section 13.3[1].

13.1 Stability of Steady-state Driving Conditions

We understand steady-state driving conditions to mean steady-state cornering ($\dot{v}_v = 0$, $\ddot{\beta} = 0$, $\ddot{\psi} = 0$, $\dot{\delta}_s = 0$) and steady-state driving in a straight line (for which $\beta = 0$ and $\dot{\psi} = 0$ additionally holds). We omit the index v in the following, thus $v = v_v$, etc. The differences between steady-state cornering and steady-state straight motion are irrelevant for the following considerations; the difference between the two types of motion is that ρ_{cc} is finite for steady-state cornering, whereas it is infinite for straight-line motion.

To assess the stability of these steady-state driving conditions, we assume the linear equations of motion (11.49) and (11.50) without air forces:

$$mv\dot{\beta} + (c'_{\alpha 1} + c_{\alpha 2})\beta + (mv^2 - (c_{\alpha 2}\ell_2 - c'_{\alpha 1}\ell_1))\frac{\dot{\psi}}{v} = c'_{\alpha 1}\frac{\delta_s}{i_s}, \qquad (13.1)$$

$$J_z\ddot{\psi} + (c'_{\alpha 1}\ell_1^2 + c_{\alpha 2}\ell_2^2)\frac{\dot{\psi}}{v} - (c_{\alpha 2}\ell_2 - c'_{\alpha 1}\ell_1)\beta = c'_{\alpha 1}\ell_1\frac{\delta_s}{i_s}. \qquad (13.2)$$

In order to assess the stability of a steady-state solution, we substitute β and $\dot{\psi}$ in Equations (13.1) and (13.2) using the following $e^{\lambda_e t}$ approach:

$$\beta = \beta_{\text{stat}} + \hat{\beta}e^{\lambda_e t}, \qquad (13.3)$$

$$\dot{\psi} = \dot{\psi}_{\text{stat}} + \hat{\dot{\psi}}e^{\lambda_e t}. \qquad (13.4)$$

[1] The derivations of the formulas in this chapter closely follow the monograph of Mitschke and Wallentowitz 2004.

Vehicle Dynamics, First Edition. Martin Meywerk.
© 2015 John Wiley & Sons, Ltd. Published 2015 by John Wiley & Sons, Ltd.
Companion Website: www.wiley.com/go/meywerk/vehicle

For a steady-state solution, the steering angle $\delta_s = \delta_{s\ stat}$ is constant. The eigenvalues, λ_e, indicate whether the steady-state solution is stable or unstable: $\mathrm{Re}(\lambda_e) < 0$ means a stable solution, while $\mathrm{Re}(\lambda_e) > 0$ means an unstable solution. The case of $\mathrm{Re}(\lambda_e) = 0$ results in a so-called centre[2].

Substituting Equations (13.3) and (13.4) for β and $\dot{\psi}$ in (13.1) and (13.2), we obtain a system of two linear equations for the constants $\hat{\beta}$ and $\hat{\psi}$:

$$\underbrace{\begin{pmatrix} mv\lambda_e + (c'_{\alpha 1} + c_{\alpha 2}) & \frac{mv^2 - (c_{\alpha 2}\ell_2 - c'_{\alpha 1}\ell_1)}{v} \\ -(c_{\alpha 2}\ell_2 - c'_{\alpha 1}\ell_1) & J_z\lambda_e + \frac{c'_{\alpha 1}\ell_1^2 + c_{\alpha 2}\ell_2^2}{v} \end{pmatrix}}_{=S} \begin{pmatrix} \hat{\beta} \\ \hat{\psi} \end{pmatrix} = \begin{pmatrix} 0 \\ 0 \end{pmatrix} . \tag{13.5}$$

Based on the condition that the determinant of the coefficient matrix $\underline{\underline{S}}$ vanishes, we obtain the following equation for determining the eigenvalues λ_e:

$$0 = \det(\underline{\underline{S}})$$
$$= \lambda_e^2 + 2\sigma_f\lambda_e + \nu_f^2 = 0 , \tag{13.6}$$

where,

$$2\sigma_f = \frac{m(c'_{\alpha 1}\ell_1^2 + c_{\alpha 2}\ell_2^2) + J_z(c'_{\alpha 1} + c_{\alpha 2})}{J_z mv} , \tag{13.7}$$

$$\nu_f^2 = \frac{c'_{\alpha 1}c_{\alpha 2}\ell^2 + mv^2(c_{\alpha 2}\ell_2 - c'_{\alpha 1}\ell_1)}{J_z mv^2} . \tag{13.8}$$

The real parts of the solutions, λ_e, of the quadratic equation are smaller than zero when σ_f and ν_f^2 are greater than zero:

$$\mathrm{Re}(\lambda_{e\ i}) < 0 (i = 1, 2) \iff \sigma_f > 0 \text{ and } \nu_f^2 > 0 . \tag{13.9}$$

The eigenvalues are

$$\lambda_{e\ 1,2} = -\sigma_f \pm \sqrt{\sigma_f^2 - \nu_f^2} . \tag{13.10}$$

In the expression for σ_f, all constants are greater than zero; therefore $\sigma_f > 0$. The expression for ν_f^2 can be described using the square of the characteristic velocity, v_{ch}^2:

$$\nu_f^2 = \frac{c'_{\alpha 1}c_{\alpha 2}\ell^2}{J_z mv^2} \left(1 + \frac{v^2}{v_{\mathrm{ch}}^2}\right) . \tag{13.11}$$

---

[2] We do not wish to enter into details here; for more information on dynamic systems, we refer to Verhulst 2006. When $\mathrm{Re}(\lambda_e) = 0$, simple conclusions are not possible with the linearized equations.

The sign of the characteristic velocity

$$v_{\text{ch}}^2 = \frac{c_{\alpha 1}' c_{\alpha 2} \ell^2}{m(c_{\alpha 2} \ell_2 - c_{\alpha 1}' \ell_1)} \tag{13.12}$$

is crucial for understeer or oversteer:

$$\delta_s = \delta_{s0} \left(1 + \frac{v^2}{v_{\text{ch}}^2} \right) . \tag{13.13}$$

Here δ_s is the steering wheel angle and $\delta_{s0} = \frac{i_s \ell}{\rho_{cc}}$. If $v_{\text{ch}}^2 > 0$, then the vehicle exhibits understeering behaviour, which is then independent of the velocity:

$$v_f^2 > 0 \text{ for } v_{\text{ch}}^2 > 0 . \tag{13.14}$$

Understeering vehicles are therefore always stable. For oversteering vehicles, it holds that $v_{\text{ch}}^2 < 0$. The following applies ($v_{\text{crit}}^2 = -v_{\text{ch}}^2$):

$$v_f^2 > 0 \text{ for: } v_{\text{ch}}^2 < 0 \text{ and } v^2 < v_{\text{crit}}^2 , \tag{13.15}$$

$$v_f^2 < 0 \text{ for: } v_{\text{ch}}^2 < 0 \text{ and } v^2 > v_{\text{crit}}^2 , \tag{13.16}$$

This means: if the velocity of an oversteering vehicle is greater than the critical velocity, $v > v_{\text{crit}}$, the vehicle becomes unstable[3]. Figure 13.1 shows the steering angles for an understeering vehicle (line with positive slope) and an oversteering vehicle (line with negative slope). Generally, the vehicle behaviour is stable for positive steering angles (on the left-hand circle). However, when the oversteering vehicle exceeds the critical velocity, the behaviour becomes unstable. When exceeding the critical velocity, the steering angle changes signs; the driver then needs to countersteer.

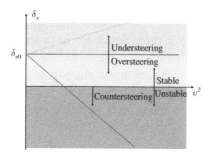

Figure 13.1   Stability of steady-state circular driving

---

[3] The critical velocity is defined for oversteering vehicles only, because in the case of understeering vehicles, $v_{\text{crit}}^2 = -v_{\text{ch}}^2 < 0$ means that the critical velocity, $v_{\text{crit}} = \sqrt{-v_{\text{ch}}^2}$, would be imaginary.

Apart from the pure consideration of stability, the decay behaviour also plays a role in the assessment of the dynamic behaviour as well as the frequency of the decaying oscillation. The decay behaviour is determined by the attenuation factor

$$D_f = \frac{\sigma_f}{\nu_f} \qquad (13.17)$$

and the dynamic behaviour is detected by the natural angular frequency of the damped system

$$\nu_{fd} = \nu_f \sqrt{1 - D_f^2} \ . \qquad (13.18)$$

However, the term natural angular frequency only makes sense for $D_f < 1$ because for $D_f > 1$ the decay behaviour is aperiodic and then ν_{fd} can no longer be referred to as frequency. For oversteering vehicles, the attenuation factor is greater than 1 above a certain speed, and the behaviour of the vehicle is then aperiodic. As long as $\nu_f^2 > 0$, there is a decay in perturbations of steady state. However, if $\nu_f^2 < 0$ (this holds for $v > v_{\mathrm{crit}}$), the real part of one eigenvalue $\lambda_{e\ i}$ is positive and perturbations increase exponentially: the driving behaviour is unstable. In understeering vehicles, the sign of $\nu_f^2 > 0$ does not change. The eigenvalues, $\lambda_{e\ i}$, will always have a negative real part. However, in the case of an understeering vehicle, perturbations may fade aperiodically.

13.2 Steering Behaviour

Section 13.1 investigates the stability of steady-state driving conditions. The steering angle was considered to be constant. In the following section, we consider the vehicle's response to a step steer input, i.e. a discontinuous change of the steering angle from the value of zero to a constant value, $\delta_{s\ \mathrm{stat}}$. This step function is also used in the testing of vehicles, (e.g. step steering angle according to ISO 7401: $v = 80$ km/h; $v^2/\rho_{\mathrm{cc}} = 4$ m/s^2; $\partial \delta_s/\partial t > 200^\circ$/s), where the step steering angle in the experiment can only be approximated by a ramp function. In order to define the step steering angle as a function of time only, the steering wheel angle velocity, $\partial \delta_s/\partial t > 200^\circ$/s, is required but not the steering wheel angle, δ_s, is prescribed. The angle δ_s must be chosen so as to achieve a lateral acceleration of $v^2/\rho_{\mathrm{cc}} = 4$ m/s^2. Since the velocity and the lateral acceleration are given, after step steering the vehicle has to drive on a circle with the radius

$$\rho_{\mathrm{cc}} = \left(\frac{80}{3,6}\right)^2 \left(\frac{\mathrm{m}}{\mathrm{s}}\right)^2 \frac{1}{4\ \mathrm{m/s}^2}$$

$$\approx 123.5\ \mathrm{m}\ . \qquad (13.19)$$

The Laplace transform is a helpful tool for investigating the response of linear systems of differential equations to non-harmonic excitations. The Laplace transform,

$\hat{f}(s)$, of a function $f(t)$ (with $f(t) = 0$ for $t < 0$) is obtained by (s is not the track but the complex argument of the Laplace transform):

$$\hat{f}(s) = \int_0^\infty e^{-st} f(t)\, \mathrm{d}t \ . \qquad (13.20)$$

The Laplace transform of the time derivative of f is

$$s\hat{f}(s) = \int_0^\infty e^{-st} \dot{f}(t)\, \mathrm{d}t \ . \qquad (13.21)$$

The reverse transformation is carried out using the following formula:

$$f(t) = \frac{1}{2\pi j} \oint \hat{f}(s) e^{st}\, \mathrm{d}s. \qquad (13.22)$$

The line integral, $\oint$, has to be chosen such that all poles of the function $\hat{f}(s)$ lie within the closed curve of integration. For simple functions, the reverse transformations can be specified:

$$\frac{1}{2\pi j} \oint \frac{1}{s+a} e^{st}\, \mathrm{d}s = e^{-at} \ , \qquad (13.23)$$

$$\frac{1}{2\pi j} \oint \frac{1}{(s+a)^k} e^{st}\, \mathrm{d}s = \frac{1}{(k-1)!} t^{k-1} e^{-at} \ , \qquad (13.24)$$

$$\frac{1}{2\pi j} \oint \frac{1}{s^k} e^{st}\, \mathrm{d}s = \frac{1}{(k-1)!} t^{k-1} \ \text{for } k \geq 2 \ . \qquad (13.25)$$

The Laplace transform of the step steering angle to the value of $\delta_{s\,\text{stat}}$ is

$$\hat{\delta}_s(s) = \frac{\delta_{s\,\text{stat}}}{s} \ . \qquad (13.26)$$

In the following, we start from the linear system of ordinary differential equations (13.1) and (13.2), transform both differential equations using the step function for the steering angle δ_s and obtain

$$(mvs + (c'_{\alpha 1} + c_{\alpha 2}))\hat{\beta}(s) + (mv^2 - (c_{\alpha 2}\ell_2 - c'_{\alpha 1}\ell_1))\frac{\hat{\dot{\psi}}(s)}{v} = c'_{\alpha 1}\frac{\delta_{s\,\text{stat}}}{i_s s} \ , \qquad (13.27)$$

$$-(c_{\alpha 2}\ell_2 - c'_{\alpha 1}\ell_1)\hat{\beta}(s) + (vJ_z s + (c'_{\alpha 1}\ell_1^2 + c_{\alpha 2}\ell_2^2))\frac{\hat{\dot{\psi}}(s)}{v} = c'_{\alpha 1}\ell_1\frac{\delta_{s\,\text{stat}}}{i_s s} \ . \qquad (13.28)$$

Here we have introduced the Laplace transform of $\dot{\psi}$ and not ψ. The two equations form an algebraic system of equations for $\hat{\beta}(s)$ and $\hat{\dot{\psi}}(s)$. In the following, we restrict ourselves to the solution for $\dot{\psi}$ referred to as size of the step steering angle $\delta_{s\,\text{stat}}$:

$$\frac{\hat{\dot{\psi}}(s)}{\delta_{s\,\text{stat}}} = \frac{1}{i_s\ell} \frac{v}{1 + v^2/v_{\text{ch}}^2} \frac{1 + T_{z1}s}{1 + \frac{2\sigma_f}{v_f^2}s + \frac{1}{v_f^2}s^2} \frac{1}{s} \ , \qquad (13.29)$$

where
$$T_{z1} = mv\ell_1/(c_{\alpha2}\ell) .$$
(13.30)

The rational function can be reduced to the following simple form using a partial fraction decomposition:
$$\frac{A}{s} + \frac{B}{s - s_1} + \frac{C}{s - s_2} .$$
(13.31)

The poles s_1 and s_2 are the zeros of the denominator polynomial $(s^2 + 2\sigma_f s + \nu_f^2)$:
$$s_{1,2} = -\sigma_f \pm \sqrt{\sigma_f^2 - \nu_f^2} .$$
(13.32)

Due to the simplified representation with the help of partial fraction decomposition, the inverse transformation can be easily performed with the help of the formula given above:
$$\frac{\dot{\psi}(t)}{\delta_{s\,\text{stat}}} = \frac{1}{i_s\ell} \frac{v}{1 + v^2/v_{\text{ch}}^2} \left(1 + \frac{s_1 + 2\sigma_f - T_{z1}\nu_f^2}{-s_1 + s_2} e^{s_1 t} \right.$$
$$\left. + \frac{-s_2 - 2\sigma_f + T_{z1}\nu_f^2}{-s_1 + s_2} e^{s_2 t} \right) .$$
(13.33)

It can be seen in the response function to the step steering that the same mathematical expressions occur as in Section 13.1. The poles s_1 and s_2 are equal to the eigenvalues of $\lambda_{e\,1}$ and $\lambda_{e\,2}$ of the stability study. For $t = 0$, we obtain the following values:
$$\dot{\psi}(0) = 0 ,$$
(13.34)
$$\frac{\ddot{\psi}(0)}{\delta_{s\,\text{stat}}} = \frac{c'_{\alpha1}\ell_1}{J_z i_s} .$$
(13.35)

We consider three cases ($\sigma_f > 0$ always applies):

1. Case: $\nu_f^2 > 0$ and $\sigma_f^2 - \nu_f^2 > 0$
This vehicle is stable according to the stability investigation, both the eigenvalues $\lambda_{e\,1}$ and $\lambda_{e\,2}$ of the stability study as well as the poles s_1 and s_2 are purely real and less than zero. In other words, the yaw angular velocity, $\dot{\psi}$, relative to the size of the step steering angle, $\delta_{s\,\text{stat}}$, asymptotically approaches the steady value
$$\lim_{t \to \infty} \frac{\dot{\psi}}{\delta_{s\,\text{stat}}} = \frac{1}{i_s\ell} \frac{v}{1 + v^2/v_{\text{ch}}^2} .$$
(13.36)

This behaviour can occur below the critical velocity in oversteering vehicles and in understeering vehicles.

2. Case: $\nu_f^2 > 0$ and $\sigma_f^2 - \nu_f^2 < 0$
This vehicle is also stable; it approaches the steady state with oscillation. This behaviour can additionally occur below the critical velocity in oversteering vehicles and in understeering vehicles.

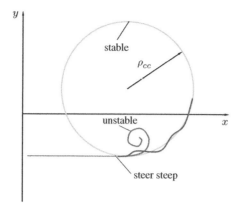

Figure 13.2   Path curves (schematic) with stable and unstable driving behaviour

3. Case: $\nu_f^2 < 0$

This has the effect that $\sigma_f^2 - \nu_f^2 > 0$ and s_1 becomes positive. The vehicle is unstable and it reacts to the step steering input with an exponentially increasing yaw rate.

Figure 13.2 shows an example. First, the vehicle runs straight. At $t = 0$, the steering wheel is turned abruptly. Vehicles that meet Cases 1 and 2 drive on a steady-state circular path. Vehicles which satisfy Case 3 exhibit instability: on a path with a decreasing radius of curvature, the yaw angular velocity increases exponentially.

We can see that stable handling and the response to a step steering input are closely linked.

13.3   Crosswind Behaviour

In addition to the aerodynamic drag forces in the longitudinal direction of the vehicle, air forces as a result of crosswinds in the lateral direction also play a role. Crosswind often does not have a sufficiently great impact on the dynamic behaviour to cause accidents, but the driver does have to countersteer in crosswinds, which means that the response of the vehicle to crosswinds is an aspect of comfort: the less the driver has to steer in crosswinds, the more comfortable the vehicle will be in this respect. The following determines the steering wheel angle that is necessary to compensate for crosswinds so that the vehicle travels in a straight line. The air forces in crosswinds affect the so-called centre of pressure, S_{pp}, which is shifted by ℓ_{pp} from the centre of mass, S_{cm}. In Figure 13.3, S_{pp} is shifted forward[4].

---

[4] In Figure 13.3, two situations are shown, in which countersteering is necessary: crosswind and split-μ braking. In the crosswind situation, a lateral force is acting on the vehicle, in the split-μ situation no lateral force is acting on the vehicle.

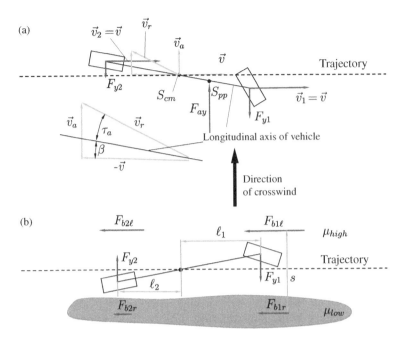

Figure 13.3 Countersteering: (a) Velocities in crosswinds on the single-track model; (b) split-μ braking (adapted from Mitschke and Wallentowitz 2004)

We consider steady-state straight-line travel. In this case

$$\ddot{\psi} = 0, \; \dot{\psi} = 0, \; \dot{\beta} = 0 \; . \tag{13.37}$$

As a result of crosswinds, it is possible for a steady-state vehicle sideslip angle to remain constant during straight-line travel. Assuming these steady-state values and by considering the crosswind term[5]

$$F_{ay} = c_y A \frac{\rho_a}{2} v_r^2 \tag{13.38}$$

in Equations (13.1) and (13.2), we then obtain

$$(c'_{\alpha1} + c_{\alpha2}) \beta = c'_{\alpha1} \frac{\delta_s}{i_s} + c_y A \frac{\rho_a}{2} v_r^2 \; , \tag{13.39}$$

$$-(c_{\alpha2} \ell_2 - c'_{\alpha1} \ell_1)\beta = c'_{\alpha1} \ell_1 \frac{\delta_s}{i_s} + c_y A \frac{\rho_a}{2} v_r^2 \ell_{pp} \; . \tag{13.40}$$

The air coefficient, c_y, depends on the angle of incidence, τ_a. For a small angle of incidence ($\tau_a < 20°$) the air coefficient, c_y, can be approximated with sufficient accuracy

---

[5] Here we are using the equations of motion for the single-track model (cf. Mitschke and Wallentowitz 2004); MacAdam 1989 adopts a simpler approach using an equilibrium of moments for this specific and easier consideration of steady-state linear motion.

by a linear function:

$$c_y \approx c_{y1} \tau_a \, , \tag{13.41}$$

such that:

$$F_{ay} \approx c_{y1} \tau_a A \frac{\rho_u}{2} v_r^2$$
$$\approx k_y v_r^2 \tau_a \, , \tag{13.42}$$

where $k_y = c_{y1} A\rho_a/2$ is a linearized air coefficient. For the special case considered here (wind direction perpendicular to the direction of movement of the centre of mass $S_{\rm cm}$) we obtain (assuming small angle τ_a and β) from

$$v_r \sin(\tau_a + \beta) = v_a \, , \tag{13.43}$$

$$v_r \cos(\tau_a + \beta) = v \tag{13.44}$$

the following equations (from Equation (13.44) it follows that $v_r = v$):

$$\tau_a = -\beta + \frac{v_a}{v} \, , \tag{13.45}$$

$$v_r^2 \tau_a = -v^2 \beta + v \, v_a \, . \tag{13.46}$$

Overall, the steady-state steering wheel angle relative to the wind speed, v_a, is

$$\frac{\delta_{s\ \text{stat}}}{v_a} = -\frac{i_s k_y v}{c_{\alpha 1}'} \frac{c_{\alpha 2}(\ell_2 + \ell_{\rm pp}) - c_{\alpha 1}'(\ell_1 - \ell_{\rm pp})}{c_{\alpha 2}\, \ell + k_y(\ell_1 - \ell_{\rm pp})\, v^2} \, . \tag{13.47}$$

The steady-state steering wheel angle is proportional to the wind speed, v_a, the vehicle speed, v, and the linearized air coefficient, k_y. It is generally smaller as tyre stiffness at the front axle and steering stiffness increase. The value $\delta_{s\ \text{stat}} = 0$ is also possible if the following holds:

$$c_{\alpha 2}(\ell_2 + \ell_{\rm pp}) - c_{\alpha 1}'(\ell_1 - \ell_{\rm pp}) = 0 \, . \tag{13.48}$$

This is equivalent to

$$\ell_{\rm pp} = \frac{c_{\alpha 1}'\, \ell_1 - c_{\alpha 2}\, \ell_2}{c_{\alpha 1}' + c_{\alpha 2}} \, . \tag{13.49}$$

For an understeering vehicle it is

$$c_{\alpha 1}'\, \ell_1 - c_{\alpha 2}\, \ell_2 < 0. \tag{13.50}$$

This means that we do not have to countersteer in an understeering vehicle when the centre of pressure is located behind the centre of mass, so $\ell_{\rm pp} < 0$, and $\ell_{\rm pp}$ takes the value of (13.49). The pressure point is often in front of the centre of gravity, so countersteering is necessary. In front-wheel-drive vehicles, the centre of mass, $S_{\rm cm}$, is usually more in front of the vehicle, i.e. more in the neighbourhood of the centre of pressure, $S_{\rm pp}$, so such vehicles are often less susceptible to side winds than rear-wheel-drive vehicles.

13.4   Questions and Exercises

Remembering

1. How does the steering wheel angle behave in relation to the square of the speed for an understeering and an oversteering vehicle?
2. Which equations are needed to obtain information on the stability behaviour of steady-state cornering?
3. Which approaches are used to obtain information on the stability behaviour of steady-state circular travel?
4. What kind of driving behaviour can lead to instability?
5. Which parameters play a role in crosswind behaviour?

Understanding

1. How is stability predicted for steady-state cornering? Explain the procedure.
2. How can a vehicle respond to a step steering angle input?
3. How does a crosswind change the behaviour of an oversteering and an understeering vehicle?
4. Explain the trends in ℓ_{pp}, $c_{\alpha 1}$, $c_{\alpha 2}$, ℓ_1, ℓ_2 on $\frac{\delta_{s\ stat}}{v_a}$, e.g. increasing ℓ_{pp} results in increasing/decreasing $\frac{\delta_{s\ stat}}{v_a}$!

Applying

The following parameters are given: $c_{\alpha 1} = 50$ kN/rad, $c_{\alpha 2} = 60$ kN/rad, $\ell_1 = 2.1$ m, $\ell_2 = 2.2$ m, $\rho_{cc} = 100$ m, $i_s = 19$, $\tilde{k}_s = 10$ kNm/rad (for a rack and pinion steering system), $m = 1350$ kg, $J_z = 3000$ kg m^2, $v = 30$ m/s, $n_c = 0.04$ m.

1. Calculate v_f^2 and σ_f^2.
2. Calculate the natural angular frequency of the undamped system.
3. Estimate v_f for an understeering vehicle at very high velocities (look at the limiting behaviour of the equation).

14

Influence of Wheel Load Transfer

In the single-track model, it was assumed that the centre of mass is on the level of the roadway. As a result, no changes would occur in the wheel loads while cornering (and also when accelerating or braking). In this chapter, we turn to the cornering of a vehicle which has a centre of mass at a value of h_{cm} above the roadway. This two-track model allows us to draw the conclusions on wheel load changes and their impact on the dynamics. In Section 14.1, we consider the wheel load changes in a simplified manner due to the centripetal acceleration. In Section 14.2, we consider the wheel load changes in more detail.

14.1 Wheel Load Transfer without Considering Vehicle Roll

We assume a vehicle travelling with a steady-state motion at a velocity v on a circle of radius ρ_{cc}. The centripetal acceleration is v^2/ρ_{cc}. The resulting moment

$$M = m\frac{v^2}{\rho_{cc}}h_{cm} \tag{14.1}$$

due to the centrifugal force, mv^2/ρ_{cc}, results in transfer of the wheel loads from the inner to the outer wheels of the vehicle. These wheel load transfers are denoted by ΔF_{zji} and ΔF_{zjo} for the wheel load transfer on the front ($j = 1$) or on the rear ($j = 2$), inside (index 'i') of the curve and outside (index 'o') of the curve of the wheel, respectively.

Forces are shown in Figure 14.1 together with the side forces F_{y1i}, F_{y1o}, F_{y2i} and F_{y2o} and the centrifugal force.

The track of the front axle, s_1, is equal to that, s_2, of the rear axle. The lateral forces on the individual wheels are dependent on the slip angles and the wheel loads. If we assume that the radius of curvature of the trajectory, ρ_{cc}, is large compared to the track width, s_1 and s_2, the slip angles of the respective inner and outer wheels are equal, too[1]:

$$\alpha_{1i} = \alpha_{1o} \text{ and } \alpha_{2i} = \alpha_{2o} . \tag{14.2}$$

[1] Different kinematics in steering geometry and suspension geometry are neglected for this simplified consideration.

Vehicle Dynamics, First Edition. Martin Meywerk.
© 2015 John Wiley & Sons, Ltd. Published 2015 by John Wiley & Sons, Ltd.
Companion Website: www.wiley.com/go/meywerk/vehicle

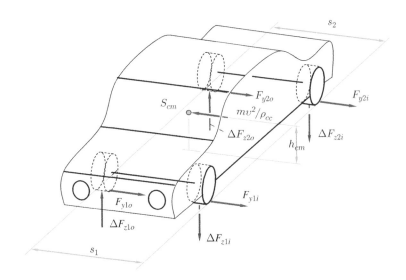

Figure 14.1 Wheel load distribution when cornering

This means that the partitioning of the cornering forces on the inner curve and outer curve wheels is not due to different slip angles but to different wheel loads. We illustrate the relationships using the diagram in Figure 14.2. This shows the curves for the lateral force F_y as a function of the slip angle for three different wheel loads, $F_{z0} + \Delta F_z$, F_{z0} and $F_{z0} - \Delta F_z$. Now it is shown that the mean slip angle must increase as a result of changes in the wheel loads. We assume that the distribution of the total load on the front and rear axles is not changed and, in the following, consider one axle, for example, the rear axle with the wheel loads F_{z0} on both wheels without considering load transfer from inner to outer side. Due to the cornering (and centrifugal force), the wheel load on the inside wheel is reduced by ΔF_z. Since the total load on the rear axle remains constant, the wheel load increases on the outer wheel when cornering by the same magnitude, ΔF_z. The lateral forces on the wheels which would arise if these wheel load changes were neglected are shown in Figure 14.2 by the arrows at $\alpha = 4°$ (for the middle curve F_{z0}). The sum of these arrows at $\alpha = 4°$ gives the side force, $F_{y \text{ tot}}$, which is necessary to drive the vehicle through the turn (without load transfer). First, the lateral forces are determined that would be obtained if we assume that the slip angles for the outer and inner wheels remain the same and only the wheel load changes. The side forces arising are thus shown in the highlighted rectangle on the left. It can be seen that the sum of the lateral forces $\tilde{F}_{y \text{ tot}}$ on the inside wheel (left, dashed small arrow) and the side force on the outer wheel (left, dashed-dotted large arrow) is below the total lateral force (sum of the arrows at $\alpha = 4°$ from the vehicle without load transfer). This means that the lateral force is not sufficient to compensate for the corresponding part of the centrifugal force. The slip angles of both wheels must therefore increase by $\Delta \alpha$. The resulting forces can be seen highlighted on the right-hand rectangle.

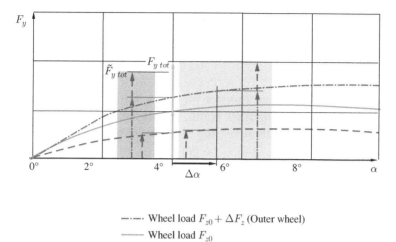

---- · Wheel load $F_{z0} + \Delta F_z$ (Outer wheel)

——— Wheel load F_{z0}

----- Wheel load $F_{z0} - \Delta F_z$ (Inner wheel)

Figure 14.2   Increase of the slip angle due to changes in wheel load distributions

When the varying wheel loads between the inner and outer wheels are taken into account, the slip angle increases: as the slip angle increases, both lateral forces move closer to the maximum and the maximum transferable lateral forces decrease with changes in the wheel load distribution. We should therefore pay attention when designing a vehicle so that the wheel load changes are small when cornering in order to make the maximum use of the adhesion area. As the whole lateral force $F_{y\,\text{tot}}$ ($=F_{y2i} + F_{y2o}$ for the rear axle) does not change, the increase in the mean slip angle $\overline{\alpha}_2 = (\alpha_{2i} + \alpha_{2o})/2$ results in a reduction of the mean cornering stiffness $\overline{c}_{\alpha2}$:

$$\underbrace{F_{y\,\text{tot}}}_{\text{constant}} = \underbrace{\overline{c}_{\alpha2}}_{\text{decreases}} \overbrace{\overline{\alpha}_2}^{\text{increases}} . \tag{14.3}$$

In the two-track model, further consequences arises from the wheel load changes, especially if we target the influence of the wheel load changes on the front and rear axles. Here the load transfer has an influence on oversteering and understeering behaviour.

To obtain an initial overview of which parameters roughly influence the load transfer, we consider a simplified case of changing wheel loads in the following. We assume that the amount of wheel load transfer on all wheels is the same ΔF_z, and that the front and rear tracks are equal to $s = s_1 = s_2$. The equilibrium (sum of the moments about an axis lying in the roadway in the middle between the wheels vanishes) results in

$$0 = \frac{mv^2}{\rho_{\text{cc}}}h_{\text{cm}} - 2\Delta F_z s . \tag{14.4}$$

The wheel load transfer when cornering under these simplified assumptions is

$$\Delta F_z = \frac{h_{cm}}{2s} \frac{mv^2}{\rho_{cc}} \ . \tag{14.5}$$

Consequently, the wheel load transfer increases with increasing centre of mass height, h_{cm}, and with decreasing track width, s. This means that the maximum transferable side force decreases with increasing centre of mass height and with decreasing track width. The ratio of centre of mass height to track width is thus a design criterion for achieving good utilization of the adhesion.

Using Equation (14.5), it is possible to evaluate an estimation for roll over. The limit case for roll over applies if

$$\Delta F = \frac{mg}{4} \ . \tag{14.6}$$

In this limit case, the wheel loads on the inner wheels become zero.

Substituting this limit case for ΔF in (14.5) yields an inequality which describes a simple roll-over condition:

$$g\frac{s}{2h_{cm}} \le \frac{v^2}{\rho_{cc}} \ . \tag{14.7}$$

This equation describes the values of v and ρ_{cc} for which a vehicle with given geometrical parameters s and h_{cm} will tilt. In a simple approach[2] lateral acceleration v^2/ρ_{cc} cannot exceed $\mu_a g$ (here μ_a is the adhesion limit). If we extend (14.7) with $v^2/\rho_{cc} \le \mu_a g$, the gravitational acceleration, g, cancels out and we obtain

$$\frac{s}{2h_{cm}} \le \mu_a \ . \tag{14.8}$$

If (14.8) holds, the vehicle will at least probably tilt at the highest possible cornering velocity. Conversely, if

$$\frac{s}{2h_{cm}} > \mu_a \ . \tag{14.9}$$

holds, then the vehicle will probably not tilt. Tilting of the vehicle is especially important for sport utility vehicles (SUV) as the height of their centre of mass is relatively large. The characteristic value $s/(2h_{cm})$ is called the static stability factor (SSF). It should be emphasized that this SSF consideration is only a rough method to estimate the danger of roll-over of a vehicle. In more complex models or in reality, a greater number of influencing factors have to be considered.

---

[2] If we assume that the total vertical load of the vehicle is mg and that the maximum lateral force is limited by the adhesion limit value, μ_a, then the centrifugal force, mv^2/ρ_{cc}, cannot exceed this limit $mg\mu_a$, which results in $v^2/\rho_{cc} \le \mu_a g$.

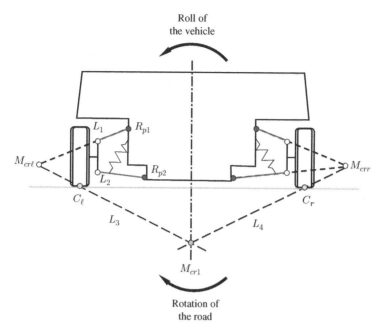

Roll of
the vehicle

$M_{cr\ell}$

L_1 R_{p1}

R_{p2}

L_2

M_{crr}

C_ℓ

C_r

L_3

L_4

M_{crl}

Rotation of
the road

Figure 14.3   Derivation of the instantaneous centre of rotation (adapted from Mitschke and Wallentowitz 2004)

14.2   Wheel Load Transfer Considering Vehicle Roll

In this section, we determine the wheel load changes resulting from the consideration of roll motion of the body. To describe the roll, we will first introduce the concept of the instantaneous axis of rotation. This is the axis about which the body of the vehicle rotates instantaneously. We first consider a cross-section (perpendicular to the x-direction) of the body through an axle, for example the front axle. We assume that the wheels are suspended by double wishbone suspensions (Figure 14.3). The position of the instantaneous axis for other wheel suspensions can be derived analogously.

To calculate the instantaneous centre of rotation (roll axis) of the body for one cross section, for example at the front axle (we call this the instantaneous centre of rotation M_{crl}), we use a modified observation layout which is designed for easier viewing or deliberation. We assume that we are fixed to the body and do not consider its rotation; consequently, the roadway has to rotate. The instantaneous centre of rotation of the road is then M_{crl}, and this is the same instantaneous centre of rotation M_{crl} of the vehicle body. The roadway rotates (horizontal line in Figure 14.3), so the wheels move up and down. Since the body is fixed, the wheels must rotate around the points of rotation R_{p1} and R_{p2}. The velocities of each point of the control arms and their extensions, L_i, $(i = 1, 2)$ are perpendicular to the corresponding control arm. The intersections $M_{cr\ell}$ and M_{crr} of the extensions therefore cannot move, and so the velocity must be zero (because the velocity vectors are perpendicular to the control arms and because the control arms are not collinear). This means that $M_{cr\ell}$ and M_{crr} yield the instantaneous centres of rotation of the wheels. The velocity of the contact point C_ℓ of the left wheel and the road can now be easily constructed by joining $M_{cr\ell}$ and C_ℓ with a line L_3: the velocity vector is perpendicular to this line L_3. Since the wheel, C_ℓ, sticks to the road, we know the speed of the road at point C_ℓ, which is also perpendicular to L_3. Similarly, we obtain the velocity direction of the road at the point C_r. On the one hand, the straight line L_3 goes through the points $M_{cr\ell}$ and C_ℓ and, on the other hand, the line L_4 goes through M_{crr} and C_r. The intersection of L_3 and L_4 is the instantaneous centre M_{cr1} (for the front axle).

We obtain the instantaneous centres M_{cr2} and M_{cr1} for the rear axle and the front axle. The axis that connects the instantaneous centres, M_{cr1} and M_{cr2}, is called the instantaneous roll axis (cf. Figure 14.4). This axis is not fixed but moves depending on the roll angle because of many non-linearities.

In the derivation of the wheel load transfer, it must be noted that the vehicle is a statically indeterminate system. We proceed as usual (for such systems) and first divide the whole system into subsystems. We then consider the deformations or deflections of the subsystems which are in turn separated as functions of the external loads and section forces. Subsequently, the actual deflections of the subsystems are determined. These deflections must be geometrically compatible, which means equal. Hence, we obtain the desired deflections from the resulting set of equations. Here we derive the wheel load changes for a vehicle with rigid axles. Figure 14.5 shows the entire vehicle together with the instantaneous roll axis. The instantaneous roll axis is fixed to the body. Figure 14.6 shows the subsystems as free-body diagrams in different views. We first turn to the body. Figure 14.6(a) indicates the points of the instantaneous roll axis, M_{cr1} and M_{cr2}, and the body rotates by the angle κ due to the centrifugal force, $m_b v^2/\rho_{cc}$. The structure is cut free from the rigid axles; the section moments M_1 and M_2 from the front axle and rear axle, respectively, are summarized in the moment $M = M_1 + M_2$. Here, we assume a small inclination angle, γ, of the instantaneous roll axis and therefore neglect trigonometric correction terms ($\cos\gamma \approx 1$).

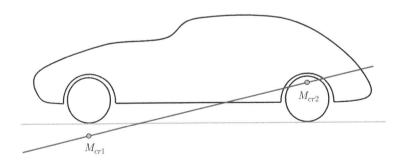

Figure 14.4   Derivation of the instantaneous axis of rotation

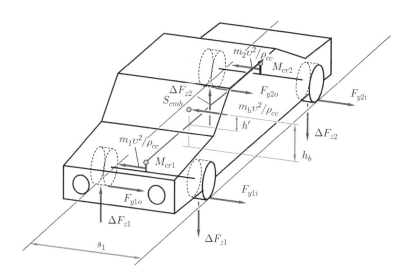

Figure 14.5 Total system with instantaneous roll axis

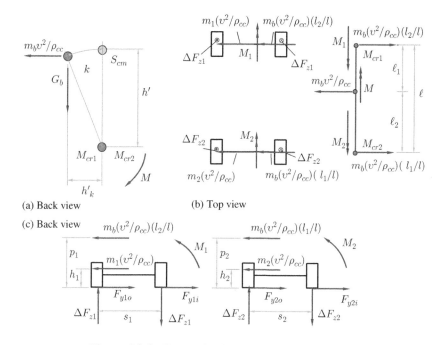

(a) Back view

(b) Top view

Figure 14.6 Determination of wheel load changes

We obtained this moment M from the free-body diagram when making the sum of the moments about the instantaneous axis

$$M = \frac{m_b v^2}{\rho_{cc}} h' + G_b h' \kappa . \tag{14.10}$$

Here, we have assumed that κ is small and therefore use $\cos \kappa \approx 1$ and $\sin \kappa \approx \kappa$. The section forces between the instantaneous roll axis and the rigid axles are not shown for the sake of clarity. However, this plays no role in the formation of the sum of the moments with respect to the instantaneous roll axis.

Figure 14.6(b) shows a line between M_{cr1} and M_{cr2} to represent the instantaneous roll axis of the structure. The instantaneous axis is cut free from the rigid axles (section moments M_1 and M_2). The sum of the moments in the longitudinal direction is as follows:

$$M = M_1 + M_2 . \tag{14.11}$$

The section moments M_1 and M_2 resulting from the roll stiffness C_1 and C_2 at the front or rear axle and the roll angle κ are given by

$$M_j = C_j \kappa , \ j = 1, 2 . \tag{14.12}$$

From Equations (14.10) and (14.12)

$$C_1 \kappa + C_2 \kappa = \frac{m_b v^2}{\rho_{cc}} h' + G_b h' \kappa . \tag{14.13}$$

we obtain the angle κ

$$\kappa = \frac{G_b h'}{C_1 + C_2 - G_b h'} \frac{v^2}{g \rho_{cc}} . \tag{14.14}$$

With the help of the angle κ, we can immediately specify the moments M_1 and M_2. The section forces $m_b (v^2/\rho_{cc})(\ell_2/\ell)$ and $m_b (v^2/\rho_{cc})(\ell_1/\ell)$ in the respective points M_{cr1} and M_{cr2} are obtained using the free-body diagram in Figure 14.6(b) from the equilibrium of forces in the lateral direction, and from the moment equilibrium about the vertical axis. (A negligible tilt of instantaneous roll axis is assumed here.)

The sum of the moments with respect to the wheel contact points (left or right) with the help of the free-body diagrams (Figure 14.6(c)) gives the wheel load changes for the two rigid axles. The static portion of the wheel loads are not shown in these free-body diagrams:

$$\Delta F_{z1} = \frac{m_b v^2}{\rho_{cc}} \left(\frac{\ell_2}{\ell} \frac{p_1}{s_1} + \frac{C_1}{C_1 + C_2 - G_b h'} \frac{h'}{s_1} + \frac{G_1}{G_b} \frac{h_1}{s_1} \right) , \tag{14.15}$$

$$\Delta F_{z2} = \frac{m_b v^2}{\rho_{cc}} \left(\frac{\ell_1}{\ell} \frac{p_2}{s_2} + \frac{C_2}{C_1 + C_2 - G_b h'} \frac{h'}{s_2} + \frac{G_2}{G_b} \frac{h_2}{s_2} \right) . \tag{14.16}$$

m_b :   mass of the body
ℓ_1, ℓ_2 :   distance: front/rear axle – centre of mass
ℓ :   distance front axle – rear axle
C_1, C_2 :   rolling stiffness front/rear axle
G_b :   weight of the body $G_b = m_b g$
h' :   distance: centre of mass – instantaneous roll axis
G_1, G_2 :   weight front/rear axle
s_1, s_2 :   track front/rear
p_1, p_2 :   distance: instantaneous roll axis – road; front/rear
h_1, h_2 :   Distance: centre of mass of front/rear axle – road

The wheel load transfer depends on partly influenced constructive variables, such as

- the position of the centre of mass ℓ_1/ℓ, ℓ_2/ℓ, h_1, h_2,
- the heights of the instantaneous centres of rotation divided by the tracks: p_1/s_1, p_2/s_2,
- the distance of the centre of mass to the instantaneous roll axis divided by the tracks h'/s_1, h'/s_2,
- the ratio of the roll spring constants $C_1/(C_1 + C_2 - G_b h')$, $C_2/(C_1 + C_2 - G_b h')$.

Section 14.1 explained that the wheel load transfer also results in an increase in the slip angles. Appropriate selection of the roll stiffness for the front and rear axles makes it possible to deliberately influence the mean slip angles and thereby the slip angle difference between front and rear axle (or the mean cornering stiffness). By influencing the wheel load transfer, we can alter the behaviour of the vehicle, for example from an oversteering to a neutral or an understeering behaviour. The roll spring stiffness, C_i, depends on the spring stiffness, k_i ($i = 1$, $i = 2$), and the geometric relationships of the suspension. As a rough approximation, we obtain the following for the roll stiffness, C_i, from suspension springs, k_i:

$$C_i = 2\left(\frac{s_i}{2}\right)^2 k_i \ . \tag{14.17}$$

However, the suspension spring stiffness, k_i, must be chosen so that, for example, comfort and safety standards are met. Consequently, these values must not be so large that they limit the body accelerations (and comfort deteriorates). One way to increase roll stiffness, C_i, and not to change spring constants, k_i, is to use a so-called anti-roll bar, Figure 14.7.

The anti-roll bar is a rod (highlighted in Figure 14.7) that is fixed at the control arms to the wheel bearing and to the body. The stiffness of the anti-roll bar is determined by its bending and torsional stiffness and the geometric parameters a_s, b_s and α_s. If $\alpha_s = 0°$, then the rod is subjected to torsional deformation , for $\alpha_s = 90°$ to bending. In the case of pure torsion $\alpha_s = 0°$, the lengths of the lever arms a_s and the torsional stiffness of the rod determine the overall stiffness of the anti-roll bar significantly,

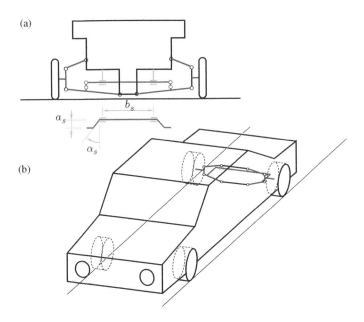

Figure 14.7 Anti-roll bar

while the bending stress of the anti-roll bar arms plays only a minor role. The greater the angle α_s, the larger the bending portion will be.

The anti-roll bar has no effect if both suspension springs are compressed in the same manner, because no torsional moments are transmitted in the connection between anti-roll bar and body of subframe. The anti-roll bar achieves its greatest effect when the structure rolls. An obstacle at one side of the vehicle and a pothole on the other side has the same effect as the rolling of the structure, stimulating the left and right wheels with a phase-shift of $180°$; the anti-roll bar also starts to act against this excitation. The comfort of the vehicle drops in these situations. An anti-roll bar is often installed at the front axle, because this increases the roll stiffness, C_1. This in turn results in an increase in the wheel load changes, ΔF_{z1}, resulting in an increase in the mean slip angle, $\bar{\alpha}_1$, at the front axle. This leads to a reduction in the mean cornering stiffness, $\bar{c}_{\alpha1}$. If we consider the steering wheel angle, δ_s, in Equation (14.18), which depends on the centripetal acceleration and the coefficient and therefore determines the oversteer and understeer behaviour:

$$\delta_s = \frac{i_s\ell}{\rho_{cc}} + mi_s\frac{c_{\alpha2}\ell_2 - c'_{\alpha1}\ell_1}{c'_{\alpha1}c_{\alpha2}\ell}\frac{v^2}{\rho_{cc}} \; , \tag{14.18}$$

it can be seen, with respect to

$$c'_{\alpha1} = \frac{c_{\alpha1}}{1 + \frac{c_{\alpha1}n_c}{k_s}} \; , \tag{14.19}$$

that a reduction in $\bar{c}_{\alpha1}$ changes the behaviour of the vehicle towards understeering.

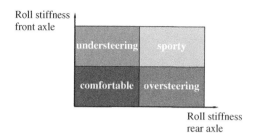

Figure 14.8   Influence of the stiffness of anti-roll bars at front and rear axles

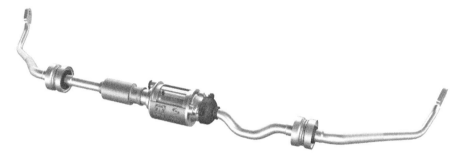

Figure 14.9   Active roll stabilization (ARS): active anti-roll bar (reproduced with permissions of ZF Friedrichshafen AG)

An anti-roll bar at the front axle thus amplifies any understeer driving behaviour (or it reduces oversteer driving behaviour), whereas an anti-roll bar at the rear axle does the opposite. In addition to a deterioration in ride comfort, a further disadvantage of the anti-roll bar occurs: an increase in the mean slip angle of the axle on which the anti-roll bar is used means that the wheels of this axle are closer to the limit of adhesion.

Active anti-roll bars can be used to actively affect the dynamic behaviour of a vehicle (cf. Figure 16.8 for a front axle with an active anti-roll bar). This is a conventional anti-roll bar that is split into two parts; these two parts are connected by an hydraulic rotary actuator. It is possible to adjust the dynamic behaviour of the car according to the lateral acceleration in order to prevent roll movement (up to a certain lateral acceleration) and thus improve comfort when driving straight ahead and improve the behaviour during off-road driving by deactivating the coupling between the two parts of the anti-roll bar. Figure 14.8 shows the main influence of the anti-roll bar stiffness. If both stiffness values are small, the vehicle is more comfortable. If they are stiff, the vehicle has a sportier performance. Increasing the stiffness at the front axle raises the tendency of the vehicle to understeer, and increasing it at the rear axle raises the tendency to oversteer.

Figure 14.9 shows an example of an active anti-roll bar. The active element can be seen in the middle.

14.3   Questions and Exercises

Remembering

1. What causes the changes in wheel load?
2. What do the wheel load changes depend on when cornering if we do not take the roll into account?

Understanding

1. What are the consequences of changes in wheel loads when cornering, with respect to the slip angle and with respect to the cornering stiffness of a single-track model?
2. What do the wheel load changes depend on when cornering if we take the roll into account?
3. Why is an anti-roll bar often used on the front axle? Explain the relation.
4. Using the roll stiffness, explain how the vehicle behaviour can be influenced in terms of oversteer and understeer.

Applying

1. What happens with the vehicle, if the anti-roll bar stiffness at the front axle is increased during cornering: Will the vehicle turn to the inside or the outside of the curve?
2. What happens with the vehicle, if the anti-roll bar stiffness at the rear axle is increased during cornering: Will the vehicle turn to the inside or the outside of the curve?
3. How do the parameters p_1 and p_2 influence oversteering or understeering behaviour?

Analysing

1. How does the anti-roll bar stiffness at the front and rear axles influence the vehicle sideslip angle for steady-state driving of the single-track model at constant speed on a circle?
2. How does the anti-roll bar stiffness at the front and rear axles influence the vehicle sideslip angle gradient for steady-state driving of the single-track model at constant speed on a circle?

15

Toe-in/Toe-out, Camber and Self-steering Coefficient

In this chapter, Section 15.1 examines the influence of toe-in/toe-out and camber.

15.1 Toe-in/Toe-out, Camber

In this section, we concentrate on the toe-in/toe-out and camber of the wheels and their influence on the lateral forces. Both variables are expressed in terms of an angle and give the position of the wheels (see Figure 15.1).

Toe: The toe angle describes a static rotation of the wheel about the $\vec{e}_{wz}$-axis. We refer to toe-in when the wheels are turned inwards (cf. Figure 15.1(a)), and toe-out, when the wheels are turned outwards (Figure 15.1(b)). The angle δ_{10} is positive for toe-in and negative for toe-out.

Camber: The camber is the angle between the wheel $\vec{e}_{wx}$–$\vec{e}_{wz}$ plane and the vertical $\vec{e}_{iz}$-axis. The constructive camber angle, γ, is positive when the wheel is inclined towards the outside of the vehicle and negative if it is inclined to its inside.

Both toe-in/toe-out and camber affect the driving behaviour. We consider the cornering of a vehicle with and without the toe-in angle, δ_{10}, on the front axle. As shown in Figure 15.2, we start with the case of no changes in wheel loads when cornering. The total lateral force arises from the sum of the two forces (left arrows in Figure 15.2 at $\alpha = 4°$) when the slip angle is $\alpha_0 = 4°$. If we consider the changes in wheel load, the slip angle increases by a value of $\Delta\alpha$, so that the whole cornering force (sum of the arrows at $\alpha = 4°$) is achieved as the sum of the lateral forces (large solid arrow

Vehicle Dynamics, First Edition. Martin Meywerk.
© 2015 John Wiley & Sons, Ltd. Published 2015 by John Wiley & Sons, Ltd.
Companion Website: www.wiley.com/go/meywerk/vehicle

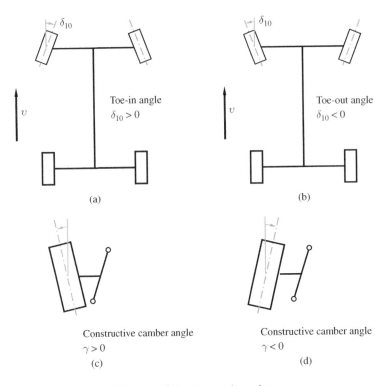

Figure 15.1   Toe and camber

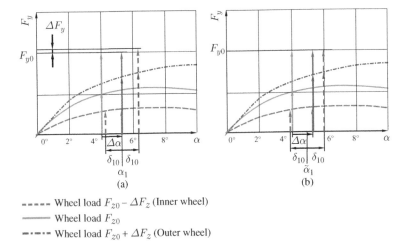

- - - - - Wheel load $F_{z0} - \Delta F_z$ (Inner wheel)
———— Wheel load F_{z0}
- - - - - Wheel load $F_{z0} + \Delta F_z$ (Outer wheel)

Figure 15.2   Influence of toe-in on the slip angle

and small solid arrow at $\alpha_1 \approx 5.2°$). In the following case, we consider a vehicle in which the front wheels have a toe-in; hence, the front wheels are inclined inwards by an angle of δ_{10} (see Figure 15.1(a)).

The toe-in angle, δ_{10}, alters the slip angle. The slip angle of the inside wheel decreases, and the slip angle of the outside wheel increases. By decreasing and increasing the slip angle, α_1, as shown in Figure 15.1(a) to δ_{10}, we obtain the lateral forces with a sum greater than F_{y0} by the magnitude ΔF_y as a result. Consequently, in order to achieve the sum of F_{y0}, the mean slip angle can be reduced to the value of $\tilde{\alpha}_1$ (see Figure 15.2(b)). The actual slip angles $\tilde{\alpha}_1 - \delta_{10}$ and $\tilde{\alpha}_1 + \delta_{10}$ then result in lateral forces that yield just F_{y0} when added up. The following applies to the angle $\tilde{\alpha}_1$

$$\tilde{\alpha}_1 + \delta_{10} > \alpha_1 \; . \tag{15.1}$$

The value of $\tilde{\alpha}_1$ must be determined iteratively from the curves; it is not possible to specify this angle directly using the curves because of non-linearities. By reducing the mean slip angle to the value of $\tilde{\alpha}_1$, we obtain an increased distance to the maximum transferable lateral force on the inside wheel and thus an increased cornering force reserve. On the outside of the curve of the wheel, the reserve is reduced, but since this reserve is already larger, we obtain a greater distance overall (with respect to the slip angle) from the maximum possible cornering force. The toe-in angle increases the maximum possible centripetal acceleration v^2/ρ_{cc} of a vehicle and thus, for a given curve radius ρ_{cc}, the toe-in angle increases the maximum allowable velocities.

In the following, we consider the influence of the camber angle on the lateral forces. In order to describe this dependency by a formula, we have to deviate from the adopted sign convention of the constructive camber angle. The camber angle of a wheel is positive when the wheel is inclined to the outer side and negative when the wheel is inclined towards the inner side of the curve. A negative camber increases the maximum transferable lateral force (constant slip angle assumed). At lower values of slip angle and smaller camber angle, the lateral force can be approximated by a linear relationship:

$$F_y = c_\alpha \alpha - c_\gamma \gamma \; . \tag{15.2}$$

Both the lateral force slip angle coefficient, c_α, and the lateral force camber coefficient, c_γ, increase with the wheel load. Figure 15.3 shows the lateral forces, F_y, as a function of the slip angle, α. The solid, middle curve shows the lateral force at the inner and outer wheels, again without camber and without the wheel load changes. For this case, the lateral forces on the two wheels are the same; the total lateral force is the sum of the individual forces of the same magnitude (solid arrows in the centre).

We compare this case with a vehicle with negative constructive camber angle. This means that the wheel camber angle, γ, in Equation (15.2) on the outer wheel is less than zero and on the inner wheel is greater than zero. Without considering the wheel load changes, these camber angles yield the dashed, lower and the dotted, upper curve for the determination of the lateral forces. Through the wheel load transfer, we obtain the lateral forces from the dashed, bottom curve and the dashed–dotted top curve.

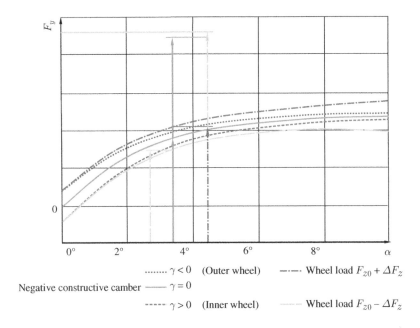

Figure 15.3   Influence of camber on the slip angle

The sum of the resulting lateral forces results in a higher lateral force. Consequently, the slip angle decreases. A constructive negative camber will therefore reduce the slip angle or increase the cornering forces at the same slip angle. If the absolute value of the negative constructive camber increases (at the same cornering force), the mean slip angle decreases and hence the mean cornering stiffness increases.

The quotient,

$$\frac{c_{\alpha 2}\ell_2 - c_{\alpha 1}\ell_1}{c_{\alpha 1}c_{\alpha 2}\ell} , \tag{15.3}$$

which determines the sign in the formula for the steering angle of the front wheels given by

$$\delta_1 = \frac{\ell}{\rho_{cc}} + \frac{c_{\alpha 2}\ell_2 - c_{\alpha 1}\ell_1}{c_{\alpha 1}c_{\alpha 2}\ell}m\frac{v^2}{\rho_{cc}} , \tag{15.4}$$

is crucial for an understeer or oversteer behaviour (or similarly the equation for δ_s). In the case of an increase in the absolute value of a negative constructive camber angle at the rear axle, the rising mean cornering stiffness, $\bar{c}_{\alpha 2}$, results in more understeering vehicle behaviour. A positive constructive front camber results in an increased slip angle and thus in a decreasing mean front cornering stiffness, $\bar{c}_{\alpha 1}$, which also affects the behaviour of the vehicle in the direction of understeering behaviour.

15.2 Questions and Exercises

Remembering

1. How do we define toe-in/toe-out?
2. How do we define camber?

Understanding

1. Explain the influence of toe-in on cornering.
2. Explain the influence of the camber angle on cornering.
3. Explain how different vehicle parameters influence the vehicle behaviour, and specifically the self-steering behaviour.

16

Suspension Systems

In this chapter, we discuss suspension systems. Examples include independent suspensions and solid-axle suspensions. Examples of independent suspensions include general multi-link, McPherson or double wishbone (or upper and lower A-arm) suspensions. The six degrees of freedom of the wheel carrier (these are the general six degrees of freedom of a rigid body; the wheel carrier includes the hub and the wheel bearing) will be locked by the suspension except for one, i.e. the vertical translational degree of freedom. Independent wheel suspensions can achieve this by an appropriate arrangement of five arms (or links). A link is a rod which is pivot-mounted on the wheel carrier on one side and on the body (or subframe) on the other side, and so one degree of freedom is locked (see Figure 16.1). Two arms can be combined into one suspension arm or A-arm. With a suitable arrangement of five arms, five degrees of freedom are disabled and an independent wheel suspension is obtained in this way. Figure 16.1 shows the principles of the various types of suspension. The actual wheel carrier is shown in darkgrey. If the wheel is steerable, then one of the arms is a tie rod. An example of a double wishbone suspension is shown in Figure 16.2.

The end points of the arms move on spherical surfaces. This has the result that the movement of the wheel carrier can be very complicated. In general, it will not act in a motion of a pure translation in the vertical direction, but will instead exhibit a spatial motion that also contains rotatory portions.

Although the arms are elastic, their deformation is small because they are loaded by axial forces. To better isolate the structure from the shocks and bumps resulting from the uneven road surface which would be forwarded by the wheel suspension, the arms are often not rigidly mounted on the structure, but are mounted with the aid of elastic rubber bushings. These rubber bushings deform to keep the shocks from the body and hence also from the vehicle interior. The deformation shift of the wheel carrier is no longer on the kinematically predetermined path but, instead, on a certain path

Vehicle Dynamics, First Edition. Martin Meywerk.
© 2015 John Wiley & Sons, Ltd. Published 2015 by John Wiley & Sons, Ltd.
Companion Website: www.wiley.com/go/meywerk/vehicle

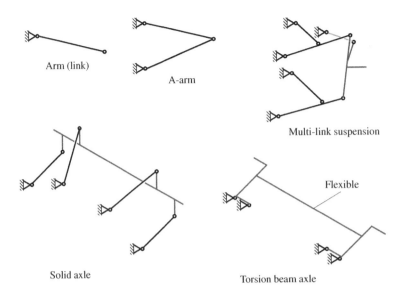

Figure 16.1   Principles of wheel suspension systems

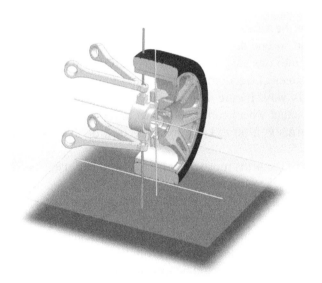

Figure 16.2   Double wishbone (or A-arm) suspension

influenced by the forces. The kinematics that arise from both the purely geometric kinematics and the movement from elastic deformation as a result of forces are called elastokinematics[1].

---

[1] As the kinematic behaviour of the wheel carrier due to the simultaneously acting forces and moments is an important characteristic which has a significant influence on the dynamic behaviour of a vehicle, this behaviour is tested in special kinematics and compliance (K&C) test facilities.

The kinematics cause changes in important geometrical parameters such as toe and camber when the carrier moves vertically. Figure 16.3 shows examples of the changes[2] as a function of the spring deflection: Figure 16.3(a) and (b) shows the toe-in/toe-out changes, $\Delta\delta$, on the front and rear wheels, while Figure 16.3(c) and (d) shows the changes, γ, in the camber. The nominal toe-in/toe-out for vanishing bounce is not included in Figure 16.3(a) and (b), but the variation is included. Due to the changes, it is possible to significantly influence the sizes of the toe and camber for driving dynamics. With the use of rubber bushing, however, no curves arise in the diagrams, but areas (grey in Figure 16.3) in which the variables $\Delta\delta$ and γ lie as functions of the bounce, z, and the forces and moment. In the diagram, the order of magnitude of the deflection, Δz, is at ± 100 mm, for the toe change, $\Delta\delta$, at $\pm 40'$ and for the camber change, γ, at $\pm 2'$.

The change of the toe as represented by the curves in Figure 16.3(a) and (b) can influence the driving behaviour during cornering as follows: the suspension of the outer wheel at the rear axle compresses (or rebounds) $\Delta z > 0$, which increases the toe-in angle of the outer wheel at the rear axle and thus the slip angle, α; the situation is reversed for toe-in at the inner wheel of the rear axle, i.e the slip angle decreases. Both effects together yield an increase in the cornering forces. This leads to a reduction in the mean slip angle, and thus an increase in mean cornering stiffness $\overline{c}_{\alpha 2}$. At the front axle, the situation is exactly reversed; the mean cornering stiffness, $\overline{c}_{\alpha 1}$, thus decreases. This leads to amplification in the expression $c_{\alpha 2}\ell_2 - c_{\alpha 1}\ell_1$, an increase in understeering behaviour of the vehicle or a decrease in oversteering behaviour.

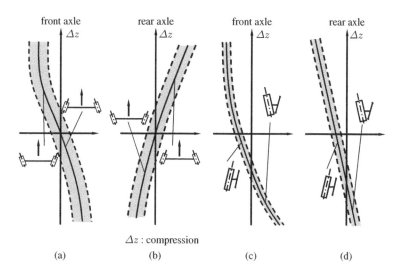

Figure 16.3 Changes of toe and camber during compression and rebound of the suspension (adapted from Mitschke and Wallentowitz 2004)

---

[2] These are typical changes from an investigation of vehicles (cf. Mitschke and Wallentowitz 2004); other characteristics are of course also possible.

Figure 16.4   Solid axle

The second form of the suspension, the so-called solid (or rigid) axle, is shown in principle in Figure 16.1; some more details are depicted in Figure 16.4. The two wheel carriers are connected by a rigid axle. This rigid axle must have two degrees of freedom: one for vertical translation and the other for rotation about the longitudinal axis. For this reason, the rigid axle is connected to the body by means of four arms, which represents one possible way of linking the axle and the body. Another, simpler method is the application of leaf springs. The vertical motion of one wheel is transmitted to the other wheel, which is a disadvantage of this kind of suspension. The camber does not change during cornering. Further disadvantages are tramp and shimmy vibrations, the latter only occur in the case of a steered axle. Rigid axles are unusual in passenger cars but can be used in heavy goods vehicles.

If the wheel carriers are not rigidly connected to each other, on the one hand, and are unable to move completely independently, on the other, the system is then called a torsion beam axle.

In order to link the vertical motion of the wheel elastically with the body, additional coil springs are often used, which are connected to the wheel carrier or to a suspension arm on one side and to the body on the other side. The shock absorber is used to damp the vertical motions of the wheel carrier. The shock absorber can be combined with the coil spring to form a strut, but it is possible to place shock absorber and spring separately. If this solution is employed, different lever arms can be introduced for the shock absorber and the spring.

The following shows some suspension systems.

Figure 16.5 shows the McPherson front axle system of a Mercedes B-Class. The constraints from the strut prevent the wheel carrier from rotating about two axes (this means that two degrees of freedom are locked). From the point of view of a multi-body system (MBS), the upper part of a McPherson suspension is a rigid body (the piston

Figure 16.5   McPherson front axle of a Mercedes B-Class (reproduced with permissions of Daimler AG)

rod) linked with the body of the vehicle by a cardanic joint[3]. The piston rod and the tube of the shock absorber are linked to each other by a cylindrical joint. The tube of the shock absorber and the wheel carrier form, in the sense of an MBS, one rigid body. When all of these factors are taken together, the wheel carrier has four degrees of freedom (two from the cardanic joint and two from the cylindrical joints). Consequently, three additional links constraining three degrees of freedom are necessary. Two links, usually one mainly in the longitudinal direction and the other mainly in the lateral direction, provide two additional constraints. The last of the five constraints comes from the steering system, i.e. the tie rod, as the McPherson suspension is mainly used for front suspensions. The longitudinal and lateral links can be substituted by one A-arm, as shown in Figure 16.5. Furthermore, this figure depicts the anti-roll bar, the differential with the drive shafts, the steering system with velocity-dependent steering ratio and the subframe.

The principle components are depicted in Figure 16.6. The advantages of McPherson struts are a reduction in the number of components and therefore a simpler design

---

[3] Strictly speaking, it is a bushing that prevents transfer of shocks and vibrations; this bushing does not allow free rotations about the two radial axes, but the compliances for these rotations are small; the axial degree of freedom is irrelevant because there is nearly no resistance in relative rotation of piston rod and tube of the shock absorber.

Figure 16.6   Principal components of a McPherson front axle suspension

and less demand for design space. As the shock absorber locks two rotational degrees of freedom, it has to be able to transmit moments. This is the reason why the piston of a McPherson shock absorber has a larger diameter than the piston rod of a conventional shock absorber. This may be a disadvantage because the shock absorber is more expensive. Another disadvantage is that the required height of the vehicle for McPherson suspension is larger than that for a spring and shock absorber linked to a lower A-arm. The McPherson strut is usually linked to the upper part of the wheel carrier; as there must be space for the jounce travel, the height of the vehicle at the upper point of connection of the strut to the body has to be large enough.

Figure 16.7 shows the rear axle of the Mercedes B-Class. It is a four-link suspension. There are three lateral links and one longitudinal. An anti-roll bar is also depicted. The shock absorber and the spring are connected at different points to the wheel carrier and to the body of the vehicle. This leads to greater independence in choosing the spring rate and the damper characteristic.

Figure 16.8 shows the front suspension of a Mercedes M-Class. It is a double wishbone suspension with an air spring with an integrated adaptive damping system (the control is performed by the skyhook algorithm). Another special component is the active anti-roll bar: this is a conventional anti-roll bar that is split into two parts; these two parts are connected by a hydraulic rotary actuator. In accordance with the lateral acceleration v^2/ρ_{cc}, it is possible to adjust the dynamic behaviour of the car to prevent roll movement (up to a certain lateral acceleration). In order to improve comfort when driving straight ahead and to improve the behaviour during off-road driving the coupling between the two parts of the anti-roll bar can be deactivated.

Figure 16.7 Rear axle of a Mercedes B-Class (reproduced with permissions of Daimler AG)

Figure 16.8 McPherson front axle of a Mercedes M-Class (reproduced with permissions of Daimler AG)

Figure 16.9   Rear axle of a Mercedes M-Class (reproduced with permissions of Daimler AG)

Figure 16.9 shows the rear suspension of the M-Class. Here, too, an active anti-roll bar (the actuator is placed behind the differential) is embedded. The air springs and the dampers are separated.

Suspensions with five links allow for the greatest design variety, as each link can be individually designed with respect to comfort, safety and dynamic behaviour. However, this kind of suspension is usually more expensive than simpler designs. Figure 16.10 shows the multi-link rear axle of a Mercedes C-Class. Three links in the front can be easily recognized, while the others are partially hidden by other components.

Figure 16.11 shows the design principle of a five-link suspension; all five links can be clearly seen in this figure. This figure additionally depicts the steering axis (nearly vertical cylinder which intersects the four extensions of the four links). It is obvious that both the scrub radius, r_k, and the disturbing force lever arm radius, r_σ, are very small.

In the last example we look at two phenomena which can be explained with the so-called elasto-kinematic axis, the circumferential steering and the steering resulting from lateral forces. First we look at the elasto-kinematic axis. Before we start with the elasto-kinematic axis, we explain the mechanical analogue of an elasto-kinematic point using Figure 16.12. In this figure a body (the square) is connected by three revolute joints to three links, which are connected to the subframe by three elastic bushings (two of them are stiff, one is weak). The inertia properties of the body are described by J and m. The translational degrees of freedom of the body are captured by x and y, the angle of rotation about the point S is φ. We want to establish the equations of motion for small angles φ. If the distance between S and the revolute

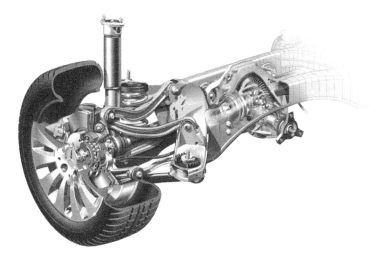

Figure 16.10 Multi-link rear axle of a Mercedes C-Class (reproduced with permissions of Daimler AG)

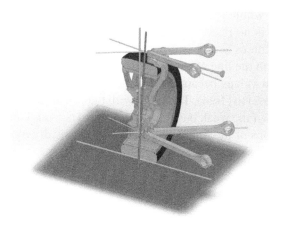

Figure 16.11 Design principle of a five-link suspension

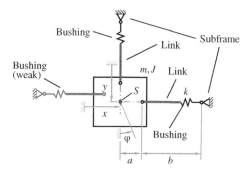

Figure 16.12 Explanation of an elastokinematic point

joints at the body is a, and the distance between the revolute joint at the body and the corresponding revolute joint at the subframe is b (both a and b are only depicted for the horizontal link; for the vertical link the distances are a and b as well), then the elongation, Δs, of the springs (bushing) for small angles φ are (we use a Taylor expansion of the square root and substitute the cosine function by its Taylor expansion $\cos \varphi = 1 - \varphi^2/2 + \cdots$)

$$
\begin{aligned}
\Delta s &= \sqrt{a^2 \sin^2\varphi + (b + a(1 - \cos \varphi))^2} - b \\
&= \sqrt{2a^2 + b^2 + 2ab(1 - \cos \varphi) - 2a^2 \cos \varphi} - b \\
&= \sqrt{b^2 + a^2\varphi^2 + ab\varphi^2 + \cdots} - b \\
&= \frac{b}{2}\left(\frac{a^2}{b^2} + \frac{a}{b}\right)\varphi^2 \cdots .
\end{aligned}
\tag{16.1}
$$

If we assume that the angle φ of rotation of the body about S is small and quadratic terms can be neglected, then the changes in length of the bushing springs during pure rotation of the body can be neglected, too. If we then look at the equations of motion (16.2)–(16.4) of the whole mechanical system (the weak bushing is neglected), we recognize that there is no restoring moment in the last equation (16.4). This means that, for even small forces acting on the body, there are significant angles of rotation if the point S is not on the line of action of the forces. Of course, the angle φ will not increase to very high values because it will be limited by the nonlinearities that we have neglected in (16.2) to (16.4). The point with the possibility about which the body can be easily rotated (if the weak bushing is neglected) is S, and this is the intersection of the linear extensions of the links, which are stiff elastically mounted

$$
m\ddot{x} + kx = 0 , \tag{16.2}
$$

$$
m\ddot{y} + ky = 0 , \tag{16.3}
$$

$$
J\ddot{\varphi} = 0 . \tag{16.4}
$$

After the explanation of the elasto-kinematic points, we continue with the elasto-kinematic axis. This axis is depicted as the cylinder with the largest diameter in Figure 16.13. The suspension is a five-link suspension. Four of the five arms are connected to the subframe or body by stiff bushings (that means elastic elements), the fifth arm (one of the arms of the lower A-arm) is weakly mounted to the subframe. Consequently, we have a similar situation to that of the elastokinematic point, where the point of easy rotation, S, is the intersection of the linear extensions of the elastically mounted arms. If we look at the linear extensions of the elastically mounted arms of Figure 16.13, we recognize, that all four extensions intersect in one straight line, which is the elastokinematic axis (the three-dimensional analogue to the elastokinematic point).

In the next paragraph, we look at the effect of cornering forces and of braking forces. To do this, we assume that the depicted suspension is a rear axle suspension. We start

Figure 16.13   A five-link suspension similar to the LSA rear suspension of the Porsche 911 Carrera

with a braking manoeuvre during cornering (this can be slight braking, or if the rear axle is driven it can be a deceleration due to the drag torque of the engine). The deceleration causes load transfer from rear to front axle. This results in a decrease in lateral forces at the rear axle, and an increase in these forces at the front axle, and, all together, it ends with a moment which tries to rotate the vehicle to the inner side of the curve. The braking forces at the rear axle on both wheels create a torque with respect to the elastokinematic axis, because the intersection between this axis and the road is situated at the outside of the vehicle (the lever arm is n, the lever arm for tractive forces is m). This means that the torques from the braking forces will easily rotate both wheels to toe-in. Furthermore, the lateral force at the inner wheel decreases (because the slip angle of the wheel decreases), while the lateral force at the outer wheel increases. As the absolute value of the increase at the outer wheel is larger than the decrease at the inner wheel (because the wheel load at the outer wheel is higher than at the inner wheel), the sum of both increases, and the above-mentioned moment, which tries to rotate the vehicle to the inner side of the curve, is partially compensated for as well.

Similar lateral forces act on the toe-in and toe-out. As can be seen in Figure 16.13, the intersection is situated behind the centre of the contact patch (the lever arm is s). This means that lateral forces cause toe-in (this in turn means an increase in the slip angle) at the outer wheel and toe-out (which is a decrease of the slip angle) at the inner wheel. This results in higher lateral forces.

16.1   Questions and Exercises

Remembering

1. How many degrees of freedom have to be locked by a suspension of one wheel?

2. How many degrees of freedom are locked by a simple link?
3. How many degrees of freedom are locked by an A-arm?
4. Which is the main degree of freedom of the wheel carrier?
5. What is the principle of an independent suspension?
6. What basic forms of wheel suspensions exist in addition to the independent suspension?
7. What quantities change during compression and rebound?
8. How is the elastokinematic axis defined?

Understanding

1. Explain the effect of changing the compression and rebound quantities with respect to the understeering or oversteering driving behaviour of a vehicle.
2. What influence does the position of the elastokinematic axis have on the driving behaviour?
3. Explain the circumferential steer forces and the lateral steer forces for different positions of the elastokinematic axis.

17

Torque and Speed Converters

In this chapter, we discuss examples of some types of speed converters (clutches) and torque converters (transmissions).

17.1 Speed Converters, Clutches

The engine speed convertor or clutch connects the engine to the torque converter (transmission). The tasks are:

- to transmit the torque from the engine to the transmission if the angular velocities of engine and transmission are not the same; this, for example, is the case when the vehicle is standing still because an internal combustion engine needs a minimum angular velocity to keep running;
- to damp torsional vibration in the powertrain;
- to enable a soft and smooth start-up and
- to enable fast shifting of the gears.

Three types of speed converters are common in vehicles: dry and wet running disc clutches (with one or more discs) and hydrodynamic power transmissions (Föttinger units, hydrokinetic fluid transmissions).

A dual dry clutch is shown in Figure 17.1. The large gear, which is usually mounted between the engine and the clutch, is for applying the starter of the combustion engine. The basic components of a clutch can be seen in the figure twice. The two release forks (in the lower left part) release the diaphragm springs. The upper fork can be clearly seen while the second is almost completely hidden. Also visible are the clutch discs with the torsional springs (one is visible), the linings and the pressure plates. Dual clutches, as shown here, consist of up to 500 single parts.

Vehicle Dynamics, First Edition. Martin Meywerk.
© 2015 John Wiley & Sons, Ltd. Published 2015 by John Wiley & Sons, Ltd.
Companion Website: www.wiley.com/go/meywerk/vehicle

Figure 17.1   Dual dry clutch (reproduced with permissions of Schaeffler)

Dual clutches are used in transmissions to transmit the torque from the engine without interruption. The dual-clutch transmission has two input and two output shafts. These two input shafts are joined to the two output sides of the dual clutch by the two discs and the pressure plates; the input sides of the dual clutch are connected to the engine. During a gear shift, for example from the first to second gear, the clutch for the first gear is disengaged at the same time as the clutch for the second gear is engaged. The principal dependences of the moments and angular velocities are depicted in Figure 17.2. The diagram at the top shows the demand of the moment, M_d, at the driven wheels; here we assume a constant moment, M_d and a constant velocity of the vehicle. The change of transmission path of the moment takes place between t_1 and t_2. The middle diagram depicts the moment from the engine (solid line). Due to the change in the gear ratio during a gear shift, the moment of the engine has to increase (we assume $i_d = 4$, $i_1 = 3$ and $i_2 = 2$ for the diagrams). During engagement and disengagement the moment, M_{c1}, transmitted by the first clutch decreases from the engine moment at $t = t_1$ to zero, whereas the moment, M_{c2}, transmitted by the second clutch increases from zero to the moment of the engine at $t = t_2$.

Figure 17.3 shows an example of a dual-clutch transmission (without the clutches). The two clutches would be attached to the two input shafts; one of the input shafts (the shorter one) is a hollow shaft. In this example, the two output shafts are not positioned axially (in the example in Figures 17.4 and 17.5 the two output shafts are axial, too). The input shafts are axial because a small volume was chosen as the necessary design space for this configuration with the two clutches. If the moment transmitted by the dual-clutch transmission is not to be interrupted, it is essential to spread neighbouring

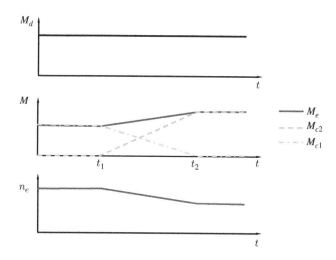

Figure 17.2   Operating principle of a dual-clutch transmission

Figure 17.3   Dual-clutch transmission

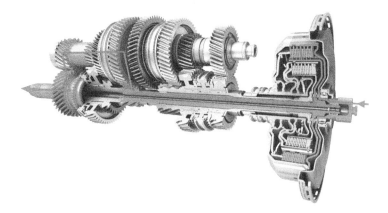

Figure 17.4 Dual clutch transmission with two axial countershafts; first gear engaged (reproduced with permissions of Dr. Ing. h. c. F. Porsche AG)

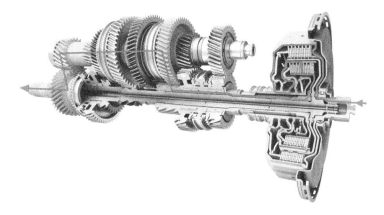

Figure 17.5 Dual clutch transmission with two axial countershafts; second gear engaged (reproduced with permissions of Dr. Ing. h. c. F. Porsche AG)

gears on different input and output shafts. This is necessary for the continuous change from one gear to another by simultaneously engaging one clutch and disengaging the other clutch. A disadvantage of an uninterrupted dual-clutch transmission is that, it is slightly larger and has a slightly higher weight than a manual shift (the efficiencies of a manual and a dual-clutch transmission are similar) but, with respect to shifting comfort, a dual clutch system is comparable to an automatic transmission. Another advantage is that the acceleration ability is higher than that for a manual transmission with interruption of tractive forces at the wheels.

In Figures 17.4 and 17.5, another design with two axial input shafts and two axial countershafts are shown. In Figure 17.4, the first gear is engaged and in Figure 17.5 the second gear. This design uses two multi-disc wet clutches; the outer for the first, third and fifth gears, while the inner clutch for the second, fourth and sixth gears. The seventh gear is not engaged using the countershaft, but by directly connecting input and output shafts.

An essential part of a single disc clutch is the clutch disc. An example is shown in Figure 17.6. Besides the basic parts, the figure also shows some additional features. First of all, a centrifugal pendulum vibration absorber (cf. Section 9.2.2) is attached to the disc. A total of four masses are used for the pendulum masses. Furthermore, the clutch lining is not rigidly connected to the output of the clutch, but connected by a four-stage torsional spring-damping device. The four stages are implemented by 16 springs, where eight of them are nested (the inner springs cannot be clearly seen in the figure). There is a clearance between the inner springs and the disc, which means that the inner springs are only active beyond a certain torsion. This results in a piecewise linear moment-angle function $M = M(\alpha)$, as depicted in Figure 17.7. The effective stiffness, C_i^* ($i = 1, \ldots, 4$), in the stages and the switching points, α_i^*, between the stages in the stiffness curve differ from the stiffness, C_i, and the clearances, α_i, given on the right-hand part of Figure 17.7. For example, the following holds:

$$C_1^* = \frac{C_1 C_2}{C_1 + C_2} \text{ and} \tag{17.1}$$

$$\alpha_1^* = \alpha_1 \frac{C_1}{C_1^*}. \tag{17.2}$$

Figure 17.6 Clutch disc with torsional damper and centrifugal pendulum absorber (reproduced with permissions of Schaeffler)

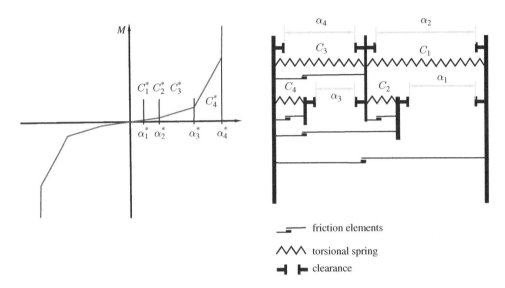

Figure 17.7   Operating principle of a clutch disc with multi-stage torsional damper

Similar relations hold for the other values of C_i^* and α_i^*. The stiffness values C_1 and C_2 have the role of a predamping device. The Coulomb friction elements are responsible for energy dissipation and hysteresis in the system. This means that the characteristic line depicted on the left-hand side of Figure 17.7 should be expanded to a moment angle area around the curve.

Another device to reduce vibration in the powertrain, especially to reduce oscillations in moment and angular velocity, which are induced by the internal combustion engine, is a flywheel. A rigid flywheel can reduce these oscillations because of its inertia. Better efficiency in reducing these oscillations can be achieved by means of a dual-mass flywheel, an example of which is shown in Figure 17.8. In this device, the mass is divided into two parts. These parts are connected by bow springs, with friction occurring between the parts. This means that the dual-mass flywheel is a vibration absorber (cf. Section 9.2.1). Furthermore, centrifugal pendulum absorber masses can be seen in Figure 17.8. These are attached on the engine side of the dual mass flywheel. The aim is a velocity-dependent absorbing frequency (cf. Section 9.2.2).

17.2   Transmission

Various transmission designs are used in passenger cars. Figure 17.9 provides an overview of the usual designs (cf. Naunheimer 2011).

In manual transmissions both gear shift and clutch (dis)engagement are carried out manually. Two countershaft designs are mainly used in passenger cars: single-stage and two-stage countershaft transmission. In single-stage transmissions the moment is transferred from the input to the countershaft, which is also the output shaft.

Figure 17.8 Dual-mass flywheel with centrifugal pendulum absorber (reproduced with permissions of Schaeffler)

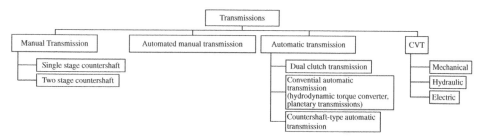

Figure 17.9 Overview of passenger car transmissions

In two-stage countershaft transmissions the input and output shaft are coaxial, and the countershaft and input or output shaft are joined by one fixed pair of gears. Single-stage transmissions are mainly used in front-wheel-drive vehicles with front-mounted engines. In this configuration the final drive with differential is integrated into the transmission. The two-stage transmission is often used for rear-wheel-drive vehicles with front-mounted engines. In this configuration the differential is located at the rear axle.

A simple two-stage countershaft transmission is shown in Figure 17.10. The gears are joined by synchromeshes with the shafts. The different gear ratios are achieved by

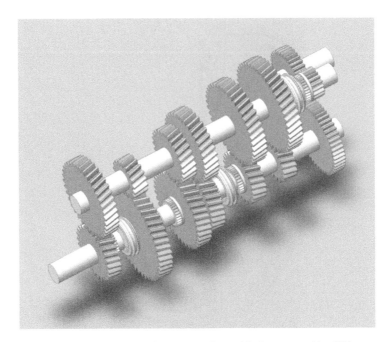

Figure 17.10   Two-stage countershaft transmission with five gears (the fifth gear with direct transmission)

joining the gears with the output shaft or by directly joining input and output shaft. In the latter case, the gear ratio of the transmission is one. The reverse gear requires an additional shaft.

In automated manual transmissions engaging or disengaging the clutch or shifting the gears can be automated. As only one process is automated in this type of transmission, these transmissions are called semi-automatic. One of the first examples came from VW in the year 1967, in which the three pairs of gears were complemented by a hydrodynamic torque converter. In this automated transmission, the gear shifting was performed manually. Semiautomatic transmissions are not used in many cars.

Automatic transmissions can be divided into the conventional automatic transmission with hydrodynamics torque converter and planetary gears, and those with dual clutches and conventional transmissions with countershafts.

The first type consists of a hydrodynamic torque converter with a lock-up clutch (in modern vehicles). The latter one locks up the turbine and the impeller in non-conversion mode of the hydrodynamics converter for increasing efficiency. The planetary gears are complemented by brakes and clutches in order to achieve the different gear ratios.

An example of an automatic transmission is shown in Figure 17.11. This transmission consists of four planetary gear sets and five brakes and clutches. The aim of transmissions with a high number of gear ratios is firstly to achieve a balanced supply

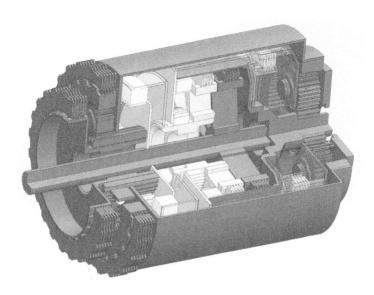

Figure 17.11   Automatic transmission, 8 gears, first gear engaged

of power and tractive force and secondly to achieve high efficiency values across a broad range of velocities and power demand (or tractive force demand). The different gear ratios are achieved by combinations of brake and/or clutch activations.

One crucial element of an automatic transmission is the hydrodynamic torque converter (or Trilok converter), which is explained below.

These hydrokinetic fluid transmissions operate as both a clutch and as a transmission, thus, in terms of the differences in angular velocities, the Trilok converter is both a clutch and a transmission.

One characteristic of this converter is that it consists of three rotating elements: the impeller (or pump wheel), the turbine and the stator. The central part of the Trilok converter, which enables a torque and a velocity conversion, is the stator, which is mounted by means of a one-way clutch to the housing and therefore to the body of the vehicle. The impeller is joined to input shaft (to the engine) and the turbine is joined to the output shaft (to the transmission). Figure 17.12 depicts an example of a Trilok converter.

There are two principal modes in which the converter runs.

In the first mode, the torque and angular velocity conversion mode, the angular velocity of the impeller is clearly greater than the angular velocity of the turbine (the angular velocity may even be zero for starting the car). In this mode, the fluid flows to the rear of the stator blades, and the one-way clutch therefore locks and prevents rotation of the stator. The locking action means that there is a moment between the housing and the stator.

In the second mode, the stator rotates freely and there is no moment between stator and housing.

Figure 17.12 Trilok converter with centrifugal absorber and clutch (reproduced with permissions of Schaeffler)

If the torque of the impeller is M_i, that of the turbine M_t and that of the stator is M_s, then these torques are in equilibrium:

$$M_i + M_t + M_s = 0 . \tag{17.3}$$

When the stator is locked, we then have

$$M_s \neq 0 \rightarrow |M_t| = |M_i + M_s| > |M_i| \tag{17.4}$$

and when the stator is not locked:

$$M_s = 0 \rightarrow |M_t| = |M_i| . \tag{17.5}$$

In the conversion mode, both, the torques and the angular velocities of input and output shaft of the converter are not the same. The torque is amplified as shown in Equation (17.4). The characteristic conversion curve is shown in Figure 17.13. Here, the input and output torque T_{in} and T_{out} are divided by an nominal torque T_o.

A different type of transmission or torque converter is the so-called continuously variable transmission, abbreviated as CVT. There are different possibilities for CVTs. One form is described here which consists of a steel chain (no belt) between two variable speed pulleys (cf. Figure 17.14). The pulleys are adjustable, which means that

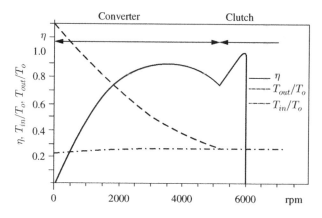

Figure 17.13   Characteristic curve of a Trilok converter

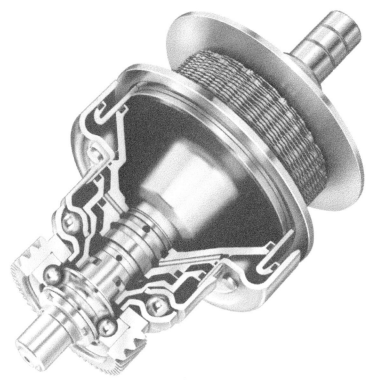

Figure 17.14   One pulley and a part of chain of a CVT (reproduced with permissions of Schaeffler)

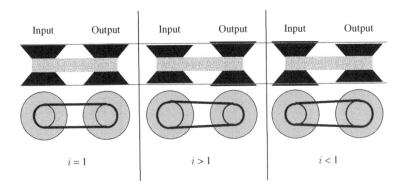

Figure 17.15   Operating principle of a CVT

stepless variable gear ratios can be achieved. In order to transfer high moments, the tension in the chain must be high. The operating principle is visualized in Figure 17.15. Depending on the relative displacements of the two parts of a pulley, different radii of the pulley take effect, which results in the variable gear ratio. With a CVT, the engine, especially an internal combustion engine, can be operated in regions of high efficiency, a change of the engine speed is not necessary. Some aspects of CVTs have been improved in recent years, with examples being the maximum transferable moment and the efficiency. An alternative to a conventional, mechanical CVT is an electronic CVT (eCVT), where the torque is transferred by means of electromagnetic forces.

17.3   Questions and Exercises

Remembering

1. What is the function of a speed converter?
2. What possibilities are there for the vibration damping in powertrains?
3. How can uninterruptible transmissions be achieved?
4. Identify different types of torque converter.

18

Shock Absorbers, Springs and Brakes

This chapter explains the construction of shock absorbers in Section 18.1, active vertical systems in Section 18.2, suspension springs in Section 18.3 and brakes in Section 18.4.

18.1 Shock Absorbers

The automotive industry most commonly uses single and twin-tube shock absorbers.

Figure 18.1 shows a schematic diagram of a monotube shock absorber. The gas is under a pressure of up to 25 bar. The high pressure in the interior of the shock absorber is intended to prevent cavitation and foaming. The movement of the piston rod causes the shock absorber piston to move up and down. During these movements, the oil flows through either the compression valve (2) or the rebound valve (1), with the two valves usually having different characteristics for compression and rebound. The separating piston between the oil and the gas seals off the gas from the oil. The gas volume is necessary to compensate for the piston rod volume, which moves inside the working cylinder during compression. Due to the high internal pressure, special demands are placed on the seal. The flow through the valves for compression is shown in Figure 18.2 and for rebound in Figure 18.3.

One advantage of the monotube shock absorber is that it can be installed in any orientation. The disadvantages are the additional costs compared to a twin-tube shock absorber due to the higher manufacturing precision and the necessary tightness. A further advantage is the efficient cooling of the oil. The static pressure acting on the piston rod means that a static force acts permanently. If we assume a pressure of 25 bar and a radius of the piston rod of 6 mm, this force is about 280 N.

Vehicle Dynamics, First Edition. Martin Meywerk.
© 2015 John Wiley & Sons, Ltd. Published 2015 by John Wiley & Sons, Ltd.
Companion Website: www.wiley.com/go/meywerk/vehicle

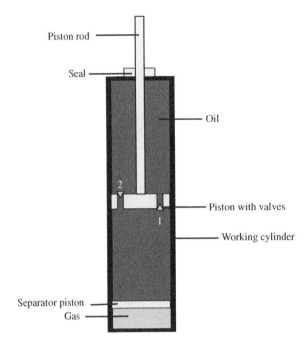

Piston rod

Seal

Oil

Piston with valves

Working cylinder

Separator piston

Gas

Figure 18.1   Monotube shock absorber

Figure 18.2   Monotube shock absorber: details of flow for compression (reproduced with permissions of ZF Friedrichshafen AG)

Figure 18.3   Monotube shock absorber: details of flow for rebound (reproduced with permissions of ZF Friedrichshafen AG)

Figure 18.4 shows the basic structure of a twin-tube shock absorber with the compression valve (1) and the rebound valve (2). Here the oil is not under a static pressure. Consequently, the demands placed on the seals and manufacturing precision are not as high as for the monotube shock absorber. Between the outer and inner cylinders is the compensation cylinder. The compensation cylinder is necessary for volume compensation of the rod. The oil flows through either the compression valve (4) or the rebound valve (3) into or out of the compensation cylinder, which is about half-filled with oil. The remaining part of the compensation cylinder is used for absorption of oil during expansion (temperatures of up to 120 °C are possible). Twin-tube shock absorbers may not be be installed in any arbitrary orientation; otherwise, air from the compensating cylinder would be drawn into the working chamber during rebound.

The flow through the valves for compression is shown in Figure 18.5 and for rebound in Figure 18.6.

Some special types of shock absorbers are in use. One is the suspension strut module, which comprises the suspension spring and the shock absorber. This strut is used in the so-called McPherson wheel suspension. As the coil spring cannot significantly transfer forces or moments in a radial direction, the shock absorber has to do this. As a result, the diameter of the piston rod has to be larger than that of a conventional shock absorber.

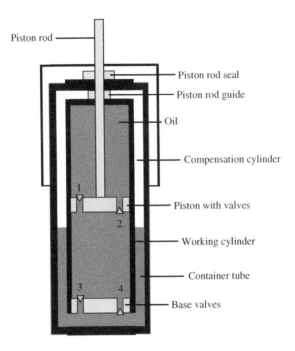

Figure 18.4   Twin-tube shock absorber

Figure 18.5   Twin-tube shock absorber: details of flow for compression (reproduced with permissions of ZF Friedrichshafen AG)

Continuous damping control (CDC) is a special form of adaptive damping system. Besides the piston valve and the base valve, a third, variable proportional valve is present in this shock absorber. This additional and adjustable valve controls the flow in a shunt flow path which is achieved by means of an additional control tube. The proportional valve can be electronically adjusted according to factors such as the road

Figure 18.6   Twin-tube shock absorber: details of flow for rebound (reproduced with permissions of ZF Friedrichshafen AG)

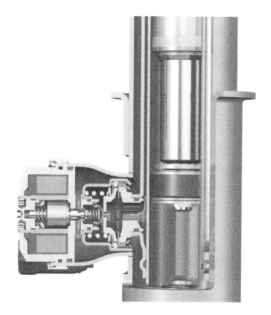

Figure 18.7   Proportional valve in a CDC (reproduced with permissions of ZF Friedrichshafen AG)

or the oscillations. The flow through this variable valve is illustrated in Figure 18.7. This type of damping is called adaptive because only the damping characteristics can be controlled, but there is no way of actively influencing the system, which means that this method is not able to bring energy or power into the system. The CDC shock absorber is used in a so-called skyhook damping system which is described in Section 18.2.2. This system controls the damping in order to isolate oscillations of the wheel from the body. This isolation is only partly possible because the shock absorber is dissipative, i.e. the shock absorber does not convert electrical into mechanical energy. The shock absorber, therefore, remains a passive system, but the characteristics are adaptable.

18.2   Ideal Active Suspension and Skyhook Damping

This section looks at some details of active body control. We first start by considering an ideal active body control and in the second subsection we look at the so-called skyhook damping.

18.2.1   Ideal Active Suspension

Ideal means that we assume an actuator which can deliver arbitrary forces or displacements. This actuator acts between the wheel and the body (cf. Figure 18.8; we prefer indices 1 and 2 instead of w and b in this section).

The force of the actuator is F; seat and the driver are not included in this consideration of the operating principle. The mechanical system is described by the following system of ordinary differential equations:

$$m_2 \ddot{z}_2 = F \ , \tag{18.1}$$

$$m_1 \ddot{z}_1 + k_1 z_1 = -F + k_1 h \ . \tag{18.2}$$

We assume that this system is excited by a harmonically uneven road. The uneven road is described by a harmonic complex function, h, with only the real part being relevant. As we are mainly concerned with the amplitudes at the end of the derivation, this consideration is sufficient. The excitation in the time domain is

$$h = \hat{h} \, e^{i\omega t}. \tag{18.3}$$

If $\kappa = 2\pi/L$ is the wavenumber of a harmonically uneven road and v is the velocity of the vehicle, then the time angular frequency is $\omega = v\kappa$. As the excitation of the system of the linear differential equations is harmonic, the time-dependent variables are harmonic too:

$$z_1 = \hat{z}_1 \, e^{i\omega t} \ , \tag{18.4}$$

$$z_2 = \hat{z}_2 \, e^{i\omega t} \ , \tag{18.5}$$

$$F = \hat{F} \, e^{i\omega t} \ , \tag{18.6}$$

$$F_z = \hat{F}_z \, e^{i\omega t} \ . \tag{18.7}$$

The last variable, F_z, is the dynamic wheel load, which means only the harmonic part, not the static load. The wheel load is a result of the deformation of the wheel.

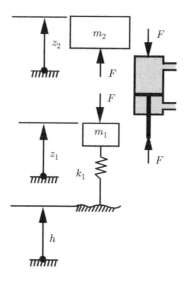

Figure 18.8 Ideal active body control (the actuator is repositioned by its forces)

The stiffness of the wheel is k_1. We then obtain

$$\hat{F}_z = k_1(\hat{h} - \hat{z}_1). \tag{18.8}$$

Solving this equation for $\hat{z}_1$ yields

$$\hat{z}_1 = \hat{h} - \frac{1}{k_1}\hat{F}_z . \tag{18.9}$$

The result of substituting $\hat{z}_1$ in Equation (18.2) (we substitute $\ddot{z}_1 = -\omega^2\hat{z}_1 e^{i\omega t}$ and divide by $e^{i\omega t} \neq 0$)

$$-m_1\omega^2\hat{z}_1 + k_1\hat{z}_1 = -\hat{F} + k_1\hat{h} \tag{18.10}$$

is

$$\hat{F} = \hat{F}_z\left(1 - \frac{m_1}{k_1}\omega^2\right) + m_1\,\omega^2\,\hat{h} . \tag{18.11}$$

Equation (18.11) contains the complex amplitudes of the actuator force, $\hat{F}$, the wheel load, $\hat{F}_z$, and the uneven road, $\hat{h}$. We introduce a frequency ratio ($\omega_w = \sqrt{k_1/m_1}$ is the natural frequency of the wheel):

$$\eta_1 = \frac{\omega}{\omega_w}, \tag{18.12}$$

This ratio can be used to rewrite Equation (18.11):

$$\hat{F} = \hat{F}_z(1 - \eta_1^2) + k_1\,\eta_1^2\,\hat{h}. \tag{18.13}$$

In this equation, it is apparent that the phase between the three complex amplitudes can be chosen as advantageous or disadvantageous. We consider two extreme cases.

First. the case $\hat{F} = 0$. In this case, the actuator is controlled so that there is no dynamic force between body and wheel (the static force does of course act between body and wheel). This vanishing force $\hat{F} = 0$ results in vanishing acceleration of the body, which results in a comfort value of zero and therefore in an optimum level of comfort. The consequence for the complex amplitude of the wheel load can be seen in the following equation, which we obtain from (18.11):

$$\hat{F}_z = -k_1 \frac{\eta_1^2}{1 - \eta_1^2} \, \hat{h} \ . \tag{18.14}$$

The denominator $1 - \eta_1^2$ becomes zero if the excitation frequency is $\omega = \sqrt{k_1/m_1}$, which means that state in which the system is excited with the natural frequency of the wheel. This results in a division by zero; if damping is considered, this results in very high values of the wheel load amplitude $\hat{F}_z$. The consequence of optimum comfort is very bad (or infinitely bad) safety values.

The second extreme case is optimum safety, which means vanishing dynamic wheel loads. From (18.11) with $\hat{F}_z = 0$

$$\hat{F} = k_1 \eta_1^2 \hat{h} \ . \tag{18.15}$$

If the excitation frequency is equal to the natural frequency of the wheel ($\eta_1 = 1$), then the uneven road affects the body so that only the wheel stiffness acts between body and wheel. As the wheel stiffness is very high, this case leads to very high forces and therefore to poor comfort values. If we look at the amplitudes $\hat{h}$, we know from the stochastic description of uneven road, that the spectral density

$$\Phi_h(\Omega) = \Phi(\Omega_0) \left(\frac{\Omega}{\Omega_0} \right)^{-w} . \tag{18.16}$$

decreases with Ω^{-w}, where the magnitude of w is 2. It follows that the amplitudes $\hat{h}$ behave like $\sqrt{\Omega^{-w}} = \Omega^{-w/2}$, or as functions of the time angular frequency, $\omega = v\Omega$ the amplitudes $\hat{h}$ decrease in the same way as $\omega^{-w/2}$. As $\eta_1^2 = \omega^2/\omega_w^2$ and as $w \approx 2$, the amplitude, $\hat{F} = k_1 \eta_1^2 \hat{h}$, increases approximately linearly with $\omega^{2-w/2}$, hence comfort will become infinitely bad for ideal safety.

Only a compromise is possible. Choosing the appropriate phases between the three complex amplitudes enables the following equation to be obtained:

$$|\hat{F}| + |\hat{F}_z(1 - \eta_1^2)| = |k_1 \eta_1^2 \hat{h}| \ , \tag{18.17}$$

In general, the following equation holds:

$$|\hat{F}| + |\hat{F}_z(1 - \eta_1^2)| \geq |k_1 \eta_1^2 \hat{h}| \ . \tag{18.18}$$

Figure 18.9   Active vertical system of Mercedes S-Class (reproduced with permissions of Daimler AG)

This means that, in the case of optimal phases (18.17), the distribution of the forces $\hat{F}$ and $\hat{F}_z$ can only be chosen within the limit $|k_1 \eta_1^2 \hat{h}|$. All together, it is obvious that the conflict between comfort and safety persists, even for an ideal active actuator.

Figure 18.9 shows an active spring system for the Mercedes S-Class. The active elements (such active body control from Mercedes) can use hydraulic actuators, but electric actuators are also possible. With this kind of actuator, it is possible to approximate the ideal phases of Equation (18.17).

18.2.2   Skyhook Dampers

The damping properties of skyhook dampers can be changed continuously[1]. These properties are made possibly by the use of proportional valves to achieve a hydraulic, controllable shunt. Accelerations of the body and wheels enter into the control algorithm, so that every wheel can be controlled individually. If, in addition, pitch and roll movements are to be influenced, the individual controllers of the wheels have to be connected to a central controller for body control.

The crucial role in the conflict between safety and comfort is played by the damper between wheel and body: the damper forces influence both comfort and safety.

---

[1] The derivations of the formulas in this subsection closely follow the monograph of Mitschke and Wallentowitz 2004.

One way out of this conflict would be a damper for the body which does not act on the wheel. As there is no possibility of fastening the damper to the body and a point of the environment, we assume that we can fasten one end of the damper to the body and the other end to an imaginary hook in the sky (which is not possible in reality, but we aim to examine whether the comfort can be improved). This skyhook damper can be adjusted to achieve good values for comfort, and the choice of the damping constant will not influence safety because the damping force will not act on the wheel.

This kind of skyhook damper is depicted in Figure 18.10(a). The skyhook damper is placed between the body and the sky, but there is no damper between the wheel and the body. The damping constant can be chosen independently of safety requirements. In the following sections, we answer the question as to whether it is possible to place a damper between wheel and body which behaves approximately or even exactly like the skyhook system. This requirement is fulfilled if the force of the skyhook damper, F_{dsky}, is the same as the force of a conventional damper, F_d, where

$$F_{dsky} = b_{sky}\dot{z}_2 \,, \tag{18.19}$$

$$F_d = b_2(\dot{z}_2 - \dot{z}_1) \,. \tag{18.20}$$

From $F_{dsky} = F_d$, we obtain a conventional damper which behaves like a skyhook damper. To achieve the skyhook behaviour, it is necessary for the damping constant, b_2, to depend on the velocities:

$$b_2 = b_{sky}\frac{\dot{z}_2}{\dot{z}_2 - \dot{z}_1} \,. \tag{18.21}$$

To evaluate whether this kind of damper is possible, we look at the power, P_{dsky}, of the damper (the power, P_{dsky}, can also be calculated if we start from

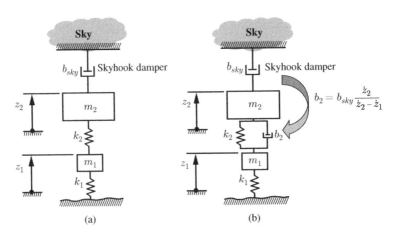

Figure 18.10   Skyhook damper (adapted from Mitschke and Wallentowitz 2004)

$P_{\mathrm{dsky}} = F_d(\dot{z}_2 - \dot{z}_1)$ and substituting b_2 by (18.21)):

$$P_{\mathrm{dsky}} = F_{\mathrm{dsky}}(\dot{z}_2 - \dot{z}_1) \tag{18.22}$$

$$= b_{\mathrm{sky}}\dot{z}_2(\dot{z}_2 - \dot{z}_1) \ . \tag{18.23}$$

As the damper is a passive element, the power always has to be positive (the damper is only able to convert mechanical energy into heat), which is fulfilled for

$$\dot{z}_2(\dot{z}_2 - \dot{z}_1) > 0 \ . \tag{18.24}$$

This requirement is equivalent to the condition that the velocity-dependent damping constant, b_2 (cf. Equation (18.21)), should be positive (if we multiply (18.21) by $(\dot{z}_2 - \dot{z}_1)^2 \geq 0$, the equivalence is obvious; division by zero is not considered here). A negative value is not possible for a passive damper. This means that a skyhook damper can only be achieved for velocities which fulfil the condition

$$\dot{z}_2(\dot{z}_2 - \dot{z}_1) > 0 \ . \tag{18.25}$$

Coming back to Figure 18.10(a), we can clearly see that the wheel is not damped at all, which could result in high values of wheel load fluctuations. To avoid these fluctuations, an additional, conventional damper (damping constant $\tilde{b}_2$) is introduced. The damping forces of the two dampers are then

$$F_{d \ \mathrm{tot}} = b_{\mathrm{sky}}\dot{z}_2 + \tilde{b}_2(\dot{z}_2 - \dot{z}_1) \tag{18.26}$$

$$= \left(b_{\mathrm{sky}}\frac{\dot{z}_2}{\dot{z}_2 - \dot{z}_1} + \tilde{b}_2 \right)(\dot{z}_2 - \dot{z}_1) \text{ for } \dot{z}_2(\dot{z}_2 - \dot{z}_1) > 0$$

$$F_{d \ \mathrm{tot}} = \tilde{b}_2(\dot{z}_2 - \dot{z}_1) \text{ for } \dot{z}_2(\dot{z}_2 - \dot{z}_1) < 0 \ . \tag{18.27}$$

With the mean of such a combination of dampers, it is now possible to adjust the damping in order to obtain good comfort values and low wheel load fluctuations. To control the skyhook part of the damper, the velocities $\dot{z}_2$ of the body and $\dot{z}_1$ of the wheel are necessary. This means that two sensors for the control of one wheel have to be built into the vehicle.

Figure 18.11 shows the hardware of a skyhook damping system (CDC): it comprises the acceleration sensor for the wheel and for the body acceleration, the control unit and the proportional valve. Figure 18.12 shows a skyhook damper. The valve for controlling the damping characteristics is visible at the lower left-hand side.

18.3   Suspension Springs

Different kinds of springs are used in automobiles for connecting the wheel carrier and body.

In passenger vehicles, a spring travel ± 100 mm may occur. Coil springs (seldom leaf springs) are used mainly, but air springs and torsion bars can also be found.

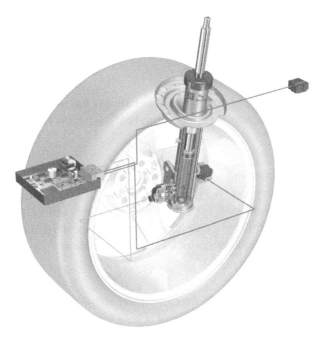

Figure 18.11 CDC: Continous damping control (reproduced with permissions of ZF Friedrichshafen AG)

Leaf springs and torsion bars play a special role because they can also be used as suspension links. Leaf spring or torsion bar suspensions can be found in heavy-duty vehicles, but seldom in new passenger cars. Coil springs and air springs are not suitable for replacing suspension links. In addition to these springs (air, leaf, coil and torsion bar), rubber bushings are also used in the suspension in order to reduce noise and vibrations.

Leaf springs (cf. spring 6 in Figure 18.13) are bending beams with a very low second moment of area, in which the deflections are large. Laminated leaf springs of different lengths or leaf springs with a variable cross-section (parabolic profile) reduce the bending stresses at the clamping end points.

Figure 18.13 shows a leaf spring and coil springs.

The coil spring is a coiled torsion bar. Figure 18.13 shows some different designs. The basic shape is cylindrical (1) with constant wire diameter and mean spring diameter. The working characteristic (i.e. the force vs. deflection) is linear.

A non-linear, progressive characteristic is achieved with constant wire and coil diameters but variable pitch. When the spring is deflected, the number of active coils decreases due to variable pitch, and hence a progressive characteristic is obtained (cf. spring 2 in Figure 18.13). A progressive characteristic curve also results from a series arrangement of two different springs (spring 3). The largest number of degrees of freedom for creating a spring characteristic is provided by a barrel spring, where the pitch, the wire diameter and the mean spring diameter are not constant (spring 4).

Figure 18.12   CDC shock absorber (reproduced with permissions of ZF Friedrichshafen AG)

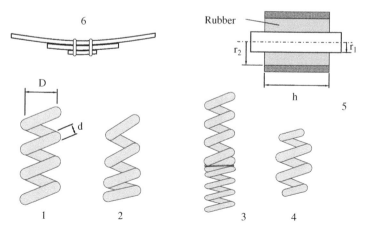

Figure 18.13   Suspension springs

The spring stiffness of a spring with constant wire diameter d, mean spring diameter D and n active coils is (where G is the shear modulus)

$$c = \frac{Gd^4}{8nD^3} \, .$$ (18.28)

Rubber bushings (cf. spring 5 in Figure 18.13) can be found as additional elastic components in vehicles, with examples being in the joints between the suspension arms and the body or subframe, in engine or transmission mounts, in anti-roll bar mounts or shock absorber mounts (e.g. for McPherson struts). The main stiffnesses of rubber bushings are axial c_a, radial c_r and torsional c_φ:

$$c_a = \frac{2\pi hG}{\ln\,(r_2/r_1)} \, ,$$ (18.29)

$$c_r = \frac{k7.5\pi hG}{\ln\,(r_2/r_1)} \, ,$$ (18.30)

$$c_\varphi = \frac{4\pi hG}{(1/r_1^2 - 1/r_2^2)} \, .$$ (18.31)

The correction factor, k, depends on the ratio between the height, h, and the thickness, $s = r_2 - r_1$, of the rubber. For $h/s = 0$ the factor is 1, and k increases progressively up to 2.1 for $h/s = 5$. The shear modulus G increases with Shore A hardness H (G in N/mm^2, H in Sh A; cf. Battermann and Koehler 1982):

$$G = 0.086 \times 1.045^H \, .$$ (18.32)

In hydropneumatic and air springs, the stiffness is determined by a fixed or variable amount of gas (nitrogen or air). The volume of the gas varies during varying jounce travel. The functioning principle is shown in Figure 18.14.

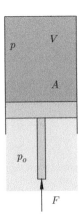

Figure 18.14 Functioning principle of an air spring

The force F acting on the piston is

$$F = A(p - p_o) \, . \tag{18.33}$$

Here, p is the pressure inside the bellows, p_o the pressure outside and A is the cross-sectional area. The pressure, p, inside the gas spring increases when the volume becomes smaller and decreases when the volume becomes larger.

According to the velocity of change in volume, the temperature stays nearly constant (which is called the isothermal change of state) or it varies. An isothermal change of state can be described by

$$pV = \text{const} \, . \tag{18.34}$$

If the volume is changed very quickly, there is nearly no exchange of heat with the environment. Consequently, this change of state is called adiabatic, which is described by the following equation: ($\kappa \approx 1.4$ for air):

$$pV^\kappa = \text{const} \, . \tag{18.35}$$

In general, the relationship between pressure, p, and volume, V, is described by the polytropic equation

$$pV^n = \text{const} \, . \tag{18.36}$$

Here, n is the polytropic exponent which is between 1 (isothermal) and 1.4 (adiabatic).

The polytropic exponent rises with increasing velocities.

The characteristic of a gas spring is not linear, but the gradient, dF/ds, of the spring force, F, can be derived as a function of the jounce travel, s. Starting from

$$F = A \left(\frac{p_0 V_0^n}{(V_0 - As)^n} - p_o \right) \tag{18.37}$$

we obtain

$$\frac{dF}{ds} = A^2 n \frac{p_0 V_0^n}{(V_0 - As)^{n+1}} \, . \tag{18.38}$$

As the gas in the bellows diffuses, a compressor, which can be combined with a bounce control, is mandatory. If the velocities are small this bounce control can compensate for changes in the bounce.

There are two types of gas springs. Those with the constant volume of gas and those with a constant mass of gas.

The principal components of a gas spring with constant volume are shown in Figure 18.15, while those for an air spring with constant mass of gas are depicted in Figure 18.16.

The piston of gas springs with constant volume presses against the bellows in which the gas is located. If the vehicle is loaded, the bounce deviates from the reference value, and the three-way valve connects the gas spring to the compressor; the latter

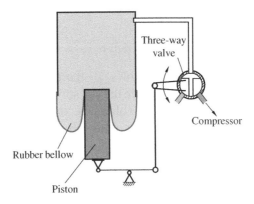

Figure 18.15 Functioning principle of an air spring with constant volume

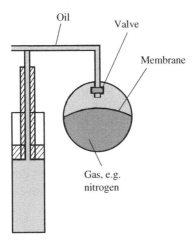

Figure 18.16 Functioning principle of an air spring with constant mass of gas (hydropneumatic spring)

increases the pressure till the bounce reference value is re-established. Unloading the vehicle causes the air spring to deflate.

Rearranging the equation for the gradient (18.38), we obtain ($p_0 V_0^n = p V^n$ and $(V_0 - As)^{n+1} = V^{n+1}$)

$$\frac{\mathrm{d}F}{\mathrm{d}s} = A^2 n \frac{p}{V} \; . \tag{18.39}$$

As the volume, V, remains constant and the static pressure increases in proportion to the load, the first natural frequency, f_1, does not change when the vehicle is loaded or unloaded. The first eigenfrequency is that of the quarter-vehicle model. If we assume that the weight, mg, is uniformly distributed over the four wheels, the pressure in one gas spring is $p = mg/(4A)$. The first natural frequency of a linear single mass oscillator (mass m and stiffness c) is $\sqrt{c/m}$) and therefore

$$f_e = \frac{1}{2\pi}\sqrt{\frac{\mathrm{d}F/\mathrm{d}s}{m/4}} \tag{18.40}$$

$$= \frac{1}{2\pi}\sqrt{\frac{Ang}{V}}, \tag{18.41}$$

which is independent of the mass and therefore of the load.

Since the volume is in the denominator of (18.41), the first natural frequency of the quarter-vehicle model with a gas spring with constant volume is independent of the load, whereas this natural frequency increases with rising load in the case of the gas spring with constant mass (because the volume becomes smaller).

The volume inside the bellows is sometimes too small to attain a low first natural frequency (cf. Equation (18.41)) so that, in some gas springs, an additional, external volume is attached to the main volume (cf. Figure 18.17). The connection between the main and the external volume can be achieved by an adaptive valve. This adaptation can be used to control the damping characteristics. An important aspect of an air spring is the shape of the bellows because the rubber becomes very stiff at high frequencies. The maximum pressure in air springs lies in the region of 15 bar.

An air spring without any additional volume is shown in Figure 18.18. Figure 18.19 shows an air spring in the wheel suspension of a truck. In passenger cars, air springs

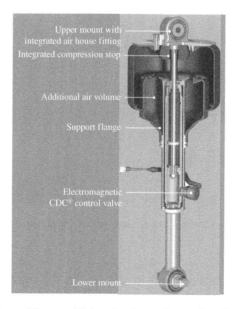

Figure 18.17   Air spring with an additional volume (reproduced with permissions of ZF Friedrichshafen AG)

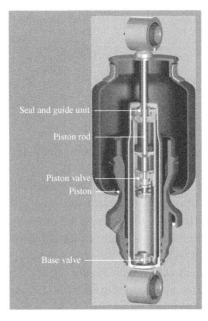

Figure 18.18 Air spring without an additional volume (reproduced with permissions of ZF Friedrichshafen AG)

Figure 18.19 Air spring in a truck wheel suspension (reproduced with permissions of ZF Friedrichshafen AG)

with constant volume can be found in vehicles in the upper segments and heavy vehicles, whereas this technology is not common in compact cars.

The second form of air springs, those with a constant mass of gas, is older than the gas spring with constant volume. This form of gas spring is generally called the hydropneumatic air spring and was introduced as a broad industrial standard by Citroen in the 1950s.

The basic structure can be seen in Figure 18.16. The idea is that the gas (usually nitrogen in passenger cars) is separated by a rubber membrane from the hydraulic oil. The compliance of the spring is a result of the gas, while the hydraulic oil is very stiff. There are some advantages of this form of suspension spring. The spring characteristic can be adjusted to a very weak setting, which yields a high comfort. The damping can be implemented by a valve for the oil. When the valve is variable, an adaptive damping characteristic is available. As the cross-sectional area in the hydraulic part is usually smaller than that in air springs with a constant volume, the gas pressure of hydropneumatic springs has to be higher (approx. 15–20 bar) than the pressure in air springs with constant volume. There are some disadvantages to the hydropneumatic spring. As the spring characteristic is weak, these systems usually have to be complemented by a hydraulic pump to compensate for decreasing chassis clearance when the vehicle is loaded. Although this involves extra effort, it is comparable to the pneumatic system necessary in air springs with constant volume.

18.4 Brake Systems

Brakes are necessary in a vehicle to reduce the velocity, to allow the driver to stop the vehicle, to limit the velocity when the vehicle is travelling downhill or to hold the vehicle at a standstill.

Many legal requirements apply, such as the following from Europe:

- A vehicle has to have two independent brake systems, which are usually achieved by two brake circuits.
- If one circuit fails, the other circuit has to be able to brake at least two wheels, which must not be located on the same side of the vehicle.
- One service brake system and one parking brake system are necessary; the latter has to hold the vehicle at a standstill on an inclined road with a gradient of $p = 25\%$.

This means that the brakes have the function of reducing the kinetic energy of the vehicle in order to reduce the velocity and they also have to keep the vehicle at a standstill.

Various brake systems exist and differ according to the type of energy conversion:

- friction brake
- engine brake or exhaust brake
- electromagnetic or eddy-current brake
- hydrodynamic brake or hydrodynamic retarder according to the Föttinger principle.

Friction brakes are most commonly used in passenger cars. These brakes are designed as disc brakes (standard in modern vehicles) or drum brakes. Figure 18.20

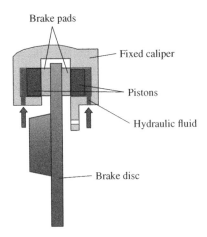

Figure 18.20   Disc brake with a fixed caliper

shows a so-called fixed caliper brake. This brake is characterized by a caliper which is not able to move perpendicularly to the disc. There are pistons on both sides of the disc. The piston on the outside (that means between disc and wheel) limits the possibility to minimize the distance between the brake and the wheel.

Figure 18.21 shows a different type of brake, namely a floating caliper brake. The caliper is able to move perpendicularly with respect to the disc, while the piston between the disc and the wheel is missing. This brake can therefore be positioned near to the wheel. Another advantage is that these brakes are cheaper and that they are self-adjusting and self-centring.

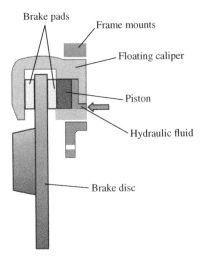

Figure 18.21   Disc brake with a floating caliper

The distance between the brake and the wheel is an important geometric factor which influences the disturbing force lever arm radius r_σ. The development of the floating caliper brake made it possible to reduce the disturbing force lever arm radius r_σ and also to reduce the scrub radius r_k to zero or even to negative values. A negative scrub radius has positive effects on the dynamics of the vehicle, but in vehicles with anti-lock braking system (ABS) a zero or small scrub radius is preferable.

Characteristics of disc brakes are:

- good cooling of the disc; therefore little fading is observed (fading is the reduced effect of braking due to heating of the disc);
- steady braking;
- no self-amplification.

The brake system is usually hydraulic, with the tandem master cylinder being the central part, see Figure 18.22. The tandem master cylinder comprises two pistons, one for each of the two braking subsystems. Pressing the brake pedal moves a push rod against the primary piston. Under normal operating conditions, the fluid (and a spring with a small contribution to the force) between the primary and the secondary piston transmits the force from the primary to the secondary piston. The pressure builds up when the primary cups of the pistons cover the bypass.

If the primary hydraulic subsystem fails (e.g. in the event of leakage) the force from the pedal to the secondary piston is transmitted mechanically by a rod at the front of the primary piston without any hydraulic pressure build-up. If the secondary hydraulic subsystem fails, the secondary piston is moved to the end of the cylinder, where it is stopped. After this, the pressure in the primary subsystem can be built up; the travelling distance of the primary piston is larger in this failure scenario. A spring is located between the pistons and between the cylinder and the secondary piston in order to expand the system after the pedal force decreases. If needed, fluid from the

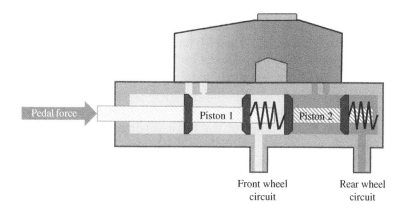

Figure 18.22   Brake circuit configurations

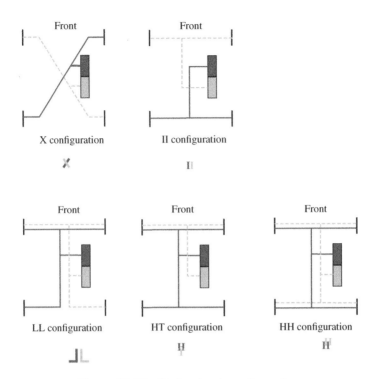

Figure 18.23 Brake circuit configurations

reservoir can refill the system. The chambers and the reservoir are connected by two bypasses, which are closed by cup seals during braking.

There are different designs of brake circuit configuration, see Figure 18.23. Their names have been chosen to match the letters that best describe their appearance: II (or TT or black-and-white), X (or diagonal), HI (or HT), LL and HH. The simplest and the most commonly used configurations today are the II and the X designs. In these two designs the brake for each wheel has only one set of pistons for one circuit (in the case of a floating caliper). The other designs (HI, LL and HH) have more than one piston sets at some of the wheels: HI and LL have two piston sets at each front wheel, and for HH two piston sets are necessary at each wheel.

In vans or heavy-duty vehicles, other, additional brakes are used, such as exhaust brakes (in which the exhaust pipe is partially closed by a valve), eddy-current brakes (in which a rotor is attached to the axles, a stator to the chassis; braking occurs when an electric current in the windings of the stator generates a magnetic fields in the rotor, which induces eddy currents in the rotor) or hydraulic retarder, in which a fluid generates the braking torque between a rotor and a stator.

18.5   Questions and Exercises

Remembering

1. Explain the different types of shock absorbers and suspension springs.
2. Explain the difference between the two types of gas spring with respect to the first vertical eigenfrequency.

Understanding

1. What is the reason that the brakes of one brake circuit are not located on the same side of the vehicle? Explain.
2. Are the pressures in the two circuits equal?
3. How different pressures at front and rear wheels can be achieved?

Applying

1. Explain the development of yaw moments when one circuit fails for the different brake circuit configurations of Figure 18.23 for straight ahead motion and for cornering. Take into account the sign of the scrub radius as well as the elastokinematic axis.

19

Active Longitudinal and Lateral Systems

This chapter explains active systems which influence lateral or longitudinal dynamic behaviour of a vehicle. The systems discussed are anti-lock braking system (ABS), anti-slip regulation (ASR) and electronic stability programme (ESP).

19.1 Main Components of ABS

In critical driving conditions where full braking is applied, one or more wheels may be locked if the vehicle does not have the anti-lock braking system (ABS). This locking of the front wheels may mean that the vehicle can no longer be steered, or the locking of the rear wheels may cause instabilities in the driving conditions. For this reason, it is helpful to prevent locking of the wheels.

The main components of an ABS are shown in Figure 19.1. A classical brake system is composed of a brake pedal, brake power unit, a master cylinder with a reservoir, brake lines and brake hoses and wheel brakes with cylinders. Additional components of an ABS are wheel-speed sensors at all four wheels, a hydraulic unit (hydraulic modulator with magnetic valves) and a controller. The wheel speed sensors detect the rotational speed of all four wheels and are necessary in order to detect the locking tendency of individual wheels. The acceleration of the wheels and the slip are used as essential parameters for the tendency to lock. There are variants of ABS that operate with only three speed sensors, as shown by the first ABS from Bosch in Figure 19.2, in which the rotational speeds of both driven rear wheels have been detected by a sensor on the Cardan shaft. The reason for the use of two control values (wheel acceleration and slip) is based on the fact that some manoeuvres may only be identified as critical by considering one of these two quantities. Consequently, the slip cannot be readily adjusted by only the slip-based control during panic braking or a sudden

Vehicle Dynamics, First Edition. Martin Meywerk.
© 2015 John Wiley & Sons, Ltd. Published 2015 by John Wiley & Sons, Ltd.
Companion Website: www.wiley.com/go/meywerk/vehicle

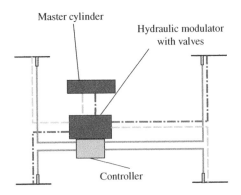

⫿ Sensors for the angular velocities of the wheels

--·· Brake lines

—— Electric wires

Figure 19.1 Components of an ABS

change in adhesion. Pure control over the wheel deceleration may provide no satisfactory solution for gentle braking to high slip values, because the high slip values are difficult or impossible to detect. The final adhesion limit is not often exceeded during a rapid build-up of acceleration, but a reliable detection is, nevertheless, simpler with the additional aid of slip. For example, the velocities of the non-driven wheels can be used for determining the velocity for calculation of the slip.

We can recognize the limited possibilities of a purely slip-based control in the following example. If we assume that all four wheels tend to lock alike, then the circumferential speed, v_{ci}, $i = 1, \ldots, 4$, tends to zero in a similar manner (we assume for this explanation that all four wheels decrease their velocity identically): $v_{ci} = at$ (here a is an acceleration value and t is the time) and we continue to assume that the driving speed $v_v = \frac{1}{4}\sum_{k=1}^{4} v_{ck}$ is calculated from averaging the four circumferential speeds v_{ci}. As a result for the slip values S_i, $i = 1, \ldots, 4$, on the four wheels we obtain

$$S_i = \frac{v_v - v_{ci}}{v_v} \tag{19.1}$$

$$= \frac{\frac{1}{4}\overbrace{\sum_{k=1}^{4} v_{ck}}^{=at} - \overbrace{v_{ci}}^{=at}}{\frac{1}{4}\sum_{k=1}^{4} v_{ck}} \tag{19.2}$$

$$= 0 \, . \tag{19.3}$$

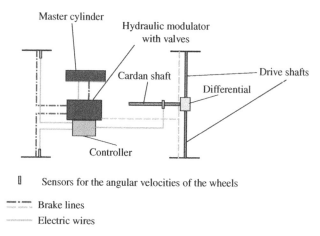

Figure 19.2 Topological principle of the first ABS from Bosch (1978) with one velocity sensor for the Cardan velocity

Of no relevance for this observation is the fact that the circumferential speeds decrease linearly with time. ABS systems use both inductive wheel speed sensors (in older systems) and wheel speed sensors based on the Hall effects. The difficulty in inductive wheel speed sensors is that they produce a small measured signal at low speeds which cannot accurately undergo further processing under certain circumstances. The Hall sensors provide a measurement signal of the same quality which is independent of the speed.

The hydraulic unit controls the brake pressures in the four wheel brake cylinders. It is essential here (see Figure 19.3) that the pressure in the brake circuit should not be passed directly to the wheel brake cylinder, but that two 2/2 solenoid valves should be provided for each wheel brake. One of these solenoid valves, the so-called inlet valve (IV), constitutes the connection between the master cylinder and the brake cylinder, the second, called the outlet valve (OV), constitutes a connection to the feedback circuit. The low-pressure storage takes the brake fluid in the brake decompression; the return pump supports the brake fluid return.

A constant pressure is maintained in the wheel brake cylinders provided the IV is closed, and the OV is also closed. This is one way in which the control device can influence the braking pressure. If the IV is closed and the OV is opened further, the braking pressure and hence the braking torque on the concerned wheel decreases. This is a second way in which the control device can influence the braking torque. After the brake pressure has been reduced in such a way, it can be increased again by closing the OV and re-opening the IV. Opening the IV briefly enables the pressure to be increased in steps. Likewise, the pressure can also be reduced in steps with the IV closed and the OV being opened briefly.

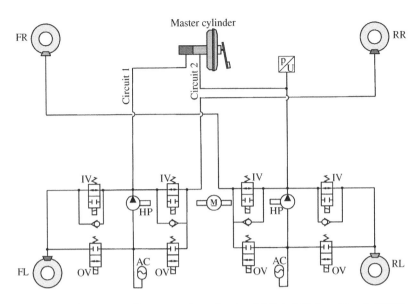

Figure 19.3   Hydraulic connections for the ABS (cf. Robert Bosch 2007)

Figure 19.3 shows the complete hydraulic circuit diagram of an ABS system. The requirements to be met by an ABS device are numerous, which include the following:

- Driving stability and steer ability should be guaranteed under different road conditions (dry, wet, icy road surfaces).
- The coefficient of adhesion should always be fully utilized on as many roadways as possible.
- In rapidly changing road conditions, the intervention of the ABS must be limited to a short period so that the braking distance is not prolonged unnecessarily. The longer intervention time of the ABS during such varying road conditions means that the good coefficients of adhesive of dry, non-icy road would not be utilized, thus prolonging the braking distance unnecessarily.
- If the road surfaces for the left and the right wheels are of a different nature (this is known as a split-μ road), then the yaw moments caused by the different braking forces on the left and right sides should rise slowly so that a normally skilled driver can countersteer.
- Even when cornering, the ABS should enable shorter braking distances, bearing in mind that the limit speed plays an essential role here during cornering. In this context, it is important that the total transmittable forces through the tyres are limited (Kamm's circle).
- Uneven roads lead to wheel load fluctuation. ABS should also be able to control these varying conditions.
- ABS should also be able to control aquaplaning.

19.2   ABS Operations

The areas in the longitudinal force coefficient–slip curve (μ-S curve) in which the ABS engages are determined substantially by adhesion limits for straight-line driving. The diagram in Figure 19.4 shows the curves for different road surfaces. The longitudinal force coefficient, μ, is the ratio of tangential longitudinal force F_x to the wheel load F_z:

$$\mu = \frac{F_x}{F_z} . \tag{19.4}$$

Similarly, for the cornering force, F_y, a cornering force coefficient $\mu_s = F_y/F_z$ can be defined. The cornering force depends on the slip angle α; for small slip angles this dependence can be linearized: $F_y = c_\alpha \alpha$. Both forces, F_x and F_y, act in the contact area and influence each other. For this reason, the longitudinal force coefficient, μ, and the cornering force coefficient, μ_s, depend on the slip angle, α (cf. Figure 19.5).

The areas in which the ABS is engaged are located at the maximum of the curves near the coefficient of adhesion, μ_a. The maximum also corresponds to the optimal range for the minimum stopping distance. The only exception is the curve for snow. This occurs because the loose snow forms a wedge in front of a locking wheel (this

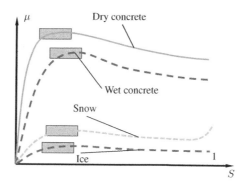

Figure 19.4   Operating areas of ABS (adapted from Robert Bosch 2007)

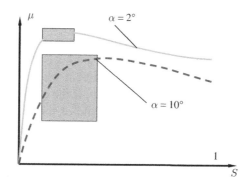

Figure 19.5   Operating areas of ABS with slip angle (adapted from Robert Bosch 2007)

is indicated by the slope of the curve for very large slip values), thus increasing the braking effect. Similar behaviour occurs when braking on gravel or crushed stone. Since these are unusual conditions for passenger cars, their significance is low, yet braking on these roads with ABS leads to an extension of the braking distance. If a driver is driving on a roadway with conditions such as these, then switching off the ABS leads to shortened braking distances. However, it is important to note that this is more common in off-road vehicles. These situations (snow, gravel) may lead to significantly longer braking distances because of the absence of wedge formation due to the ABS intervention; under certain circumstances, it is not possible to stop the vehicle on an inclined road. In commercial vehicles, it should therefore be possible to switch off the ABS when the driving conditions are as described.

Figure 19.5 shows the conditions during cornering. The main difference while cornering when compared to the straight-ahead driving is that the wheels must not only transmit longitudinal but also lateral forces. This means that the wheels need to be controlled during cornering in different ways with respect to the ABS than when driving in a straight line. For small lateral slip angles (as shown in Figure 19.5 for $\alpha = 2°$) the operating area of the ABS is almost identical to the area of straight-ahead driving.

For larger lateral slip angles ($\alpha = 10°$), however, the ABS system initially operates with very small slip values, S, and a small longitudinal force coefficient, μ. Braking reduces the speed so that the lateral forces decrease rapidly. The quadratic dependence of the centrifugal forces, $F_c = \frac{mv^2}{\rho_{cc}}$, plays an important role here. Due to the quadratically decreasing lateral forces as braking continues, the ABS braking can be active towards higher slip values and thus be active at a higher longitudinal force coefficient, μ. Consequently, it can then operate at higher braking forces.

Control variables used for the ABS are the circumferential velocity, v_c, of the wheel (which is used to calculate the circumferential acceleration), the wheel slip, the reference speed and the vehicle deceleration. These quantities are calculated or estimated from the wheel speed sensor signals. One means of determining the vehicle speed is to calculate the mean value of the speed of the diagonally opposite wheels. This possibility of vehicle speed estimation is no longer available if ABS emergency braking is engaged. Then the speed can only be roughly estimated. The control variable of wheel circumferential velocity, v_c (or acceleration) must be assessed in different ways for driven and non-driven wheels. For non-driven wheels, the inertial properties are known; hence, the increase in wheel circumferential deceleration can be assessed relatively easily in an emergency stop. For driven wheels, the increase in wheel circumferential deceleration drops in a slightly different way, depending on whether it is braked with the engaged or non-engaged clutch. If the clutch is engaged in the first or second gear, the inertia values of the wheel and of the drive train, including the transmission and the motor and the drag torque of the combustion engine, are considerably greater than those in the non-engaged state. One result of the different moments of mass (or different mass correction factors) under emergency braking conditions is therefore a slower increase in the wheel circumferential acceleration, making the detection of wheel lock difficult.

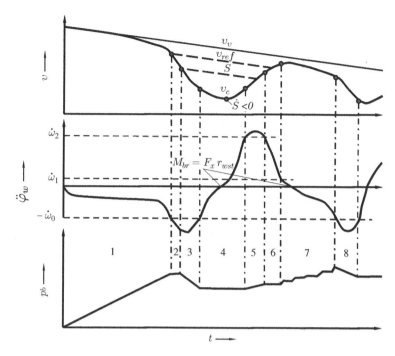

Figure 19.6 Control cycle of an ABS (adapted from Robert Bosch 2007; discontinuities are added in the time derivative of $\ddot{\varphi}_w$ at the switching points of the valves, since a discontinuity in the time derivative of the braking pressure results in a discontinuity of $\dddot{\varphi}_w$: $J_w \dddot{\varphi}_w = \dot{M}_{br} - \dot{F}_x r_{wst}$. At the point with $\dot{v}_c = 0$ the slip rate is smaller than zero, $\dot{S} = \dot{v}_v v_c / v_v^2 < 0$, the slip is greater than $\mu^{-1}(\mu_h)$; at the points with $\ddot{\varphi}_w = 0$ the braking moment $|M_{br}|$ and the moment from the tangential force $|F_x r_{wst}|$ are equal: $M_{br} - F_x r_{wst} = 0$)

In the following, a typical control cycle during braking on a dry road surface is explained. The essential variables are shown in Figure 19.6, as functions of time (brake pressure, wheel circumferential acceleration and vehicle speed and wheel circumferential speed). The entire control cycle comprises phases 2–7; phase 8 marks the start of a new operating cycle.

First, in phase 1, the brake pressure is built up. This reduces the vehicle driving velocity, v_v, and the circumferential wheel velocity, v_c. Comparison of the two velocities reveals that the wheel velocity, v_c, decreases faster than the vehicle velocity, v_v. Consequently, the slip increases and the longitudinal force coefficient, μ, approaches the adhesion coefficient, μ_a. Exceeding the adhesion coefficient, the wheel goes into the non-stationary region, whereby it is greatly accelerated. This high acceleration is detected by the ABS control unit. As shown in Figure 19.6, the wheel circumferential acceleration thus exceeds the negative value $-\dot{\omega}_0$ (this is a fixed parameter of the ABS), and the ABS control unit intervenes. This involves the inlet valve closing, whereby the pressure in stage 2 is maintained at a constant value. The velocity at the start of phase 2 is the so-called reference velocity, which continues to decrease following a specific target(e.g., extrapolation of the deceleration at the beginning of braking). The assumption behind the extrapolation is that the slip, S, is in an area where a constant longitudinal force coefficient, $\mu \approx \mu_a$, predominates, so that we can assume that the vehicle deceleration is nearly constant. This decreasing reference speed is reduced to the so-called slip threshold (denoted by S in the velocity diagram at the top of Figure 19.6). If the wheel circumferential velocity falls below this threshold, the third phase is initiated, in which the brake pressure is reduced by the OV opening. Reduction of the brake pressure ends when the absolute value of the wheel circumferential acceleration falls below the critical value of $-\dot{\omega}_0$; it is followed by phase 4 at constant pressure (inlet and outlet valves are closed). This phase ends when the circumferential wheel acceleration exceeds the positive value of $\dot{\omega}_2$. This point clearly represents an acceleration of the wheel and an approach of the wheel circumferential velocity, v_c, to the vehicle velocity, v_v. If the circumferential acceleration of the wheel exceeds the threshold $\dot{\omega}_2$, the brake pressure is increased by the intake valve opening until the threshold $\dot{\omega}_2$ is attained for a second time, but in this case for decreasing circumferential acceleration. At this point, the inlet valve is closed (phase 6) and the pressure is constant until the circumferential acceleration falls below the threshold $\dot{\omega}_1$. Thereafter, the brake pressure is increased in steps in phase 7 until the deceleration is again below the negative threshold $-\dot{\omega}_0$.

Afterwards, the second cycle starts with phase 8, in which the brake pressure is immediately reduced; a phase with constant pressure comparable to phase 2 does not take place in this second braking cycle.

The control cycles of the ABS braking on road surfaces with small adhesion coefficients differ from those with large adhesion coefficients. For roadways with a small coefficient of adhesion, phases 1 and 3 take the same course for both road conditions, in phase 4, a further reduction of the pressure is necessary, otherwise the time for accelerating the wheel again would be too long, the control cycles would become too long and the ABS would not meet the requirement for short intervention times.

19.3   Build-up Delay of Yaw Moment

If road characteristics for the left wheels are different from those for the right (different adhesion limits, μ_{high} and μ_{low}), a deceleration of the vehicle causes a yaw moment due to the different braking forces. This yaw moment has different effects depending on the vehicle class and requires different behaviour from the driver. The building-up of yaw moment depends not only on the adhesion conditions but also on the track of the vehicle. For a large track, the product of force and lever arm is greater than for a vehicle with a smaller track. If s is the track, then the maximum moment, M_{max} (with respect to the centre of mass of the vehicle) is (G is the weight of the vehicle)

$$M_{\mathrm{max}} = \frac{s}{2}(\mu_{\mathrm{high}} - \mu_{\mathrm{low}})\frac{G}{2} \ . \tag{19.5}$$

The impact of this moment will differ depending on the wheelbase ℓ (distance from the front to the rear axle) and the mass moment of inertia of the vehicle for rotation about the vertical axis. First, the moment $M = M(t)$ results in an angular momentum $L = J_z \dot{\psi}$ (J_z is the mass moment of inertia for a rotation about the vertical axis and $\dot{\psi}$ is the yaw rate). The angular momentum can be calculated by the integral

$$L = \int_0^{\Delta t} M(t) \, dt \, . \tag{19.6}$$

This results in a yaw rate after the time Δt:

$$\dot{\psi} = \frac{1}{J_z} \int_0^{\Delta t} M(t) \, dt \, . \tag{19.7}$$

The yaw moment must be compensated for by cornering forces, which means that the driver has to countersteer (cf. Figure 13.3). The moment of countersteering with respect to the centre of mass is (assuming that the centre of mass is in the geometric centre of the four contact patches; $c_{\alpha 1}$ is the cornering stiffness of one wheel at the front axle and $c_{\alpha 2} = c_{\alpha 1}$ at the rear axle, and δ_{counter} is the angle of countersteering at the front wheels)[1]

$$M_{\text{counter}} = 2 c_{\alpha 1} \delta_{\text{counter}} \frac{\ell}{2} \, . \tag{19.8}$$

The necessary compensation forces are lower for a large wheelbase than those for a small wheelbase. In large heavy vehicles, the mass moment of inertia J_z is large, so that the yaw moment only leads to a slow increase in the yaw rate of the overall vehicle. With the slow increase and the large wheelbase, it is easier to countersteer in a large vehicle than in a small vehicle. Countersteering is necessary for both vehicles, and, to make this easier, the ABS build-up of pressure at the μ_{high} side is delayed. The delay is smaller for the large car than for the smaller car.

Figure 19.7 shows the pressure in the master cylinder, and the pressures at the μ_{high} and the μ_{low} side as examples. The ABS cycle starts nearly simultaneously on both sides. The marked area stands for the integral of the difference of the two pressure curves, and this integral stands for the change in angular momentum, because the pressure is proportional to the braking force and therefore the pressure difference is proportional to the difference in braking forces; with this difference in braking forces the yaw moment can be calculated.

Figure 19.8 shows the pressure delay for a small vehicle, while Figure 19.9 shows that for a large car. It is evident that the pressure for the small vehicle at the μ_{high} side increases very slowly, which results in a small integral and therefore in a small change in angular momentum, whereas the pressure build-up delay for the larger vehicle is smaller, while the integral and the change in angular momentum are greater.

---

[1] Similar to the consideration of the effect of crosswind, a vehicle sideslip angle $\beta \neq 0$ and slip angles $\alpha_1 \neq 0$ and $\alpha_2 \neq 0$ occur at the front and the rear wheel of a single track model. In the described situation only a moment acts on the vehicle, no lateral forces occur. This means that the lateral forces at the front and the rear axle have opposite signs and the same absolute values: $\alpha_2 = \delta_{\text{counter}}/2$ and $\alpha_2 = -\delta_{\text{counter}}/2$ or vice versa; $|\beta| = \delta_{\text{counter}}/2$. In this consideration we neglect the geometric non-linearities from $\beta \neq 0$.

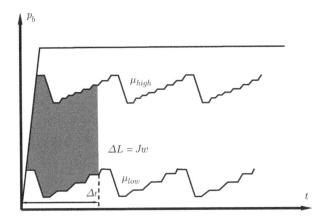

Figure 19.7   Pressure build-up for split-μ without delayed pressure build-up

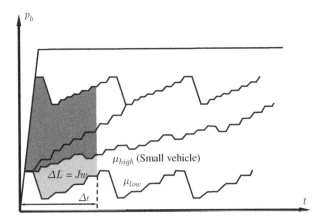

Figure 19.8   Delayed pressure build-up for a small vehicle

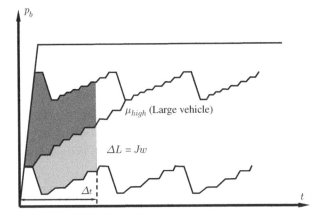

Figure 19.9   Delayed pressure build-up for a large vehicle

   The delay of the yaw moment plays an important role in cornering. When braking takes place during cornering, an increase in the wheel load on the front axle and a reduction in the wheel load on the rear axle occur. Due to the larger wheel loads, the lateral forces on the front axle are increased and the lateral forces on the rear axle are reduced. This results in a yaw moment which turns the vehicle to the inside of the curve. If we consider the wheel load, it is apparent that the wheel loads differ due to the moment of the centrifugal force: on the outer wheels the loads are larger than those on the inside wheels. If in this situation, the build-up delay for the pressures from ABS takes effect, then the braking force would have a delayed build-up on the outer wheels. However, this is not desirable because a rapid build-up of the longitudinal braking forces on the outer wheels counteracts the inward curve turning yaw moment from load transfer. Consequently, the yaw moment build-up delay must not be active during cornering as it is necessary for split-μ braking.

19.4   Traction Control System

In some situations, the wheels may spin when accelerating during unfavourable road conditions. One problem with spinning wheels is that not only do the longitudinal forces decrease due to the transition from adhesion to sliding, but the cornering forces are also no longer transferable. Anti-slip regulation (ASR) systems are used to avoid this. There are essentially two reasons for large slip. One reason is a low coefficient of adhesion for both wheels of a driven axle. In this case, both wheels will spin, causing the lateral forces to break and thus the lateral stability will decrease. The side force loss on rear-wheel-drive vehicles is critical because this can pose the threat of the rear breaking out. The other reason is a split-μ road surface. The wheels of an axle on the vehicle are connected with each other by a differential gear. This differential gear is required because it helps to drive the vehicle around the curve, because the wheels should be able to spin relatively to each other during cornering (the radii or the outer and inner wheels differ). The disadvantage of an axle differential, however, is that the overall transferable torque is determined by the wheel that rolls on the part of the roadway with the lower adhesion coefficient μ_{low}. If the coefficient of adhesion is very low, then a low drive torque will be transmitted even on the wheel that rolls on the part of the roadway with a high adhesion coefficient, μ_{high}. Conventional bevel gear differentials distribute the drive torque in equal parts between the left and right wheels. When the vehicle starts off on the split-μ road surface so that the μ_{low} wheel spins, the same torque transferred to the road via the μ_{low} -wheel will also be transmitted to the μ_{high}-wheel.

   This is the point at which ASR is engaged. ASR has two tasks: firstly, it influences the total torque of the engine and, secondly, it controls the torque distribution between left and right wheels. If we consider the total torque, M_{tot}, on a wheel, we can see that it comprises half of the moment of the Cardan shaft, $M_{car}/2$, a braking torque, M_b, and, due to circumstances defined by the roadway, braking torque as a result of rolling resistance, M_{road}:

$$M_{\text{tot}} = M_{\text{car}}/2 + M_b + M_{\text{road}} \ . \tag{19.9}$$

In this formula, the braking torque, M_b, and the road moment, M_{road}, are taken as negative. The moment, M_{tot}, can be influenced by $M_{\text{car}}/2$ and M_b. The moment from the Cardan shaft, $M_{\text{car}}/2$, can be adjusted by varying the parameters of the engine (electric, combustion or hybrid), and the moment M_b can be adjusted via the individual braking moment from the brakes.

This second option is that ASR can be engaged by torque control, based on the individual braking torques for the left and right wheels. In order to brake the left and right wheels during acceleration of a vehicle, in addition to the conventional ABS system, there should be a possibility to provide for a pressure build-up in the braking system without brake pedal operation. For this reason, the return pump in the ASR system must be a self-priming pump in order to bring about the pressure build-up. However, no pressure is taken from the brake cylinders through this self-priming pump, a special non-return valve between the return pump and the respective OVs of the ABS must be additionally provided.

The ASR includes two controls: first, the Cardan controller, which regulates the torque at the Cardan shaft, and the second the differential lock control, which controls the torque difference between the wheels. The Cardan speed regulator responds to a rapid increase in Cardan shaft speed, which prevents the corresponding two wheels of the driven axle from spinning. The differential lock controller responds to differential speeds between the wheels and brakes the fast-spinning wheel by applying a braking torque. This braking torque primarily affects the overall torque balance on the corresponding wheel. However, a larger torque can be transferred to the non-spinning wheel on the μ_{high} side by the increased torque on the spinning wheel.

19.5 Lateral Stability Systems

The ESP (electronic stability programme)[2] helps the driver in critical road scenarios involving lateral dynamics. The ESP prevents accidents due to skidding and reduces the driver's steering effort. The ESP is intended to perform when certain driving situations deteriorate. In today's systems, the ESP is often secondary to ABS and ASR. In contrast to ABS, the ESP is based on the control variables such as sideslip angle, β, and yaw rate, $\dot{\psi}$. The sensors used for the ESP system are the angular velocity sensors on all four wheels. In addition to ABS, a steering wheel angle sensor, δ_s, a yaw rate sensor, $\dot{\psi}$, an acceleration sensor for the lateral acceleration, a_c, and a pressure sensor for detecting the pressure, p_b, in the master cylinder are also used (cf. Figure 19.10). The idea of the ESP systems is to apply a corrective torque about the z-axis to the vehicle by braking individual wheels in order to stabilize the vehicle. The effect which is

---

[2] Other names for similar systems are used: DSC (dynamic stability control), VSC (vehicle stability control) as well as many other names and abbreviations.

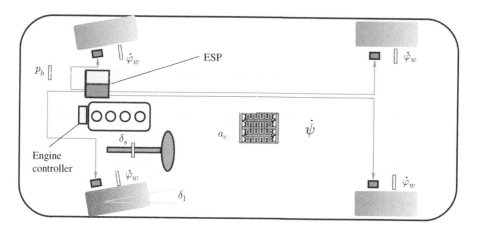

Figure 19.10 Components of an ESP system (adapted from Robert Bosch 2007)

obtained through the braking of individual wheels is an additional yaw moment due to brake longitudinal forces, and a reduction in the cornering forces, especially in the area of the limits of Kamm's circle.

Using the measured variables such as lateral acceleration, a_c, steering wheel angle, δ_s, yaw rate, $\dot{\psi}$, angular velocities of the wheels $\dot{\varphi}_{wi}$, $(i = 1, \ldots, 4)$, as well as the estimated value of the vehicle longitudinal speed, v_x, it is possible to estimate other values such as the braking forces, F_{xi}, on the four wheels, the sideslip angle, β, the slip angles at the four wheels, α_i, the lateral speed, v_y, as well as the cornering forces on the wheels, F_{yi}, the wheel loads, F_{zi}, and the resulting forces in the contact patches, F_{ri} $(i = 1, \ldots, 4)$.

In order to estimate the vehicle sideslip angle, we proceed from the following equation:

$$\dot{\beta} = -\dot{\psi} + \frac{1}{v_x}(a_c \cos \beta - a_x \sin \beta) . \tag{19.10}$$

From the relationship between centripetal acceleration $a_c = \frac{v_x^2}{\rho_{cc}}$ (strictly speaking, the total velocity has to enter in the equation, but the vehicle sideslip angle is usually small, and then $v \approx v_x$) and from the angular velocities of $\dot{\beta}$ and $\dot{\psi}$, we obtain

$$\frac{v_x^2}{\rho_{cc}} = v_x(\dot{\beta} + \dot{\psi}) . \tag{19.11}$$

Here we utilize the fact that the velocity, v, of a point on a circular trajectory with radius R defines the angular velocity, $\omega = v/R$. Substituting $R = \rho_{cc}$, $v = v_x$ and $\omega = \dot{\beta} + \dot{\psi}$ yields Equation (19.11).

If we assume small longitudinal accelerations, a_x, and small vehicle sideslip angles, β, Equation (19.10) then simplifies to

$$\dot{\beta} = \frac{a_y}{v_x} - \dot{\psi} . \tag{19.12}$$

Integrating equation (19.12) gives us a formula for calculating the vehicle sideslip angle, β, as a function of time:

$$\beta(t) = \beta_0 + \int_{t_0}^{t} \left(\frac{a_y}{v_x} - \dot\psi \right) \, \mathrm{d}t \quad . \tag{19.13}$$

Due to errors in the measurement values, a Kalman filter can be used on the basis of the differential equations for a two-track model in order to estimate the longitudinal velocity of the vehicle.

The target values for the vehicle sideslip angle, β, and the yaw rate, $\dot\psi$, are determined from the measured quantities of the lateral acceleration, a_c, steering wheel angle δ_s, brake pressure in the brake master cylinder, p_b, and the desired motor torque. These are compared with the estimated values of the vehicle; if the difference is too large, a correction torque is calculated, which is obtained by braking individual wheels.

19.6 Hydraulic Units for ABS and ESP

Figure 19.3 shows the hydraulic circuit diagram of a four-channel ABS hydraulic unit for an X-brake design. In the X-design, each diagonally opposite wheel is braked by the two brake circuits. The diagram shows that an IV and an OV exist for each wheel brake cylinder. Furthermore, there is a hydraulic pump, HP, in each brake circuit. The pump is used to transport the brake fluid when the OV is opened in order to reduce the pressure. Since the response times of the OVs are very short, a low-pressure accumulator, AC, is additionally provided as a reservoir in the return circuit to receive this very short-term accumulation of brake fluid quantities. The ABS devices of the first and second generations used 3/3-way solenoid valves in which functions such as pressure build-up, pressure reduction and pressure maintenance were achieved by only one valve. Since these valves were very expensive in terms of electrical activation and complicated in terms of mechanics, they were replaced by two 2/2-solenoid valves. In the hydraulic circuit diagram illustrated all wheels can be individually controlled. There are other arrangements in which the wheels of the rear axle are controlled as a whole. In these systems, the select-low principle is applied, this means that the slip of the wheel which rolls on the road surface with a low coefficient of adhesion, μ_{low}, determines the intervention of the ABS for the two rear wheels.

If a vehicle is also equipped with an ASR system, the return pump must be able to independently build up pressure. A self-priming pump is therefore used in this case. Furthermore, two additional valves must be provided for each circuit of an X-brake layout, whereby an ASR system is equipped with a total of twelve valves. Figure 19.11 shows the hydraulic circuit diagram of an ESP hydraulic unit (also for an X-brake layout). This diagram shows the two additional valves in each brake circuit. In contrast to the ASR system, it may be necessary to increase the brake pressure that is applied by the driver in a brake circuit in the ESP system. For this reason, the HSV (high-pressure

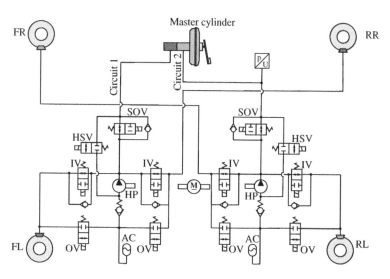

Figure 19.11   Complete ESP hydraulics (cf. Robert Bosch 2007 or Bauser and Gawlik 2013)

selector valve) in an ESP system is designed so that it can switch against higher differential pressures. During active braking intervention for both ASR and ESP systems, the HSV is open and the SOV (switch over valve) is closed. This allows the pump, HP, to increase the pressure in the respective brake circuit, with brake fluid being pumped into the respective circuit from the reservoir. Consequently, the self-priming pump does not take in the brake fluid from the brake cylinders unintentionally, and a non-return valve is provided for each pump.

19.7   Active Steering System

In the field of steering systems, active steering systems are being used increasingly in both the upper and in the medium and compact classes (active front steering, AFS). This technology does not rely on pure steer-by-wire systems, but often uses so-called superimposed steering systems. The mechanical feed-through from the steering wheel to the rack is present at all times, and is implemented by means of a planetary gearbox (see Figure 19.12). The steering wheel is connected to the steering column by the planetary gear, while the sun gear forms the connection to the rack. An electric motor with a screw drive is used as active element, acting on the ring gear member. In the case of failure of the electric motor or the control electronics, an electromagnetic lock ensures that the ring gear member is locked and so the mechanical feed-through is also guaranteed by the steering wheel to the rack.

The active steering (superimposed steering) system can perform different tasks. Firstly, a dependence of the steering ratio i_s on the velocity of the vehicle can be achieved. This example is shown in the lower part of Figure 19.12. For low speeds,

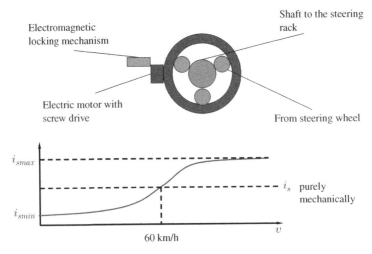

Figure 19.12 Schematic diagram of active steering with planetary gear; variable steering ratio

the steering ratio is small, then the steering is direct and small steering wheel angles lead to large steering movements. At high speeds, however, the gear ratio is large, the steering is indirect and small steering movements lead to small effects on the front wheels. This behaviour is useful at low speeds because it makes the vehicle easier to manoeuvre. At high speeds, the system provides a gain in vehicle comfort due to enhanced controllability.

In addition to this variable steering ratio, the system can actively intervene in the driving process to make course corrections in a similar way to ESP. Compared to the ESP, a steering intervention by an active steering system is quicker, less noticeable than a braking intervention by the ESP, but the stabilizing effect of the active steering is not as great as in the case of the ESP. A possible steering intervention can be advantageous for split-μ braking.

19.8   Questions and Exercises

Remembering

1. What components is ABS composed of?
2. What additional hydraulic components are necessary for ASR compared to ABS?

Understanding

1. Explain how a hydraulic ABS operates.
2. Explain the intervention strategy of ABS in the brake pressure.
3. Explain the requirements for ABS.

4. Explain the differences between ABS for small and large vehicles with regard to split-μ braking and explain appropriate corrective measures.
5. Explain why the build-up delay of yaw moments when braking in curves is generally undesirable.
6. Explain the operation of ASR when starting off on split-μ road surfaces.
7. Explain the operation of the two built-in ASR controls for the speed of the Cardan shaft and differential speeds between the driven wheels.

20

Multi-body Systems

In the virtual development process of new vehicles, dynamic behaviour is calculated by so-called multi-body systems (MBS). The simple models described in the preceding chapters are not sufficient to obtain details of the behaviour, and, instead, more precise models have to be used, for example models that capture geometric non-linearities or which capture more precisely the behaviour of the tyre. In this chapter we therefore present a brief introduction to MBS[1]. However, the basics from the preceding chapters may be helpful in understanding and interpreting the results calculated with the aid of such MBS models.

The main components of MBS are rigid bodies, which are connected by joints and/or force elements such as springs.

One characteristic of the bodies is that they are rigid, but modern software for MBS is also able to consider flexible bodies approximately.

We start with some sample applications.

Example 20.1 One broad area of application is cars and trains, but robots are also investigated using MBS. Figure 20.13 shows an MBS model of a front axle. Typically, these models consist of one central rigid body, which is the body of the car with additional masses for features such as seats and other interior equipment. Other rigid bodies are chassis subframe, suspensions (trailing arms, wishbones, etc.), wheel carriers and wheels as well as the different parts of the powertrain, e.g. engine, clutch, transmission, Cardan shaft, differential and drive shafts. Although it is typical for the bodies to be rigid in MBS, some of them are approximated by means of flexible algorithms in order to capture their compliances.

Example 20.2 In order to investigate vehicle safety in the automotive industry, dummies are employed in crashtests. The behaviour of these dummies can be simulated by means of MBS. For instance, the extremities are several rigid bodies which are joined, for example, by revolute joints (elbow joint) or by spherical joints (hip joint).

---

[1] Further recommended reading can be found in Blundell and Harty 2004 or Roberson and Schwertassek 1988.

Since flexibility is important for other parts of the human body such as the abdomen or thorax, these parts are approximated by several rigid bodies which are connected by springs and dampers.

The goals of the MBS investigations are to reduce the risk of several injuries to occupants. This goal can be achieved by optimizing occupant safety systems such as safety belts or airbags.

Example 20.3 Other typical examples for MBS are robots in which the different arms are rigid bodies. These arms are connected by joints. The difference between these robot examples and the dummies is that active elements (electric motors) have to be provided in a robot for each joint.

20.1   Kinematics of Rigid Bodies

Generally a rigid body has six degrees of freedom in three-dimensional space. For these six degrees of freedom on the one hand, six variables are necessary on the other hand to describe both the position of a rigid body and its orientation.

Three coordinates are necessary in order to define the position of a point P in three-dimensional space. If we start from an inertial frame $(O, \vec{e}_{ix}, \vec{e}_{iy}, \vec{e}_{iz})$ (cf. Figure 20.1), the vector from the origin of the frame O to the point P, $\overrightarrow{OP} = \vec{r}_p$, can be described by using three coordinates x, y, z (Cartesian coordinates or rectangular coordinates):

$$\vec{r}_p = x\vec{e}_{ix} + y\vec{e}_{iy} + z\vec{e}_{iz} . \tag{20.1}$$

We prefer the notation using tuples and scalar products:

$$\vec{r}_p = x\,\vec{e}_{ix} + y\,\vec{e}_{iy} + z\,\vec{e}_{iz}$$

$$= \underbrace{(x, y, z)}_{=\underline{r}_p^T} \underbrace{\begin{pmatrix} \vec{e}_{ix} \\ \vec{e}_{iy} \\ \vec{e}_{iz} \end{pmatrix}}_{=\vec{\underline{e}}_i} \tag{20.2}$$

$$= \underline{r}_p^T \vec{\underline{e}}_i . \tag{20.3}$$

If one point of a rigid body is fixed, the body can rotate about this point; hence, the orientation of the body is not fixed. As we need a total of six variables for six degrees of freedom, there are three additional variables apart from the so-called translational variables x, y, z (corresponding to the three translational degrees of freedom). From Euler's theorem (cf. Roberson and Schwertassek 1988), the motion of a rigid body can be divided into a translation and a rotation, and the rotation can be described by three parameters. There are several ways of using these three parameters.

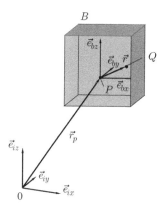

Figure 20.1   Frames and coordinate systems

One easy way is to introduce a so-called body frame or body-fixed frame $(P, \vec{e}_{bx}, \vec{e}_{by}, \vec{e}_{bz})$ (cf. Figure 20.1). The orientation of the body can be described by the orientation of the axis system $(\vec{e}_{bx}, \vec{e}_{by}, \vec{e}_{bz})$ with respect to the inertial axis system $(\vec{e}_{ix}, \vec{e}_{iy}, \vec{e}_{iz})$. Since both systems are dextral and orthonormal systems (following the right-hand rule: thumb $= \vec{e}_x$, forefinger $= \vec{e}_y$, middle finger $= \vec{e}_z$), it is known (from mathematics) that a rotation matrix $\underline{\underline{R}}$ exists, which maps the inertial axis system to the body-fixed system:

$$\underline{\vec{e}}_b = \underline{\underline{R}}\, \underline{\vec{e}}_i \ . \tag{20.4}$$

We will come back to the properties of the matrix later. At this point, we need only the inverse of $\underline{\underline{R}}$, which is simply the transposed:

$$\underline{\underline{R}}^{-1} = \underline{\underline{R}}^T \ . \tag{20.5}$$

We now consider a body B (see Figure 20.1) which is translated and rotated and we want to describe one point Q in this body. The vector $\vec{r}$ from P to Q can be written with the aid of the body-fixed frame:

$$\vec{r} = \overrightarrow{PQ} = (r, s, t)\, \underline{\vec{e}}_b \ . \tag{20.6}$$

Altogether, we thus derive the vector $\vec{r}_Q = \overrightarrow{OQ}$ by adding $\overrightarrow{OP}$ and $\overrightarrow{PQ}$:

$$\vec{r}_Q = \vec{r}_p + \vec{r} \ . \tag{20.7}$$

Now we want to derive the velocity of Q (in order to obtain the kinetic energy of the body B). Consequently, we have to differentiate the vector $\vec{r}_Q$ with respect to time. For this differentiation, it is important to notice that the coordinates (x, y, z) of P depend on the time, whereas the coordinates (s, r, t) are independent of time, because the point Q is fixed within the body B. We obtain

$$\dot{\vec{r}}_Q = \dot{\vec{r}}_P + \frac{d}{dt}(\vec{r})$$

$$= \dot{\vec{r}}_P + \frac{d}{dt}((r,s,t)\,\underline{\underline{R}}\,\vec{e}_i)$$

$$= \dot{\vec{r}}_P + (r,s,t)\,\underline{\underline{\dot{R}}}\,\vec{e}_i \qquad \text{substitute} \qquad \vec{e}_i = \underline{\underline{R}}^T\,\vec{e}_b$$

$$= \dot{\vec{r}}_P + (r,s,t)\,\underbrace{\underline{\underline{\dot{R}}}\,\underline{\underline{R}}^T}_{=\underline{\underline{\Omega}}}\,\vec{e}_b$$

$$= (\dot{x},\dot{y},\dot{z})\,\vec{e}_i + (r,s,t)\,\underline{\underline{\Omega}}\,\vec{e}_b\ . \tag{20.8}$$

Let us first consider the matrix $\underline{\underline{\Omega}} = \underline{\underline{\dot{R}}}\,\underline{\underline{R}}^T$. To obtain the properties of this matrix, we compute the derivate of the identity matrix

$$\underline{\underline{I}} = \begin{pmatrix} 1 & 0 & 0 \\ 0 & 1 & 0 \\ 0 & 0 & 1 \end{pmatrix}\ . \tag{20.9}$$

As

$$\underline{\underline{I}} = \underline{\underline{R}}\,\underline{\underline{R}}^T \tag{20.10}$$

is independent of time, the derivative with respect to time is

$$0 = \underline{\underline{\dot{R}}}\,\underline{\underline{R}}^T + \underline{\underline{R}}\,\underline{\underline{\dot{R}}}^T$$

$$= \underbrace{\underline{\underline{\dot{R}}}\,\underline{\underline{R}}^T}_{\underline{\underline{\Omega}}} + \underbrace{\left(\underline{\underline{\dot{R}}}\,\underline{\underline{R}}^T\right)^T}_{\underline{\underline{\Omega}}^T}\ . \tag{20.11}$$

Thus

$$\underline{\underline{\Omega}} = -\underline{\underline{\Omega}}^T\ , \tag{20.12}$$

which means that $\underline{\underline{\Omega}}$ is an antisymmetric matrix (or antisymmetric tensor) of the angular velocities. The diagonal components of $\underline{\underline{\Omega}}$ are zero, and for the off-diagonal components, we have only three independent variables:

$$\underline{\underline{\Omega}} = \begin{pmatrix} 0 & \omega_3 & -\omega_2 \\ -\omega_3 & 0 & \omega_1 \\ \omega_2 & -\omega_1 & 0 \end{pmatrix}\ . \tag{20.13}$$

Sometimes the body-fixed axis system should be changed; then we obtain the new tensor of angular velocities $\underline{\underline{\tilde{\Omega}}}$ ($\underline{\underline{\hat{R}}}$ is the rotation matrix transforming from one body-fixed axis system to another):

$$\underline{\underline{\tilde{\Omega}}} = \underline{\underline{\hat{R}}}\,\underline{\underline{\Omega}}\,\underline{\underline{\hat{R}}}^T\ . \tag{20.14}$$

The part of the velocity of the point Q which results from the rotation can be calculated by means of a vector product or cross product:

$$(r, s, t) \underline{\underline{\Omega}} = (w_1, w_2, w_3) \times (r, s, t) . \tag{20.15}$$

20.2 Kinetic Energy of a Rigid Body

With the equation for the velocity of the point Q in the rigid body B, we are now able to compute the kinetic energy of the whole rigid body B. This kinetic energy can be used, for example, in order to derive the equations of motion with Langrange's or Hamilton's equations. In this consideration, we will assume the point P of the body-fixed frame to be the centre of mass of the body. If we denote the coordinates of the points Q within B as (r, s, t), then we have (with P being the centre of mass)

$$0 = \int_V r\rho \, dV \quad ,$$

$$0 = \int_V s\rho \, dV \quad , \tag{20.16}$$

$$0 = \int_V t\rho \, dV \quad ,$$

where V is the region of B in the three-dimensional space and ρ the mass density.
The kinetic energy, T, is

$$T = \frac{1}{2} \int_V \rho |\dot{\vec{r}}_Q|^2 \, dV , \tag{20.17}$$

where the square of the velocity can be calculated as follows:

$$|\dot{\vec{r}}_Q|^2 = (\dot{x}, \dot{y}, \dot{z})(\dot{x}, \dot{y}, \dot{z})^T$$

$$+ 2(\dot{x}, \dot{y}, \dot{z})((r, s, t) \underline{\underline{\Omega}} \, \underline{R})^T$$

$$+ (r, s, t) \underline{\underline{\Omega}} \, \underline{\underline{\Omega}}^T (r, s, t)^T . \tag{20.18}$$

The second summand vanishes after integration because the first moments of mass (20.16) are zero (because the origin of the body-fixed frame, P, is the centre of mass).

We consider the other two terms separately. The first term is the translational part of kinetic energy (M is the mass of the body B):

$$\frac{1}{2} \int_V \rho(\dot{x}, \dot{y}, \dot{z})(\dot{x}, \dot{y}, \dot{z})^T \, dV = \frac{1}{2} M |\dot{\vec{r}}_P|^2 . \tag{20.19}$$

To consider the second term, we rearrange the equation

$$(r, s, t) \underline{\underline{\Omega}} \, \underline{\underline{\Omega}}^T (r, s, t)^T$$

$$= (r, s, t) \begin{pmatrix} \omega_2^2 + \omega_3^2 & -\omega_1\omega_2 & -\omega_1\omega_3 \\ -\omega_1\omega_2 & \omega_1^2 + \omega_3^2 & -\omega_2\omega_3 \\ -\omega_1\omega_3 & -\omega_2\omega_3 & \omega_1^2 + \omega_2^2 \end{pmatrix} \begin{pmatrix} r \\ s \\ t \end{pmatrix}$$

$$= \omega_1^2(s^2 + t^2) + \omega_2^2(r^2 + t^2) + \omega_3^2(r^2 + s^2)$$

$$- 2\omega_1\omega_2 rs - 2\omega_1\omega_3 rt - 2\omega_2\omega_3 st \tag{20.20}$$

and, with the abbreviations J_{jh}, obtain the complete kinetic energy, where the last term is the rotational part of the kinetic energy.

The terms J_{jh} are the second mass moment. Here they are as follows:

$$J_{11} = \int_V \rho(s^2 + t^2)\, dV \quad , \qquad J_{22} = \int_V \rho(r^2 + t^2)\, dV \quad ,$$

$$J_{33} = \int_V \rho(r^2 + s^2)\, dV \quad , \qquad J_{12} = \int_V \rho rs\, dV \quad , \tag{20.21}$$

$$J_{13} = \int_V \rho rt\, dV \quad , \qquad J_{23} = \int_V \rho st\, dV \ .$$

Now the total kinetic energy can be rewritten as

$$T = \frac{1}{2} M |\dot{\vec{r}}_P|^2$$

$$+ \frac{1}{2}(J_{11}\omega_1^2 + J_{22}\omega_2^2 + J_{33}\omega_3^2 - 2J_{12}\omega_1\omega_2 - 2J_{13}\omega_1\omega_3 - 2J_{23}\omega_2\omega_3) \ . \tag{20.22}$$

The tensor $\underline{\underline{J}}$

$$\underline{\underline{J}} = \begin{pmatrix} J_{11} & -J_{12} & -J_{13} \\ -J_{21} & J_{22} & -J_{23} \\ -J_{31} & -J_{32} & J_{33} \end{pmatrix} \ , \tag{20.23}$$

where $J_{12} = J_{21}$, $J_{13} = J_{31}$, $J_{23} = J_{32}$, and the tensor $\underline{\underline{\Omega}}$ are formulated with respect to the body-fixed axis system $\vec{e}_b$.

If the body-fixed axis system is changed, the new tensor $\underline{\underline{\tilde{J}}}$ of mass moments can be computed:

$$\underline{\underline{\tilde{J}}} = \underline{\underline{\hat{R}}}\, \underline{\underline{J}}\, \underline{\underline{\hat{R}}}^T \ . \tag{20.24}$$

The kinetic energy can be used to derive the equations of motion, for example using Lagrange's or Hamilton's equations.

At this point, we have the kinetic energy of one rigid body. In MBS there is usually more than one body, whereas we now have to consider the case of two bodies, as depicted in Figure 20.2. The two bodies can be connected by a joint. The first of the two bodies, B_1, is described like the body B of the preceding considerations by the vector from the origin of the inertial frame and by the body-fixed frame. The joint between body B_1 and B_2 is located at J_1 in body B_1. At this point, a joint-fixed

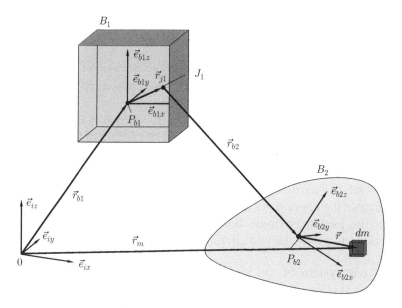

Figure 20.2   Two rigid bodies

frame is usually introduced. However, to derive the kinetic energy, we can skip this introduction of the joint-fixed coordinate system. The body-fixed axis system of $\underline{\vec{b}}_1$ can be derived by

$$(\vec{e}_{b1x}, \vec{e}_{b1y}, \vec{e}_{b1z})^T = \underline{\underline{R}}_1 (\vec{e}_{ix}, \vec{e}_{iy}, \vec{e}_{iz})^T \ . \tag{20.25}$$

where $\underline{\underline{R}}_1$ is a rotation tensor.

The vector from the centre of mass P_{b1} from body B_1 to the joint J_1 is $\vec{r}_{j1}$.

The vector from the joint J_1 to the centre of mass P_{b2} of B_2 is $\vec{r}_{b2}$.

In body B_2, we now consider the infinitesimal element of mass, dm. The vector from the origin, O, of the inertial frame to dm is

$$\vec{r}_m = \vec{r}_{b1} + \vec{r}_{j1} + \vec{r}_{b2} + \vec{r} \tag{20.26}$$

$$= \underline{r}_{b1}^T \begin{pmatrix} \vec{e}_{ix} \\ \vec{e}_{iy} \\ \vec{e}_{iz} \end{pmatrix} + \underline{r}_{j1}^T \begin{pmatrix} \vec{e}'_{b1x} \\ \vec{e}_{b1y} \\ \vec{e}_{b1z} \end{pmatrix} + \underline{r}_{b2}^T \begin{pmatrix} \vec{e}_{b1x} \\ \vec{e}_{b1y} \\ \vec{e}_{b1z} \end{pmatrix} + \underline{r}^T \begin{pmatrix} \vec{e}_{b2x} \\ \vec{e}_{b2y} \\ \vec{e}_{b2z} \end{pmatrix} \ .$$

The joint, J_1, is fixed in B_1 and the element of mass, dm, is fixed in B_2 so that the tuples $\underline{r}_{j1}$ and $\underline{r}$ of the vectors $\vec{r}_{j1}$ and $\vec{r}$ are time independent if we use the body-fixed axis systems:

$$\vec{r}_{j1} = \underline{r}_{j1}^T (\vec{e}_{b1x}, \vec{e}_{b1y}, \vec{e}_{b1z})^T \quad , \tag{20.27}$$

$$\vec{r} = \underline{r}^T (\vec{e}_{b2x}, \vec{e}_{b2y}, \vec{e}_{b2z})^T \ . \tag{20.28}$$

At this point, we have to distinguish between different kinds of joints; we give two examples:

1. If J_1 is a revolute joint, the body B_2 rotates about the axis of rotation of the joint. If we formulate the vector $\vec{r}_{b2}$ with respect to the body-fixed axis system of body B_2, the coordinates of $\underline{r}_{b2}$ are independent of time.
2. If J_1 is a prismatic joint, it is advantageous to introduce the above-mentioned joint-fixed frame $(J_1, \vec{e}_{jx}, \vec{e}_{jy}, \vec{e}_{jz})$. In this case, the origin of this joint-fixed frame is, of course, fixed to the joint J_1. The axis system $(\vec{e}_{jx}, \vec{e}_{jy}, \vec{e}_{jz})$ is oriented in such a way that one axis (e.g. the $\vec{e}_{jx}$-axis) is the axis of the revolute joint, which means the axis of relative motion coincides with this axis of the joint axis system.

   In this case, the coordinates of the vector from J_1 to P_{b2} can be expressed with respect to this joint-fixed frame. If the first axis $\vec{e}_{jx}$ is the direction of the prismatic joint motion, then only the first coordinate depends on time.

Only the axis system of the inertial frame is independent of time; the other body-fixed axis systems depend on time. We now replace the time-dependent axis systems with the inertial axis systems by introducing the time-dependent rotation matrices $\underline{R}_1$ and $\underline{R}_2$:

$$(\vec{e}_{b1x}, \vec{e}_{b1y}, \vec{e}_{b1z})^T = \underline{R}_1 (\vec{e}_{ix}, \vec{e}_{iy}, \vec{e}_{iz})^T , \tag{20.29}$$

$$(\vec{e}_{b2x}, \vec{e}_{b2y}, \vec{e}_{b2z})^T = \underline{R}_2 (\vec{e}_{b1x}, \vec{e}_{b1y}, \vec{e}_{b1z}) . \tag{20.30}$$

Then we obtain the velocity by derivation with respect to time:

$$\dot{\vec{r}}_m = \left(\dot{\underline{r}}_{b1}^T + \underline{r}_{j1}^T \dot{\underline{R}}_1 + \dot{\underline{r}}_{b2}^T \underline{R}_1 + \underline{r}_{b2}^T \dot{\underline{R}}_1 \right.$$

$$\left. + \underline{r}^T \left(\dot{\underline{R}}_2 \underline{R}_1 + \underline{R}_2 \dot{\underline{R}}_1 \right) \right) \begin{pmatrix} \vec{e}_{ix} \\ \vec{e}_{iy} \\ \vec{e}_{iz} \end{pmatrix} . \tag{20.31}$$

The first four terms yield the translational energy, and the fifth term yields the rotational energy, with both energy values being obtained after integration. Since the vectors are composed by successive rotation matrices, it is easy to derive the kinetic energy:

$$\dot{\underline{R}}_2 \underline{R}_1 + \underline{R}_2 \dot{\underline{R}}_1 = \dot{\underline{R}}_2 \underbrace{\underline{R}_2^T \underline{R}_2}_{\underline{E}} \underline{R}_1 + \underline{R}_2 \dot{\underline{R}}_1 \underbrace{\underline{R}_1^T \underline{R}_1}_{\underline{E}}$$

$$= \underbrace{\dot{\underline{R}}_2 \underline{R}_2^T}_{\underline{\Omega}_2} \underline{R}_2 \underline{R}_1 + \underline{R}_2 \underbrace{\dot{\underline{R}}_1 \underline{R}_1^T}_{\underline{\Omega}_1} \underline{R}_1$$

$$= \underline{\Omega}_2 \underline{R}_2 \underline{R}_1 + \underline{R}_2 \underline{\Omega}_1 \underbrace{\underline{R}_2^T \underline{R}_2}_{\underline{E}} \underline{R}_1$$

$$= \left(\underline{\underline{\Omega}}_2 + \hat{\underline{\underline{\Omega}}}_1 \right) \underline{\underline{R}}_2 \underline{\underline{R}}_1 \,, \tag{20.32}$$

where

$$\hat{\underline{\underline{\Omega}}}_1 = \underline{\underline{R}}_2 \underline{\underline{\Omega}}_1 \underline{\underline{R}}_2^T \tag{20.33}$$

is the tensor of angular velocities with respect to the body-fixed axis system of B_2.
Looking at the last expression in velocity:

$$\underline{r}^T \left(\underline{\underline{\dot{R}}}_2 \, \underline{\underline{R}}_1 + \underline{\underline{R}}_2 \, \underline{\underline{\dot{R}}}_1 \right) \begin{pmatrix} \vec{e}_{ix} \\ \vec{e}_{iy} \\ \vec{e}_{iz} \end{pmatrix} = \underline{r}^T \left(\underline{\underline{\Omega}}_2 + \hat{\underline{\underline{\Omega}}}_1 \right) \underline{\underline{R}}_2 \underline{\underline{R}}_1 \begin{pmatrix} \vec{e}_{ix} \\ \vec{e}_{iy} \\ \vec{e}_{iz} \end{pmatrix}$$

$$= \underline{r}^T \left(\underline{\underline{\Omega}}_2 + \hat{\underline{\underline{\Omega}}}_1 \right) \begin{pmatrix} \vec{e}_{b2x} \\ \vec{e}_{b2y} \\ \vec{e}_{b2z} \end{pmatrix} \,, \tag{20.34}$$

it is obvious that the kinetic energy can be calculated by successive multiplication of rotation tensors and their derivations with respect to time. After integration with respect to the volume of the bodies, we obtain the kinetic energy; we have omitted the translational portion of energy in the expression. The mixed terms vanish because of the vanishing first moments of mass.

20.3   Components of Multi-body Systems

Other components can be defined in addition to the rigid bodies. We describe all components including rigid bodies in the following:

Rigid bodies: The inertia properties (mass, first and second moments of mass) are necessary in order to define rigid bodies. Usually the position of a rigid body is given by its centre of mass, and this is the reference point for the definition of the mass moments (first and second); consequently, the first moments of mass are zero, and in some programmes it is not possible to define both an arbitrary point of the rigid body for the body-fixed frame and additionally the centre of mass with respect to the first point. A different way of defining the inertia properties is to define the surface of the body and the mass density. Some MBS programmes are able to compute the inertia properties by numerical integration. In this case it is advantageous to define an arbitrary point of the body, the surface (e.g. as FE mesh) and the mass density (as a constant or non-constant function with respect to space variables). It is usual to define a body-fixed frame; with this frame it is easy to define the tensor of second moments of mass and, if necessary, the surface.

Joints: In MBS, several rigid bodies interact with each other. One possibility of interaction is through joints, which are constraints because they constrain the relative motion in pairs of bodies. To simplify the definition, additional frames are located at the joint points of the rigid bodies. These joint-fixed frames define, for example, the axis of rotation for revolute joints.

Figures 20.3–20.8 depict examples of joints. Some bodies may be constrained in motion with respect to the inertial frame; as these are special constraints because they are not constraints between two bodies, some MBS programmes have special ways of defining them.

Forces: Forces can be classified in different ways. They can be subdivided into active and passive forces. Passive forces depend only on the motion (relative

Figure 20.3 Revolute joint

Figure 20.4 Translational joint or prismatic joint

Figure 20.5 Cylindrical joint

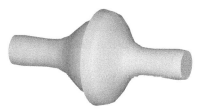

Figure 20.6   Spherical joint

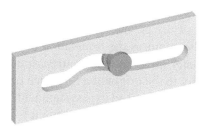

Figure 20.7   One-degree, nonlinear motion joint

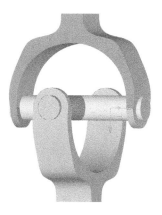

Figure 20.8   Cardanic joint

displacements and velocities) of the bodies. Active forces from actuators need power. These groups of active and passive forces are applied forces; the other group is known as constraint forces or reaction forces from the joints.

Geometry of surfaces: It is not possible in all MBS programmes to enter the surfaces of the rigid bodies. If it is possible, the surfaces can be used either for the internal computation of the inertia properties (total mass, first and second moment of inertias) or for the computation of contact forces if two bodies come into contact. The first possibility is easy to implement in the programmes. The second is challenging because additional constraint forces occur in the case of contact, and the degrees of freedom for the whole system decrease. The additional constraint forces depend on the condition in the contact point: in the case of sliding, there is one normal constraint force perpendicular to the tangential plane to both bodies at the point of contact and a tangential force which, for example, depends on the sliding velocity and on the normal force. In the case of rolling, all reaction forces between the two bodies in contact are constraint forces.

20.4   Orientation of Rigid Bodies

In addition to one point P (e.g. the centre of mass), describing a body in three-dimensional space also requires the orientation of the body. Both the point P and the orientation are usually given by relative descriptions, which means that the location of P is given relative to the origin of a reference frame, and the orientation is given by the orientation of the body-fixed axis system with respect to the axis system of the reference frame (or relative to a body-fixed frame).

As described at the beginning of this chapter, the relative displacement is given by three (Cartesian) coordinates. The orientation can be described by a rotation matrix $\underline{\underline{R}}$. The matrix has three free parameters, and, in MBS programmes, there are several ways of entering the matrix in the programme.

Rotation matrix by numerical values: One possibility is to enter all nine components. Rotation matrices defined by numerical values would not be appropriate. Consider, for example, the matrix

$$\underline{\underline{R}} = \begin{pmatrix} \frac{\sqrt{6}}{4} & \frac{\sqrt{2}}{4} & \frac{\sqrt{2}}{2} \\ -\frac{\sqrt{6}}{4} & -\frac{\sqrt{2}}{4} & -\frac{\sqrt{2}}{2} \\ \frac{1}{2} & -\frac{\sqrt{3}}{2} & 0 \end{pmatrix}, \tag{20.35}$$

which is the result of a sequence of three rotations ($30°$, $90°$ and $45°$). It is obvious that there are many square roots, and it is known that these roots (of 2 or 3 for example) can only be approximated by numerical values. This could result in the violation of a necessary condition, for example $\det(\underline{\underline{R}}) = 1$.

Euler angles: Every rotation matrix can be represented by a sequence of three simple rotations. Often the Euler convention is used for the simple rotations. Using Euler convention, the first rotation is about the $\vec{e}_3$-axis, then about the new $\tilde{\vec{e}}_1$-axis and the last rotation is about the new $\hat{\vec{e}}_3$-axis. One essential point is that the rotations are about the new axes that result from the preceding rotation. There is one angle of rotation for each rotation, which means that there are three angles and therefore three parameters. An important point is that the first and last axes of rotation depend on each other, because the last axis of rotation is the first axis of rotation rotated by the second rotational operation. To describe the sequence for the above-mentioned rotation, we may write 3–1–3. Of course, other possibilities can be used, such as 3–2–3, 1–3–1, 1–2–1, 2–1–2, 2–3–2. One feature common to all these sequences is that the last axis of rotation is the same axis of rotation as for the first operation, but rotated by the second rotational operation. The whole rotation can be described by three angles, for example φ, ϑ, ψ. Some conventions in the literature use modified angles.

There is a singularity in this representation of rotation matrices. Consider the sequence of angles with the second angle $\vartheta = \pi = 180°$ or $\vartheta = 0$. Every pair of angles for the first rotation and for the third rotation with the same value $\varphi - \psi$ or $\varphi + \psi$, resp., yields the same rotation matrix independently of the individual values for φ and ψ. The reason for this behaviour is that the first and last axes of rotation coincide (the reason in mathematics is that only trigonometric functions with the argument for $\varphi - \psi$ or $\varphi + \psi$ occur in the total rotational matrix).

An example of a rotation matrix for Euler angles is given in the following:

$$
\begin{pmatrix}
\cos\psi\cos\varphi - \cos\vartheta\sin\varphi\sin\psi & \cos\psi\sin\varphi + \cos\vartheta\cos\varphi\sin\psi & \sin\psi\sin\vartheta \\
-\sin\psi\cos\varphi - \cos\vartheta\sin\varphi\cos\psi & -\sin\psi\sin\varphi + \cos\vartheta\cos\varphi\cos\psi & \cos\psi\sin\vartheta \\
\sin\vartheta\sin\varphi & -\sin\vartheta\cos\varphi & \cos\vartheta
\end{pmatrix}
$$

Tait–Bryan angles: The main property of Euler angles is that the first and third rotations are about the same local axis. In the definition of Tait–Bryan angles all axes are different, for example the first rotation is about the $\vec{e}_1$-axis (angle α), the second rotation about the new $\tilde{\vec{e}}_2$-axis (angle β) and the third rotation about $\hat{\vec{e}}_3$ (angle γ). The angles α, β, and γ are called Tait–Bryan angles. Sometimes, especially in the German literature, they are called Cardan angles, and in programmes they are sometimes called 1–2–3-Euler angles. Similar to the Euler angle convention, a singularity also exists here if the second angle is $\beta = \pi/2$ or $\beta = 3\pi/2$.

Euler parameter: Each rotation matrix has one eigenvalue 1. The eigenvector to this eigenvalue can be interpreted as the axis of rotation; the corresponding angle is α, which can be calculated with the trace of the rotation matrix $\underline{\underline{R}}$:

$$\mathrm{tr}(\underline{\underline{R}}) = 1 + 2\cos\alpha . \tag{20.36}$$

Quaternions: Quaternions are an extension of the complex numbers. In complex numbers, the imaginary unit j ($j^2 = -1$; sometimes called i) is introduced to extend the real numbers to complex numbers. The set of quaternions together with addition operations and noncommutative multiplication operations is called non-commutative algebra. Each quaternion can be described (in a similar way to complex numbers) by

$$\alpha_1 + \alpha_2\, i + \alpha_3\, j + \alpha_4\, k\ , \tag{20.37}$$

where i, j, k are introduced as equivalents to j in the extension of real numbers to complex numbers. The additionally introduced elements i, j, k fulfil multiplication rules such as the following (this list is not complete):

$$ijk = i^2 = j^2 = k^2 = -1; \tag{20.38}$$

$$ij = k\ . \tag{20.39}$$

All quaternions with an absolute value of 1 are equivalent to the set of rotations, and Euler angles, for example, can be mapped to the quaternions.

Fields of application for quaternions include the programming of computer graphics and robots.

Caley–Klein parameters: Another means of representing rotations involves using complex matrices of the form

$$\underline{\underline{Q}} = \begin{pmatrix} \alpha & \beta \\ \gamma & \delta \end{pmatrix}. \tag{20.40}$$

If the components fulfil the following equations (the bar is the complex conjugation)

$$\alpha = \overline{\delta}\ , \tag{20.41}$$

$$\gamma = \overline{\beta}\ , \tag{20.42}$$

the matrices are a representation of the above-mentioned quaternions. With the additional condition

$$\alpha\delta - \beta\gamma = 1, \tag{20.43}$$

the matrices are a representation of the quaternions with an absolute value of 1 and therefore a representation of rotations. The parameters α, β, γ and δ are usually called the Caley–Klein parameters.

Axis system: This possibility of defining rotations is very simple and easy for users. The idea is to define the axis system by three points: the first point, N_1, is the origin of the axis system, the second point, N_2, defines the direction of the $\vec{e}_1$-vector (cf. Figure 20.9). The third point, N_3, defines a plane with N_1 and N_2. There are two vectors in this plane, which are perpendicular to $\vec{e}_1$. The vector $\vec{e}_2$ is chosen in such a way that the point N_3 is an element of the quadrant defined by the first two vectors $\vec{e}_1$ and $\vec{e}_2$ (the dashed vector in the figure does not fulfil this condition, but the solid vector does).

The third vector is defined by the cross product $\vec{e}_3 = \vec{e}_1 \times \vec{e}_2$.

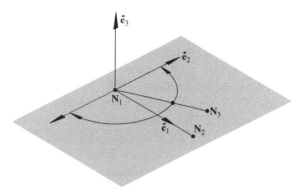

Figure 20.9   Definition of rotational matrices by axis systems

20.5   Derivation and Solution of the Equations

This section describes the principal methods for deriving the equations of motion in the first subsection, while the second subsection describes numerical algorithms for solving these equations.

20.5.1   Derivation of the Equations

In order to derive the equations of motion automatically, the computer needs a systematic description of how the different parts of an MBS are joined or connected to each other. Different possibilities of achieving this exist, with one example involving the use of graphs to describe the interconnection.

A graph is a set of vertices $V = \{v_j, j = 1, \ldots, N\}$ where each vertex represents a rigid body of the system and a set B of branches connecting the vertices. A branch is a pair of two vertices. This pair (v_j, v_k) stands for an interaction between the two bodies v_j and v_k. This interaction can be a result of a joint, but not of a force element.

An important property of the graph is whether or not a circuit exists. A circuit is a sequence of branches $\{v_{i1}, v_{i2}\}, \{v_{i2}, v_{i3}\}, \ldots, \{v_{ik-1}, v_{ik}\}$ with $v_{i1} = v_{ik}$, which means that these branches form a closed loop. The algorithms for numerical solutions of the equations of motion differ according to whether circuits exist or not. MBS without circuits can be described by a simple set of ordinary differential equations, an MBS with a circuit has to be described by a system of so-called differential algebraic equations (DAE). The numerical algorithms for solving DAE are much more complicated than those for solving a system of differential equations.

Two main approaches exist to establish the equations of motion. The first is called the Eulerian approach. This method starts with the equations of motion by Euler and Newton for each of the N rigid bodies. If Euler's and Newton's equations of motion are written in the form of first-order differential equations, 12 equations exist for each body, which means $12\,N$ equations for the whole system. One set of unknown quantities is the displacements (e.g. of the centres of gravity) and rotational angles (e.g. Euler angles) and their first derivatives with respect to the time, so that there are 12 unknowns for each rigid body. Furthermore, there are unknown forces and torques from the constraints, and, altogether, there are more unknowns than there are equations. Additional information comes from constraint equations and some principles of dynamics (Roberson and Schwertassek 1988). In one procedure of establishing equations of motion, the 12 N variables have to be reduced and the forces from the constraints have to be eliminated. This is the reason why this is called an elimination method. Another possible procedure is to introduce additional variables, the so-called Lagrange multipliers, for the unknown constraint forces and moments. As the number of unknowns grows in this procedure, this method is called an augmentation method.

Other ways of obtaining the equations of motion involve the application of Langrange's or Hamilton's equations.

One difficulty in all the methods is found in closed kinematic chains, as the variables describing the motion of the bodies depend on each other and they have to fulfil the kinematic consistency condition.

20.5.2   Solution of Equations

The equations of motion are ordinary differential equations in the case of a tree configuration of the MBS and differential algebraic equations in the case of an MBS with closed kinematic chains. The solution algorithm differs for the two types of equations.

We restrict our considerations to the easier way of ordinary differential equations without algebraic parts. The equations can be written in the simple form

$$\dot{\underline{y}} = \underline{f}(t, \underline{y}) \text{ where } \underline{y}(t_0) = \underline{y}_0 \, . \tag{20.44}$$

The tuple $\underline{y}$ comprises the state variables after elimination of constraint forces and torques (elimination method) or the tuple comprises all 12 N variables of all N bodies and, in addition, the Lagrangian multipliers for the constraint forces and torques. One of the simplest and most illustrative methods is the first-order explicit Euler method shown in Figure 20.10. In this method a polygon is computed stepwise, where the step size, h, can be variable. The local convergence rate is quadratic.

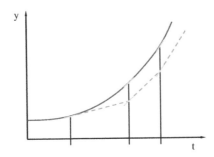

Figure 20.10   Explicit Euler method

In general, the solution procedures can be classified as explicit and implicit methods, on the one hand, and as single-step and multiple-step methods on the other hand.

The starting point is the initial value $\underline{Y}_0$ for the time $t = t_0$. From this, an iterative process yields the approximations $\underline{Y}_n$ of the function $\underline{y}$ at distinct times $t = t_n$. The general classes can be described by one formula:

$$Y_{n+1} = \psi(Y_n) \text{ explicit single-step method}, \tag{20.45}$$

$$Y_{n+1} = \psi(Y_{n+1}, Y_n) \text{ implicit single-step method}, \tag{20.46}$$

$$Y_{n+1} = \psi(Y_n, \dots, Y_{n-(k-1)}) \text{ explicit } k\text{-step method}, \tag{20.47}$$

$$Y_{n+1} = \psi(Y_{n+1}, \dots, Y_{n-(k-1)}) \text{ implicit } k\text{-step method}. \tag{20.48}$$

In all these procedures, the distances $h_n = t_n - t_{n-1}$ enters in the iteration formulae. If we assume constant step sizes $h_0 = h_1 = h_2 = \cdots$ and if we use h to denote this step size, we can then consider the local and the global truncation error of the procedures.

Assuming that $\underline{Y}_n = \underline{y}(t_n)$ is an exact solution, then the local truncation error is of the order p if

$$|Y_{n+1} - y(t_{n+1})| \leq M h^{p+1} \text{ for } p \geq 1 . \tag{20.49}$$

The global truncation error is of the order p for $\underline{Y}_n$, where $\underline{Y}_0 = \underline{y}(t_0)$ if

$$|Y_n - y(t_n)| \leq \tilde{M} h^p \text{ for } p \geq 1 . \tag{20.50}$$

Under certain conditions it is possible to deduce a global truncation error of order p from a local truncation error of order p.

20.6   Applications of MBS

Several applications for MBS exist in the area of vehicle dynamics.

One part of these applications concerns the engine and the powertrain. Figure 20.11 shows an example with a rocker arm valve drive for an internal combustion engine. In this application, for example, the forces between the cam lobe and the pushrod can be calculated using an MBS. Another example involves calculating the dynamic behaviour of the powertrain, as shown in Figure 20.12. Torques play an important role in this application, especially for dynamic manoeuvres. If compliances are introduced in the model, torsional vibrations can be investigated as well.

Vehicle dynamics investigates several aspects of the behaviour of the whole vehicle, for example

- understeering/oversteering;
- the influence of changing the engine torque during cornering;
- the influence of braking during cornering;

- the investigation of design space taking into consideration dynamic loads and relative displacements of the components;
- the kinematics and compliances of suspensions.

Figure 20.13 shows typical components of a McPherson front suspension. The rigid bodies are, for example, the wheel carrier, the wheel or the A-arm. The lower parts of the McPherson struts are not extra single bodies, because they are firmly connected to the wheel carrier. An additional rigid body is the subframe. The A-arms are connected with the subframe by rubber bushings, and the subframe itself is connected to the chassis with rubber bushings, too. Furthermore, the steering is shown in the figure as are the drive shafts. The powerplant is of course part of the MBS, but not depicted in Figure 20.13, but the two mounts for the powerplant and the roll restrictor are shown.

The main rigid body is not shown in full detail but only as a small sphere, which stands for the centre of mass and the inertia properties. Not visible, but included in the model, are the joints, for example, between the A-arm and the wheel carrier. As

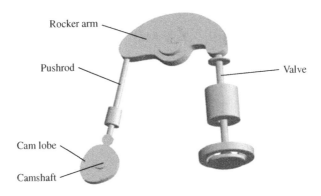

Figure 20.11   Rocker arm valve drive (example from the MBS software ADAMS)

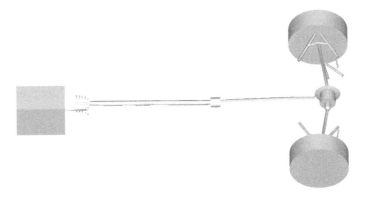

Figure 20.12   Rear-driven powertrain (example from the MBS software ADAMS)

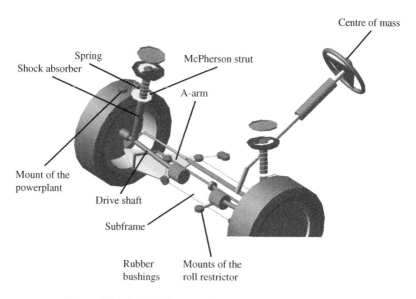

Figure 20.13   McPherson front axle in an MBS model

described above, many rigid bodies are not linked to each other by ideal rigid joints but by elastic bushings.

Some of the elements (e.g. springs, shock absorbers, bushings, mounts) are not described by a physical model but by characteristic curves. A simple example is a force–displacement curve for a spring, or a force–velocity curve for a shock absorber. The curves or maps become more complicated for multi-axial loaded mounts, such as hydromounts for the connection between the powerplant and body.

The tyres play a crucial role in MBS for automotive applications concerning the whole vehicle, for which a large number of models are available. The models can be classified using different characteristics, such as amplitude and frequency, complexity or underlying mathematical description.

Many investigations have been carried out with the aid of these models. One example is stationary cornering on a circle or step steer manoeuvre. The result of the latter is shown in Figure 20.14.

This figures shows the important dynamic quantities of the vehicle, such as yaw rate, steering angle and lateral acceleration (vertical axes from left to right). Another example is shown in Figure 20.15, in which the steering angle is shown as a function of the lateral acceleration.

The assumption that only rigid bodies are present is, of course, a simplification. In reality, bodies are never rigid. The error from assuming rigidity may be small, but in some cases deformation cannot be neglected or in some cases the deformations of rigid bodies should be calculated.

For instance, in some cases we may want to know the stresses in the chassis in order to assess reliability or durability. In other cases, the deformation may have an impact

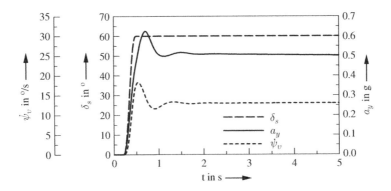

Figure 20.14 Step steer result

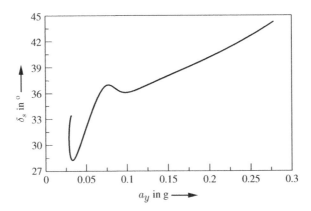

Figure 20.15 Quasi-steady-state cornering (for small lateral acceleration with some initial oscillations)

on the results; for example, the deformation of the subframe may influence kinematics and compliance of the suspension.

As stresses and strains are essential for predicting fatigue, such investigations can only be performed by considering deformations of the chassis.

Several methods exist for describing flexible bodies. Here we roughly outline one of them, similar to the Craig–Bampton method (or the fixed interface method)[2]. We briefly describe the free interface method using the example depicted in Figure 20.16, which shows a body (a rectangular plate) connected at three points by spherical joints to neighbouring rigid bodies. The spherical joints have three rotational degrees of freedom, which means that relative translations between the plate and the bodies are not possible. Describing the flexibility of the plate in an MBS first involves modelling

---

[2] The outlined method here is indicated as a free interface method and it is described in one of the first publications by MacNeal 1971; it is a generalization of the Craig–Bampton method, cf. Craig and Bampton 1968.

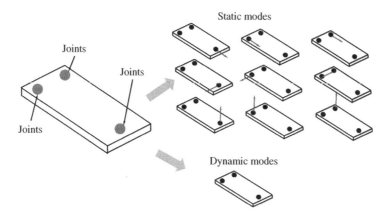

Figure 20.16   Modes for the free interface method

the plate in a finite element programme (using solid or shell elements). For each locked degree of freedom of the joints, this degree of freedom is unlocked and a unit force (for translational degrees of freedom) is introduced. If a rotational degree of freedom is unlocked, a unit moment is introduced in the FE model. With these unit quantities (forces or moments), the static deformations (or static modes) are calculated in the FE programme. In the example depicted there are nine static modes.

After this step of determining the static modes, all joints are deleted and an eigen-mode investigation is carried out for the plate. This results in a number of eigenvalues and eigenmodes (or natural modes) which we call the dynamic modes. Both the static and the dynamic modes are now used to establish the equation of motion. We call the static mode $\underline{u}_{si} = \underline{u}_{si}(x, y, z)$ and the dynamic modes $\underline{u}_{di} = \underline{u}_{di}(x, y, z)$. We can then approximate the deformation $\underline{u}$ of the plate using these functions:

$$\underline{u} = \sum_{i=1}^{N_s} \alpha_{si}\underline{u}_{si} + \sum_{i=1}^{N_d} \alpha_{di}\underline{u}_{di} . \tag{20.51}$$

This formula allows us to establish an expression for the kinetic and potential energy by integration with respect to the volume of the flexible body. As we have one static deformation mode for each locked degree of freedom, each configuration of the neigh-bouring body can be exactly described by the static modes. The dynamic modes describe the dynamic behaviour. In eigenvalue analysis, there is usually no limit to the number of eigenvalues (in FE eigenvalue analysis the limit is given by the degrees of freedom of all nodes). Consequently, the number of dynamic modes is usually restricted by a frequency bound.

The kinetic and potential energy terms contain the time-dependent coefficients α_{si} and α_{di}, which are now additional degrees of freedom of the system. The kinetic and potential energy can be used to establish the equations of motion for the whole system, for example using Lagrange's formalism.

20.7   Questions and Exercises

Remembering

1. Which components can be defined in MBS?
2. How many parameters are needed for the description of a rotation?
3. How many parameters are needed for the description of a translation?
4. Which algorithms for the solution of ordinary differential equations you know?

Understanding

1. Explain Euler and Tait–Bryan angles.
2. Explain a method for the consideration of flexible bodies in MBS.
3. Consider a system of two bodies, which are joined by a revolute (translational, cylindrical, spherical, cardanic) joint: How many degrees of freedom this system has?
4. How many static modes do you have to calculate in the free interface method for one revolute (translational, cylindrical, spherical, cardanic) joint?
5. Which static modes do you have to calculate in the free interface method for one revolute (translational, cylindrical, spherical, cardanic) joint?

Glossary

Acceleration resistance   Another resistance is due to d'Alembert's inertial forces. These inertial forces (from translational and rotational motions) are combined and are referred to as the acceleration (or inertial) resistance F_i. The acceleration resistance not only take into account the forces due to the translatory acceleration, but also take into account the forces in the longitudinal direction, which arise due to the angular acceleration of the rotating masses, 30

Ackermann steer angle   As described for the vehicle sideslip angle and the steering wheel angle, we introduce the front wheel angle δ_{10} for diminishing velocities:

$$\delta_{10} = \lim_{v \to 0} \delta_1 = \frac{\ell}{\rho_{cc}} .$$

We call this angle δ_{10} the Ackermann steer angle, 198

Aerodynamics drag force   On a vehicle with projected frontal area, A, while travelling at a speed v_v in the longitudinal direction, a longitudinal force, F_a, the so-called aerodynamic drag force, acts as follows (wind velocity $v_a = 0$):

$$F_a = c_d A \frac{\rho_a}{2} v_v^2 .$$

Here c_d is the coefficient of aerodynamic drag. The coefficient c_d of modern passenger cars is about 0.2 to 0.3. A typical size for the area A is 2 m^2, 29

Camber   This is the angle between the wheel $\vec{e}_{wx}$-$\vec{e}_{wz}$-plane and the vertical $\vec{e}_{iz}$-axis. The constructive Camber angle γ is positive when the wheel is inclined towards the outside of the vehicle and negative if it is inclined to its inside, 229

Caster   The point of application of the force F_y does not lie in the symmetry plane of the tyre, but is shifted against n_{tc} in the x_w-direction. We call n_{tc} the tyre caster trail (see Figure 11.7(b)), 177

Vehicle Dynamics, First Edition. Martin Meywerk.
© 2015 John Wiley & Sons, Ltd. Published 2015 by John Wiley & Sons, Ltd.
Companion Website: www.wiley.com/go/meywerk/vehicle

Circle of curvature The circle of curvature is a purely geometric object, which approximates the trajectory locally at one point. That is, the circle of curvature exists even when there is no vehicle moving along the trajectory; it is a characteristic of the trajectory, 170

Contact patch This is the contact area, wherein the tyre and the road are in contact. The size of the contact patch depends on the geometry and design of the tyre, the internal pressure and the wheel load. The order of magnitude for a passenger car tyre has a postcard format. (The contact area of the wheel-track contact is of the order of a thumb nail), 11

Cornering stiffness For small slip angles (approx. $\alpha < 4°$) the lateral force F_Y can approximated by a linearized law

$$F_y = c_\alpha \alpha \,.$$

The coefficient c_α is called the lateral force coefficient or the cornering stiffness, 180

Frame System A quadruple $(A, \vec{e}_x, \vec{e}_y, \vec{e}_z)$ is a frame system of an affine space. Here, A is a point (the origin) and $\vec{e}_x, \vec{e}_y, \vec{e}_z$ is a Cartesian tripod (the axis system). To describe the position of a point P with respect to A, three coordinates x, y, z are sufficient:

$$\overrightarrow{AP} = x\vec{e}_x + y\vec{e}_y + z\vec{e}_z \,.$$

The point A can be defined fixed in space (or in an inertial frame). This is called an inertial frame system (sometimes called earth or world coordinate system). If the point A and the tripod $\vec{e}_x, \vec{e}_y, \vec{e}_z$ are fixed to a body and continues to be firmly connected to the body then the result is called a body-fixed coordinate system, 6

Quarter-vehicle model The quarter-vehicle model (two-mass substitute system, Figure 10.1) is the simplest substitute system that already exhibits essential features of a vehicle in terms of vertical dynamics. The substitute system consists of the two masses, m_b (in this case m_b is one quarter of the body mass) and m_w (this is the wheel mass). The body springs and shock absorbers are located between the masses. (Spring stiffness k_b, damping constant b_b). A spring–damper system (stiffness k_w, damping constant b_w) also acts between the wheel mass, m_w, and the uneven road surface. Dividing the wheel into the components of wheel mass, m_w, wheel stiffness, k_w, and wheel damping, b_w, is a simplified model that permits a good reproduction of the wheel properties, 155

Driving performance diagrams A driving performance diagram comprises
1. the (real) supply characteristic maps of the engine converted to forces and power at the wheels as a function of the driving speed and in the same diagram
2. the required tractive effort (the driving resistances) or the effort for the power.

With the help of these diagrams, one can for example determine, the maximum speed without grading, the climbing ability in any gear and the acceleration capability, 57

Grading resistance The grading resistance (or climbing resistance) F_g is the portion of the weight of the vehicle which acts parallel to the road:

$$F_g = m_{tot}g\sin\alpha$$

, 29

Instantaneous center of rotation The instantaneous center of rotation is an imaginary point. The vehicle rotates around this point at a particular moment. If one imagines an imaginary infinite very large rigid plate which is fixed to the vehicle and which is parallel to the road, the instantaneous center or rotation is that point, which do not moves, i.e. the velocity of this point vanishes. The instantaneous center of rotation M_{cr} is the intersection of two normals of two arbitrary velocity vectors in two different points of the vehicle, 174

Progression ratio The progressive ratio α_{gz} denotes the ratio (quotient) of the transmission ratios of two adjacent gears.

$$\alpha_{gz} = \frac{i_{z-1}}{i_z} \quad z = 1, \dots, N_{z\,\text{max}}$$

, 48

Rolling resistance coefficient The rolling resistance coefficient f_r is the ratio of the rolling resistance F_r to the resulting normal force F_Z in the contact patch

$$f_r = \frac{F_r}{F_z}$$

, 16

Rolling resistance If a wheel is rolling on a road, an asymmetric normal stress distribution occurs between road and wheel in the contact patch (Figure 2.2). The line of action of the resultant force F_z of the asymmetric normal stress distribution does not intersects the center of the wheel, but is shifted in the rolling direction. The distance between the wheel center and the line of action of F_z is the eccentricity e_w. This results in a moment $M_w = e_w F_z$. To overcome this moment, a tractive torque in the case of a driven wheel or a tractive force F_r in the case of a towed wheel is necessary. This force F_r is called the rolling resistance. It can be derived by solving the sum of moment $0 = r_{wst}F_r - e_w F_z$ for F_r:

$$F_r = \frac{e_w}{r_{wst}}F_z \ .$$

In the case of a driven wheel the rolling resistance is

$$F_r = \frac{M_w}{r_{wst}}$$

, 15

Self-steering coefficient The following coefficient:

$$\frac{1}{i_s \ell} \frac{\partial(\delta_s - \delta_{s0})}{\partial(v^2/\rho_{cc})}$$

is called the self-steering coefficient of the vehicle. Likewise, the term

$$\partial(\delta_1 - \delta_{10})/(\partial(v^2/\rho))$$

is common, which is the self-steering coefficient without considering the steering stiffness, 202

Single-track model The single-track model is a key model in the lateral dynamics of a vehicle, which allows to consider important parameter dependences and to draw conclusions in the lateral dynamics. The single track model often forms the basis of simple ESP systems. Important assumption of the single track model is that the center of mass of the vehicle is on the road, which means that the distance of the center of mass to the road plane is zero: $h_{cm} = 0$. From this simplification, the limitation of the applicability of the single-track model results, 170

Slip For a driven wheel, slip is defined as the difference between the circumferential speed, $v_c = R_{w0}\omega$ and the driving speed v_v divided by the circumferential speed v_c.

$$S = \frac{v_c - v_v}{v_c} \ .$$

The slip of a braked wheel is defined as

$$S = \frac{v_v - v_c}{v_v} \ .$$

The slip is often given as a percentage, 21

Tyre Slip angle Lateral slip occurs in a tyre when the x_w-direction (i.e. the longitudinal direction in the tyre coordinate system) does not coincide with the direction of motion ($\vec{v}_w$ direction in Figure 11.7(b)). One calls this angle between the x_w-direction and $\vec{v}_w$-direction the slip angle α, 177

Toe The toe angle describes a static rotation of the wheel about the $\vec{e}_{wz}$-axis. We refer to toe-in when the wheels are turned inwards (cf. Figure 15.1(a)), and toe-out, when the wheels are turned outwards (Figure 15.1(b)). The angle δ_{10} is positive for toe-in and negative for toe-out, 229

Transmission ratio The transmission ratio i_z is the ratio (the quotient) between the input speed n_{iz} to the output speed n_{oz} of a transmission or gear:

$$i_z = \frac{n_{iz}}{n_{oz}} \ z = 1, \ \dots, N_{z\,\text{max}} \ .$$

The index z indicates the stage of transmission with $N_{z\,\text{max}}$ gears. The transmission ratio i_z is independent of the speed, 48

Tyre long. force coeff.   A tangential force F_x arises at the driven or the braked wheel, depending on the slip and the normal force F_z:

$$F_x = \mu(S)F_z \ .$$

The value μ is referred to as the tyre longitudinal force coefficient. This is a function of the slip S. The functions $\mu_b(S)$ for braking and $\mu_d(S)$ for driving is approximately equal: $\mu(S) \approx \mu_b(S) \approx \mu_d(S)$, 21

Oversteer   If $v_{ch}^2 < 0$, this means that an increase in vehicle speed, v, (on a circle with radius ρ_{cc}) requires an decrease in the steering wheel angle. We call this behaviour oversteer, 201

Understeer   If $v_{ch}^2 > 0$, this means that an increase in vehicle speed, v, (on a circle with radius ρ_{cc}) requires an increase in the steering wheel angle. We call this behaviour of the vehicle understeer, 201

Vehicle sideslip angle   The angle between the direction of motion of the vehicle's center of mass and the vehicle's longitudinal axis is called the vehicle sideslip angle β. The sum of the yaw angle and the vehicle sideslip angle is the course angle, 170

References

Abramowitz, M and Stegun A (eds.): Pocketbook of Mathematical Functions, (abridged edition of Handbook of Mathematical Functions; material selected by M. Danos and J. Rafelski), Harri Deutsch, Thun, Frankfurt/Main, 1984.

MacADAM, C C: Static Turning Analysis of Vehicles Subject to Externally Applied Forces – A Moment Arm Ratio Formulation, Vehicle Syst. Dynam., Vol. 18, pp. 345–357, 1989.

Battermann, W and Koehler, R: Elastomere Federung- Elastische Lagerung, W. Ernst u. Sohn, Berlin (in German).

Bauser D and Gawlik R: Method for controlling e.g. rear left wheel of motor car, involves determining drop in default braking pressure, and cancelling superimposed controlling of braking pressures based on drop in default braking pressure, patent DE102012008508 A1, 2013.

Blundell, M and Harty, D: The Multibody Systems Approch to Vehicle Dynamics, Elsevier, 2004.

Braess, H.-H. and Seiffert U. (Eds.): Vieweg-Handbuch Kraftfahrzeugtechnik, 2. Aufl., Braunschweig/Wiesbaden: Vieweg. 2001. (engl., Handbook of Automative Engineering SAE International, 2005).

Braess, H-H: Vom 02er zum E46 – Meilensteine der Marke BMW, in: J. Goroncy (Hrsgb.): Der neue 3er, Sonderausgabe als Beilage von ATZ 5/1998 und MTZ 5/1998, pp. 10–14, 1998.

Craig R, Bampton M: Coupling of substructures for dynamic analysis, Amer. Inst. Aero. Astro. J., 6(7), pp. 1313–1319, 1968.

Cucuz, S: Oscillation of passengers in cars: Impact of stochastics uneven roads and single obstacles of real road, PhD-Thesis, TU Braunschweig, 1993 (in German). Schwingungen von Pkw-Insassen: Auswirkungen von stochastischen Unebenheiten und Einzelhindernissen der realen Fahrbahn, Diss., TU Braunschweig, 1993.

Denman, H H: Tautochronic bifilar pendulum torsion absorbers for reciprocating engines, J. Sound Vibr., Volume 159, Issue 2, pp. 251–277, 1992.

Dresig, H and Holzweißig, F: Dynamics of Machinery, Springer, Berlin, 2010.

Dukipatti, R, Pang, J, Qazu, M, Sheng, G and Zuo, S: Road Vehicle Dynamics, SAE International, 2008.

Gillespie, T D: Fundamentals of Vehicle Dynamics, SAE International, 1992.

Harrer M, Goerich H-J, Reuter, U, and Wahl, G: 50 Jahre Porsche 911 – Die Perfektionierung des Fahrwerks (in German), Springer fuer Professionals, http://www.springerprofessional.de/50-jahre-porsche-911 – die-perfektionierung-des-fahrwerks_teil-1/4710496.html, 2013.

Heissing, B, and Ersoy, M.: Chassis Handbook, Vieweg + Teubner, 2011.

ISO 8855: International Standard Road vehicles – Vehicle dynamics and road-holding ability – Vocabulary, 2nd edn., reference number ISO 8855:2011(E), 2011.

Jazar, N J: Vehicle Dynamics, Springer New York, 2nd ed., 2014.

Mitschke, M and Wallentowitz, H: Dynamik der Kraftfahrzeuge, Springer, Berlin, 4th edn., 2004.

Naunheimer H, Bertsche B, Ryborz J, Novak W: Automotive Transmissions, Springer, Berlin, 2nd edn., 2011.

Nester, T M, Schmitz, P M, Haddow, A G, and Shaw S W: Experimental obersvations of centrifugal pendulum vibration absorber, The 10th International Symposium on Transport Phenomena and Dynamics of Rotating Machinery Honolulu, Hawaii, 7–11 March, 2004.

OECD: Key Transport and Greenhouse Gas Indicators by Country, International Transport Forum ©OECD, 2014 (http://www.internationaltransportforum.org/statistics/CO2/index.html).

Pacejka H B: Tyre and Vehicle Dynamics, 2nd edn., Butterworth and Heinemann, 2006.

Reimpell, J, Stoll, H, Betzler, J W: The Automotive Chassis: Engineering Principles, Elsevier Butterworth-Heinemann, Oxford, 2nd edn., 2001.

Reithmaier, W and Salzinger T: Determination of the state-of-the-art concerning rolling noise, rolling-resistance and safety properties of modern passenger car tyres, Research Report 201 54 112, TÜV Automotive GmbH Tire-/Wheel-Test-Center Ridlerstraße 57 D-80339 Munich, Commisioned by German Federal Environmental Agency, 2002. (http://www.umweltbundesamt.de/sites/default/files/medien/publikation/long/3163.pdf)

Richard H. MacNeal, R H: A hybrid method of component mode synthesis, Comp. Struct. Vol. 1, Iss. 4, pp. 581–601 (Special Issue on Structural Dynamics), 1971.

Roberson, R E and Schwertassek, R: Dynamics of Multibody Systems, Springer, Berlin, 1988.

Robert Bosch GmbH (eds.): Automotive electrics and electronics, in German: Autoelektrik, Autoelektronik, 5. ed., Friedr. Vieweg & Sohn Verlag/GWV Fachverlag GmbH, Wiesbaden, 2007.

Verhulst, F: Nonlinear Differential Equations and Dynamical Systems, Springer, Berlin, 2006.

Winner, H: Adaptive Cruise Control, in: Eskandarian, A, (Ed.): Handbook of Intelligent Vehicles, Springer, 2012.

Winner, H: Radar sensors, in: Eskandarian, A, (Ed.): Handbook of Intelligent Vehicles, Springer, 2012.

Index

Vehicle Dynamics, First Edition. Martin Meywerk.
© 2015 John Wiley & Sons, Ltd. Published 2015 by John Wiley & Sons, Ltd.
Companion Website: www.wiley.com/go/meywerk/vehicle

Printed and bound by CPI Group (UK) Ltd, Croydon, CR0 4YY